SECOND EDITION

LOGIC, SETS, AND RECURSION

ROBERT L. CAUSEY
University of Texas at Austin

JONES AND BARTLETT PUBLISHERS
Sudbury, Massachusetts
BOSTON   TORONTO   LONDON   SINGAPORE

World Headquarters
Jones and Bartlett Publishers
40 Tall Pine Drive
Sudbury, MA 01776
978-443-5000
info@jbpub.com
www.jbpub.com

Jones and Bartlett Publishers Canada
6339 Ormindale Way
Mississauga, Ontario L5V 1J2
CANADA

Jones and Bartlett Publishers International
Barb House, Barb Mews
London W6 7PA
UK

Jones and Bartlett's books and products are available through most bookstores and online booksellers. To contact Jones and Bartlett Publishers directly, call 800-832-0034, fax 978-443-8000, or visit our website, www.jbpub.com.

Substantial discounts on bulk quantities of Jones and Bartlett's publications are available to corporations, professional associations, and other qualified organizations. For details and specific discount information, contact the special sales department at Jones and Bartlett via the above contact information or send an email to specialsales@jbpub.com.

Production Credits
Acquisitions Editor: Timothy Anderson
Editorial Assistant: Kate Koch
Production Director: Amy Rose
Production Editor: Tracey Chapman
Marketing Manager: Andrea DeFronzo
Manufacturing Buyer: Therese Connell
Composition: Northeast Compositors, Inc.
Cover Design: Kristin E. Ohlin
Cover Image: © Photos.com
Printing and Binding: Malloy, Inc.
Cover Printing: Malloy, Inc.

Library of Congress Cataloging-in-Publication Data
Causey, Robert L.
  Logic, sets, and recursion / Robert L. Causey.-- 2nd ed.
     p. cm.
  Includes index.
  ISBN-13: 978-0-7637-3784-9 (casebound)
  ISBN-10: 0-7637-3784-4 (casebound)
  1. Logic, Symbolic and mathematical. I. Title.
  QA9.C347 2006
  511.3--dc22
                                    2005030305
6048

Printed in the United States of America
10 09 08 07 06   10 9 8 7 6 5 4 3 2

Contents

Preface

This book is an introduction to mathematical logic and related topics for undergraduates. It is designed to close a gap in mathematics literature and curriculum, and is primarily intended for students of computer science, mathematics, and philosophy. It is assumed that the reader has a good understanding of high school mathematics. No knowledge of computer science or programming is required, but the text contains many discussions and exercises applicable to those subjects.

An undergraduate student of mathematics often receives an inadequate introduction to the principles of logic, and to the basic ideas of set theory and mathematical induction. That represents a gap in the student's education. When the student enters advanced courses, he or she is expected to be familiar with those concepts and to be experienced with constructing and writing proofs. Likewise, the computer science student must be familiar with logic and related subjects and be able to construct proofs. Courses in data structures, algorithms, artificial intelligence, and many other topics require an understanding of logic and a level of mathematical sophistication beyond that imparted by calculus and "discrete mathematics" courses.

A principal goal of this book is to teach students how to construct and write informal, but rigorous, mathematical proofs using basic set theory, recursive definitions, and mathematical induction. Another major goal is to introduce formalized sentential and predicate logic. The treatment of logic emphasizes semantical concepts, and introduces formal derivations and elementary metatheory. The student should gain a basic understanding of the nature and use of sound deductive reasoning. The organization of the book is unusual; the chapters on sentential and predicate calculus are separated by informal treatments of set theory and recursion. This organization helps to relate formal and informal proofs, and it provides the mathematical background for the basic metatheory of predicate calculus. Nevertheless, some instructors may prefer to deviate from this order of presentation. Section 0.2, "Goals and Organization," describes some alternate study plans for using this text.

This text is based on a logic course that is primarily for undergraduate majors in computer science, but is also taken by many mathematics, philosophy, and engineering students. I have taught this course many times since 1982. The first edition of this text was published in 1994. Since then I have learned much about the students' engagement with the text from my own classroom experience with it and from comments of other faculty and teaching assistants. It became clear that students generally had less difficulty with formalized proofs than with standard, informal mathematical proofs. Above all, students desired more examples of problem-solving heuristics and more guidance in writing mathematical proofs.

Based on my experience with the first edition, a revised edition was published in 2001. It comprised a lightly edited version of the first edition, accompanied with a supplement containing many additional detailed examples of problem solving and proof writing, and many new exercises. This was a major improvement, but it was inconvenient to study from two books. The current, second edition merges the contents of those two books, and also incorporates much new material.

This new edition has been entirely rewritten. There are many detailed examples of problem solving. Most require a proof or counterexample, and many use recursive definition and mathematical induction. Most of these examples include a statement of the problem, a discussion of how to solve the problem, and a written proof. When appropriate, figures are used to help search for solutions, and the book contains sections devoted to proof construction heuristics. It contains a wealth of exercises; many illustrate applications of the subject matter in computing theory, and in pure and applied mathematics.

I am grateful to many people who have helped me with one or more of the editions of this book. I cannot list them all here, but especially wish to thank the following: Carl Hesler, Jr., former Vice President of Jones and Bartlett Publishers, enthusiastically supported the writing of the first edition. David Barrington, the late Jon Barwise, Herbert Enderton, Michael Gelfond, and Richard Grandy wrote useful reviews of the original manuscript of the first edition. Many students and teaching assistants, especially David Bradshaw, David Newman, Benjamin Rode, and Andrew Schwartz, made suggestions that improved the later editions. Lynn Cates proofread both the first and second editions. A number of colleagues, including Robert S. Boyer, Lynn Cates, Cory Juhl, Matt Kaufmann, and Vladimir Lifschitz, have contributed comments based on their classroom experience with the text. Several members of the excellent Jones and

Bartlett team, especially Tim Anderson and Amy Rose, have done exceptional work in the production of the book. Most of all, I thank my wife, Sandy, whose patience, careful attention, and critical eye have improved all editions of this work.

Chapter 0

Introduction

0.1   The Subject Matter

This book is an introduction to the basic concepts and methods used in modern symbolic logic and mathematics. It is primarily intended for undergraduate students studying mathematics or computing science. It will also be of interest to many students specializing in philosophy or engineering, as well as anyone seriously interested in proofs and computations. The reader is assumed to have a good understanding of the basic principles of arithmetic and algebra, and to have some general understanding of the nature of proofs as given, for instance, in elementary Euclidean geometry. I will say more about the expected background shortly. This book does not presuppose any previous study of more advanced mathematics such as calculus or modern abstract algebra. Previous study of these subjects may be of benefit to some readers of this book, but it is more likely that a good understanding of the material would be helpful in the study of advanced calculus and algebra. The concepts and methods introduced constitute the essential background for advanced mathematics and computing theory.

Logic is the study of methods and patterns of reasoning, especially reasoning that is expressed in the form of arguments. For our purposes, an *argument* consists of a set of declarative sentences called the *premises* and another sentence called the *conclusion*. An argument is intended to show that if the premises are all true, then their truth supports the truth of the conclusion. There are many different kinds and forms of arguments. We will be mainly concerned with the types of arguments used in mathematical proofs, for example, proofs of the Pythagorean theorem in elementary geometry. In the study of logic, one is not concerned with specific information about right triangles or other facts of geometry, but rather with general principles of reasoning. These general principles may be used in any mathematical proof, as well as in legal, philosophical, engineering, or scientific arguments. Most people use many of these principles without even being aware of them. In this book, the principles

1

will be explicitly stated, analyzed, and extensively used. Chapters 1 and 4 are devoted to the detailed study of principles of pure logic formulated in precise, artificial, symbolic languages (symbolic logic). Of course, general principles of logic will be used in various forms throughout the book. Typical mathematical proofs and examples are presented in what may be called "mathematical English" (or "mathematical French," German, etc.). This refers to the rigorous but somewhat informal style of presentation found in mathematics books and articles. The artificial, symbolic languages invented by logicians are not at all convenient for ordinary mathematical writing. It is my view that there are two main motivations for the introduction of artificial, symbolic languages: They possess the high degree of clarity and precision needed for a careful analysis of logical principles and the foundations of mathematics, and they are required for automated reasoning by computing machines. Historically, the first reason was the primary impetus for the invention of symbolic logic, although the second reason was also envisioned. Today it may be true that more people study symbolic logic because of the second reason rather than the first. Logical symbolisms are not only used in specialized programming languages for artificial intelligence and automated theorem proving; but they are also components in other, general-purpose programming languages.

Most students learn some facts about sets in school. Intuitively, a *set S* is a collection of distinct objects considered just as a collection, without regard to the order or manner in which these objects may be named or described. The objects that S comprises are called the *elements* or *members* of S. Our idea of a set includes the case in which there may be no elements in the collection. If there are just a few elements in a set, it may be described by listing these elements within braces. Here are some examples of simple sets:

- the set of planets in the solar system

- the set of people on Manhattan Island at noon, June 1, 1992

- $\{\,0,\,1,\,2,\,3\,\}$

The branch of mathematics that is concerned with general principles about sets is called *set theory*. A major achievement of twentieth-century mathematics is the demonstration that most, if not all, mathematics can be developed from basic concepts and assumptions of logic and set theory. The influence of set theory has been so strong that it is now nearly impossible to read moderately advanced mathematics texts without familiarity with the basic concepts and facts about sets. Chapter 2 introduces some of the most basic ideas of set theory. Our treatment of this subject is mathematically rigorous, although it uses an informal style and is based on only three of the fundamental axioms (or postulates) of set theory. This treatment is limited in scope, but it is adequate preparation for much additional study of mathematics, and it covers most of the set theory one is likely to encounter in computing science.

Chapter 3 is largely concerned with what are called *recursive definitions*. Here is a simple example: Suppose that one is learning the arithmetic of the nonnegative integers, 0, 1, 2,$\cdots$. Also suppose that one already knows how to add two such numbers together, e.g., $3 + 5 = 8$. Now we want to *define* multiplication, $\times$, in terms of addition. In fact, we just consider multiplication to be repeated addition, so we write the following, where i is a nonnegative integer:

$$1 \times i = i,$$

$$2 \times i = i + i = i + (1 \times i),$$

$$3 \times i = i + (i + i) = i + (2 \times i),$$

$$4 \times i = i + (i + (i + i)) = i + (3 \times i),$$

and so on. We might decide to summarize all of this as

$$1 \times i = i$$

and, for $k = 1, 2, 3, \cdots$,

$$(k + 1) \times i = i + (k \times i).$$

But this ignores the case $k = 0$, which we decide to handle as follows:

$$0 \times i = 0$$

and, for $k = 0, 1, 2, \cdots$,

$$(k + 1) \times i = i + (k \times i).$$

These last two equations constitute what is called a *recursive definition* of $\times$ in terms of $+$. Definitions of this kind are widely used in mathematics, in computing theory, and in practical computer programs. They are used not only for numerical operations, but also for operations on strings of characters, lists of data, and in other contexts. One may not be troubled by this simple example, but many recursive definitions are subtle and complex; if they are not properly formulated, they may not yield the intended values. Worse yet, in computer programs, bad recursive definitions can cause the program to fall into eternal (infinite) loops, so that the program produces no useful output at all, but uses up computer time indefinitely until it is stopped by some intervention. The justification of recursive definitions is closely related to the structure of the *natural numbers* (nonnegative integers), so Chapter 3 discusses the natural numbers and initially develops recursive definitions within this context.

Recursive definitions are related to a very general principle of proof called *mathematical induction*. In its simplest form, mathematical induction is used to prove that some proposition, say, ϕ, is true of all natural numbers. One proves that the proposition is true for 0. One also proves that, if ϕ is true for a number k, then ϕ is true for the next number $k + 1$. Since it is true for 0, it must therefore be true for 1, and hence also for 2, and so on. The *principle of mathematical induction* states that we may now conclude that ϕ is true for all natural numbers. Chapter 3 discusses mathematical induction and contains many examples of proofs that use this principle and related principles. The methods developed in this chapter are applied not only to numbers but also to non-numerical data. Mathematical induction proofs are used in many branches of mathematics, and they are indispensable in the advanced study of logic and computing theory.

0.2   Goals and Organization

This book naturally divides into the four chapters mentioned earlier. One goal of the book is to impart a familiarity with the basic concepts and methods of logic. Among other things, this includes the following: What is a logically sound argument? How can one prove that an argument is sound? How can one show that an unsound argument is unsound? How are logical principles used in mathematical proofs? In addition, I try to indicate, mainly by examples, how logic is useful in computing theory and in programming. Unlike most introductory logic books used in philosophy or linguistics courses, relatively little space is devoted to the relations between symbolic languages and ordinary English. I do discuss how the logical principles analyzed in the symbolic languages relate to "mathematical English," the language that mathematicians typically use in proofs and examples. Beyond this, there is no extended analysis of the structure of ordinary arguments in English. The analysis of such arguments is a large and important topic that is beyond the scope of this text. This book emphasizes general principles of reasoning. Symbolic languages are examined in Chapters 1 and 4 because they contribute to the analysis of these general principles, and because they are used in programs for automated reasoning by computers.

Another goal is to present the basic ideas of set theory and to show how these ideas can be used in a variety of theoretical and practical contexts. For example, one section of Chapter 2 is concerned with *relations*. In mathematics, one uses relations such as *less than*, as in $2 < 3$. In ordinary life, we refer to relations when saying things such as "John is a brother of Sue." In set theory, relations are studied in a general and abstract form. The same is true of functions, which are special kinds of relations. The results of these studies can be used in computer programs, especially in what are called *relational databases*. The material in Chapter 2 can help prepare one for further study of these more specialized topics.

Besides presenting basic concepts and methods of set theory, this book also presents the fundamental axioms for the natural numbers and develops and justifies the use of recursive definitions. Chapter 3 contains numerous examples of such definitions and their proofs. In addition to imparting fundamental concepts, a major goal of this book is to help one learn how to construct *proofs*. Most proofs combine specific information (e.g., about right triangles) with general principles of reasoning to move from premises to a conclusion. Our concern is with these general principles of reasoning: first, the principles of pure logic, and second, the principle of mathematical induction (in various guises). Thus, most of the text is devoted to proofs — principles of reasoning, suggestions for the construction and writing of proofs, and examples of proofs.

Mathematicians usually present their proofs in brief form; even if the proof is convincing, one is often left wondering how it was thought of in the first place. I have departed from this practice. When a new type of proof is first presented in this book, I usually try to explain how I constructed it. These proofs are not original; they all exist in the mathematical literature. Most are so routine that experienced mathematicians construct them without even seriously thinking about how each is accomplished. Yet the thought processes involved often seem utterly mysterious to beginners. It is hoped that descriptions of how I construct some of the proofs will help others to strengthen their own problem-solving and proof-construction abilities. This analysis of problem solving and proof construction is of interest for another reason. If we hope to program a computer to solve problems and construct proofs "intelligently," it is likely that some of our ways of thinking will need to be incorporated somehow into the program. Much artificial intelligence research has already been devoted to these matters, but a great deal remains to be done.

To summarize, it is hoped that this book will contribute to a better understanding of the following:

- Basic concepts of logic

- Concepts of sets, relations, and functions

- The natural number system

- Recursive definitions of functions on numbers and other types of data

In addition, this text is intended to enhance one's ability to read and understand mathematical proofs; assess alleged proofs; and invent and write proofs, including mathematical induction proofs.

This is a good place to mention the order of presentation. Many books on introductory logic begin by presenting a symbolic language and strict rules for performing deductions in this language. The symbolic logical system is interpreted only by corresponding fragments of English. It is often further assumed that, if one is to study a mathematical system such as set theory or the natural number system, then this study should be carried out within symbolic logic using strict rules of deductions. In my opinion, this is not the best way to achieve initial understanding of either logic or mathematics. The approach used here is mixed: Chapter 1 begins with a very simple symbolic language that can be adequately developed with little mathematical apparatus. The logical principles developed in Chapter 1 are then used informally in the proofs in Chapters 2 and 3. The material developed in Chapter 2, together with some of the material in Chapter 3, is then used in the development of a more powerful logical theory in Chapter 4. This mixed approach has the added advantage of illuminating the close relationships between formal and informal styles of proof.

For the majority of readers, the four chapters are best studied in order, although many sections can be skipped without seriously harming study of later sections. Readers who are already familiar with basic set theory, and who wish only to learn symbolic logic, may choose to read Chapters 1 and 4. Except for a few proofs in Section 4.3.3, such readers should be able to master the material in these two chapters. Readers who already know some symbolic logic or have some experience with mathematical proofs should be able to master Chapters 2 and 3, with perhaps an occasional glance back to Chapter 1. Chapter 3 is the longest in the book. Sections 3.1 through 3.3.3 provide a detailed introduction to recursive definitions and mathematical induction. The remainder of this chapter covers some applications to number theory and non-numerical data. The latter should be of special interest to computer science students. Even readers with no previous knowledge of the book's subjects could mix up their order of reading. One could study most of Chapter 1, followed by the first two sections of Chapter 2. This could then be followed by Sections 4.1 through 4.3. This would provide a good introduction to basic set theory and symbolic logic. If desired, one could then go back and complete Chapter 2, study part or all of Chapter 3, and then finish more of Chapter 4. Although this pattern of study is feasible, I still strongly recommend following most of the topics in the order presented.

This book contains many exercises. Some merely provide practice with the material presented in the main text. Others present extensions of this material. Still others are simple applications of this material to special kinds of problems. The exercises are numbered consecutively in each chapter. These numbers are prefixed by "E" followed by the chapter number. Thus, E 1-24 is the twenty-fourth exercise in Chapter 1; E 2-7 is the seventh exercise in Chapter 2, etc. Most of the major sections within a chapter are followed by sets of exercises. As a general rule, the later exercises within one of these sets are more difficult than the earlier ones.

Some of the exercise numbers are printed in bold font, e.g., **E 1-22**. Problems that introduce important principles, useful applications, are especially difficult, or are special in some way, are indicated by this bold print. In some cases, parenthetical comments describe the special features of the problem. If such comments are missing, the exercise is nonetheless of special value in introducing a new concept, principle, or technique. It is not necessary to work on all of the exercises, but one should try to solve a fair sample of the various kinds of problems within each exercise set. It is difficult to master this material without devoting considerable effort to this endeavor. As Euclid said to a ruler of ancient Egypt who wanted to learn mathematics in a quick, easy way, "There is no royal road to geometry." This is as true today as it was in 300 B.C.E. To ease the task a little, answers to a few, selected exercises are included at the end of the book. These answers are intended to illustrate some standard techniques and styles for proofs. Other problems may require different techniques, so one should not simply try to mimic the selected answers that are provided.

0.3   Assumed Background Knowledge

0.3.1   Sets, Numbers, and Algebra

This section briefly reviews some of the background required for understanding this book. As previously mentioned, I assume an intuitive familiarity with the concept of a set. In addition, it is assumed that one has an intuitive idea of the difference between a finite and infinite set. For example, it is assumed that sets like

$$\{\, 2,\, 4,\, 5,\, 11 \,\}$$

are finite, whereas the set of natural numbers and many other sets are infinite. We also assume an intuitive understanding of the difference between a finite sequence such as

$$2,\, 4,\, 8,\, 16,\, 32,$$

which has exactly five elements, and an infinite sequence like

$$2,\, 4,\, 8,\, \cdots,\, 2^k,\, \cdots,$$

which never ends. A precise definition of *sequence* is given in Chapter 3, where it is also shown how to give clear meanings to uses of the three-dot ellipsis notation used here.

When a set has a finite number of elements that can be named by symbols such as, α_1, α_2, $\cdots$, α_n, then the notation $\{\, \alpha_1,\, \alpha_2,\, \cdots,\, \alpha_n \,\}$ denotes the set consisting of exactly these elements. In addition to the intuitive idea of a set, I also assume an understanding of *set union*, $\cup$, and *set intersection*, $\cap$. If Γ and Δ are sets, then $\Gamma \cup \Delta$ is the set that consists of all elements that are either in Γ or in Δ (or both). Thus,

$$\{\, 0,\, 3,\, 4,\, 7 \,\} \cup \{\, 4,\, 7,\, 11 \,\} = \{\, 0,\, 3,\, 4,\, 7,\, 11 \,\}.$$

$\Gamma \cap \Delta$ is the set that consists of all elements that are both in Γ and in Δ. Thus,

$$\{\, 0,\, 3,\, 4,\, 7\,\} \cap \{\, 4,\, 7,\, 11\,\} = \{\, 4,\, 7\,\}.$$

For examples and other purposes, it will be necessary to refer to different systems of numbers. The set of *natural numbers* is the set of nonnegative integers and is denoted by *Nat*. Thus,

$$Nat = \{\, 0,\, 1,\, 2,\, \cdots \}.$$

The set of all *integers* is denoted by *Int*, so

$$Int = \{\, \cdots,\, -2,\, -1,\, 0,\, 1,\, 2,\, \cdots \}.$$

A *rational number* is a common fraction that can be expressed in the form a/b, where a and b are integers, with $b \neq 0$. Any integer i is also a rational number, since $i = i/1$. It is customary to reduce a rational number to its "simplest form," e.g., convert $-36/210$ to $-6/35$, but we will not assume that this is always done. The set of rationals is *Rat*.

It is important to understand how these different number systems are related to algebraic equations. Suppose that we seek a natural number x such that

$$5x + 3 = 13.$$

There is a solution, $x = 2$. But the equation

$$5x + 18 = 13$$

has no natural number solution. It does have an integer solution, $x = -1$.

The equation

$$5x + 19 = 13$$

has no integer solution, but it has a rational number solution, $x = -6/5$. The invention of new number systems was partly motivated by the search for solutions of algebraic equations. The preceding equations can all be put into the form

$$ax + b = 0,$$

where a, b are nonzero integers. Equations of this form always have rational number solutions. The situation is different for quadratic equations. Consider the simple case,

$$x^2 = 2.$$

From algebra, one recalls that there are two solutions,

$$x = \sqrt{2} \text{ and } x = -\sqrt{2}.$$

The ancient Greeks proved that $\sqrt{2}$ is not a rational number. I assume that the reader is generally familiar with the set of *real numbers*, which I denote by *Re*. The numbers $\sqrt{2}$ and $\pi = 3.14159\cdots$ are irrational (i.e., they are not rational numbers), but they are real numbers. For our purposes, it is sufficient to consider the real numbers to be the set of all numbers that can be expressed in infinite decimal form, including such examples as

$$5.250000\cdots ,$$

$$0.142857142857142857\cdots ,$$

and

$$1.41421356237309504880168872\mathbf{4}\cdots .$$

The real numbers can be added, subtracted, multiplied, and divided (except by 0) to yield real numbers. With certain kinds of exceptions, they can also be raised to powers. Several principles apply to addition and multiplication, among which the following are noteworthy. Let x, y, z, be any real numbers, then

$$x + y = y + x, \tag{0.1}$$

$$xy = yx, \tag{0.2}$$

$$x + (y + z) = (x + y) + z, \tag{0.3}$$

$$x(yz) = (xy)z, \tag{0.4}$$

$$x(y + z) = xy + xz, \tag{0.5}$$

The first two equations are called the *commutative laws*, or *principles of commutation*, for addition and multiplication, respectively. Equations (0.3) and (0.4) are the *associative laws*, or *principles of association*, for addition and multiplication, respectively. Equation (0.5) states the *principle of distribution*, or *distributive law*, for multiplication over addition.

I assume some familiarity with simple algebraic functions and how they can be plotted in a coordinate system. In the standard X–Y, or Cartesian, coordinate system, each point in the coordinate system corresponds to a pair of real numbers written in a particular order. The *origin* is the pair <0, 0>. It is customary to represent simple functions in the form

$$y = f(x),$$

where f is a *function* of the one real variable x, i.e., the possible values of x are real numbers. In place of $f(x)$, one often writes a particular symbolic expression that may provide a way to calculate a value of y for a given value of x. This notation is convenient when plotting a function in the coordinate system. Thus, if we plot the pairs, $< x, y >$, of real numbers that satisfy the equation

$$y = 2x + 5,$$

the plot is a straight line that passes through the point <0, 5>. Similarly, the equation

$$y = x^2 - 4$$

determines a function plotted as a parabola passing through $< 2, 0 >$ and $< -2, 0 >$. More complex relationships can also be plotted; the set of all points that satisfy this expression

$$x^2 + y^2 = 25 = 5^2$$

constitutes a circle with a radius of 5 and center at the origin. The set of points satisfying

$$x^2 + y^2 = r^2$$

is the circle of radius r with center at the origin. If $r < 5$, this will be a circle inside the circle of radius 5. So, the set of points satisfying

$$x^2 + y^2 < 25$$

is the set of *all points inside* the circle of radius 5. I will always use the term *plot* to refer to sets of points in the Cartesian coordinate system corresponding to algebraic expressions. The term *graph* will be used for a different purpose in this text. On many occasions, I will want to refer only to *positive* numbers, i.e., those greater than zero. The terms Nat^+, Rat^+, and Re^+ denote the positive natural numbers, rationals, and reals, respectively. The principal number systems that we will use are summarized in Table 0.1.

0.3.2   Functions

Consider the algebraic function $f(x) = 5x^2 - 2x + 3$. This expression is said to represent a function of *one argument*, because f may be given one "input

Symbol	Description	Examples
Nat	natural numbers	0, 1, 2, $\cdots$
Nat$^+$	positive natural numbers	1, 2, 3, $\cdots$
Int	integers	$\cdots$, -2, -1, 0, 1, 2, $\cdots$
Rat	rational numbers	$-2/3$, $-7/1$, $0/1$, $7/5$, $13/1$, etc.
Rat$^+$	positive rational numbers	$7/5$, $13/1$, $3/5$, etc.
Re	real numbers	-2.0013, 31.000, $-0.14159\cdots$, etc.
Re$^+$	positive real numbers	31.000, $0.14159\cdots$, etc.

Table 0.1 *Familiar Sets of Numbers*

value" (such as -3.046) for which it "returns" one corresponding output value (55.482580) that can be calculated by using the algebraic expression on the right-hand side of the equation. Thus, we may write $f(-3.046) = 55.482580$. When working with a function, it is important to specify the possible values that its argument may have. For instance, in this example, x might have a value that is any real number, or its possible values might be restricted to any element of *Re*$^+$ or any element of *Nat*, etc. Suppose the possible values of x are any elements of *Re*, then the values of $f(x)$ are also in *Re*. It is often said that f *maps Re to Re*, and a function is often called a *mapping*.

Functions may have more than one argument, e.g., the function represented by

$$g(x, y) = 7xy^3,$$

which has two arguments, indicated by the variables x and y. Any other pair of variables (say, u and v) could be substituted respectively for x and y, and the resulting expression,

$$g(u, v) = 7uv^3,$$

would represent the same function. "Argument," when used to refer to an argument of a function, has a very different meaning than when it is used in the study of logic to refer to a set of premises together with a conclusion. Although it is ambiguous in this way, the context of its use should enable us to avoid confusion.

Conceptually, a function is not the same as a procedure for calculating its values. In fact, there may exist more than one procedure for calculating the values of a function. For example, there are different procedures for computing square roots of positive real numbers. Therefore, we will not think of a function as a specific procedure, but rather as a special kind of abstract relationship.

Let Γ and Δ be nonempty sets. Think of a *function* from Γ to Δ as an abstract specification that associates with each element of Γ a *unique* element of Δ. When this is the case, it is also said that the function is *defined on* Γ. We may know only that the association exists in theory; it is not necessary that we be able to compute the values in Δ that f has (returns) for various elements of Γ. It is not necessary that Γ or Δ be sets of numbers in order to have a function from Γ to Δ. Let Γ be a set of people at a sporting event at noon on January 1, 2006. Let $\Delta = \{\,\mathbf{F}, \mathbf{M}\,\}$, so Δ is just a set of two boldface characters. Now define f by: For each person x in Γ, let

$f(x) = \mathbf{F}$, if x is female,

and

$f(x) = \mathbf{M}$, if x is male.

This defines a function f, which maps Γ to Δ. It is a function that maps a set of people to a set whose elements are just two boldface characters. We will use sets and functions extensively—sometimes in very abstract ways and sometimes with computational procedures. Sets and functions are treated in detail in Chapter 2. There are many questions that can be raised about the adequacy of the specification of a function and about the procedures that are supposed to compute the values of a function. We will consider some of these questions in this book, but the preceding informal discussion is adequate review for the first chapter.

0.3.3   Inequalities and Identity

Although we are not concerned with geometrical proofs, I assume familiarity with their general structure and purpose. It is further assumed that simple algebraic operations on equations and inequalities are familiar. In the case of inequalities, not much is needed; it will be helpful if one recalls a few facts such as the following, where x, y, u are real numbers:

If $0 < x$ and $0 < y$, then $0 < xy$.

If $x < y$, then $x + u < y + u$.

If $u < x$ and $x < y$, then $u < y$.

I have mentioned equations several times. Equations are just statements of identity. We may write $5 = 5$ to express the "fact" that the number 5 is identical with itself. If *bob* is the name of a person, we may also write *bob = bob* to say that *bob* is identical with himself. These examples illustrate the first of the two basic *laws of identity*, which are stated in Figure 0.1.

Identity statements are of primary interest when different expressions are on the opposite sides of the identity (equals) sign, e.g.,

$$2^3 = 8,$$

for every real number x, $7 - (5 + x) = 2 - x$.

I_1 For anything x, $x = x$.

I_2 Let α and β be symbols that denote mathematical objects, and let ϕ be a mathematical expression containing occurrences of α (and possibly also β). Suppose that the expression ψ results from replacing zero or more occurrences of α in ϕ by occurrences of β. Then: If ϕ is true, and if $\alpha = \beta$ or $\beta = \alpha$ is true, we may infer that ψ is true.

Figure 0.1 *Laws of Identity*

In school, one learns the rule that "equal things may be substituted for one another." This is vague, largely because it does not specify the kinds of expressions into which the substitutions may be made. For the present purposes, this rule is stated as I_2 in Figure 0.1. In I_2 the term *mathematical objects* refers to numbers, sets, functions, etc. The term *mathematical expression* refers to algebraic expressions, set theory statements, and other formulas used in standard mathematics. To specify exact meanings for the terms *symbol* and *occurrence*, it is necessary to specify the structure of the language one is using. Chapters 1 and 4 will give such specifications for some symbolic logic languages. When one is doing informal mathematics, it is assumed that one is familiar with what count as symbols, meaningful expressions, and occurrences of symbols within these expressions. For example, suppose that ϕ is the expression

$$x = y^5 + 2y,$$

and ψ the expression

$$x = 3^5 + 2y.$$

Let α denote the symbol 'y' and β denote the symbol '3'. Notice that ψ results from ϕ by replacing an occurrence of α (y) by an occurrence of β (3). If ϕ is true and also $y = 3$ or $3 = y$ is true, then we may infer that ψ, namely, $x = 3^5 + 2y$ is true. In this example, it is clear that we are replacing an occurrence of one symbol, y, by an occurrence of another symbol, 3. For contrast, suppose that we started with $25 = 5^2$ and $a = 5$. It would be a *mistake* to infer that $2a = 5^2$. This is not a proper use of I_2 because the 5 in 25 is not an explicit occurrence of 5 in the equation $25 = 5^2$. On the other hand, 5 does occur as an explicit term in $20 + 5 = 5^2$. Thus, we can substitute a for 5 in it and infer the true result, $20 + a = 5^2$. In informal mathematics, it is assumed that one recognizes that 25 is a *single symbol*, denoting a particular number, so we cannot substitute a for the 5 in 25.

I_2 is the *Principle of Substitutivity of Identicals. Although this principle holds for typical mathematical objects and expressions, there are exceptions to it.* An example of such an exception is presented in Chapter 2. For now we need not be concerned with this, since we will be working with simple mathematical expressions that are not exceptions to I_2. Here is a simple example of the use of the laws of identity. Suppose that we are interested in the intersection of the two straight lines corresponding to the equations

$$x + y = 10 \tag{0.6}$$

and

$$x - 2y = 4. \tag{0.7}$$

This is a routine problem in elementary algebra: If possible, find values for x and y that satisfy both equations. There are general techniques for attacking such problems, but we can do the following: First, multiply both sides of Equation (0.6) by 2, producing

$$2x + 2y = 20. \tag{0.8}$$

Why is this justified? By I_1, we have

$$2(x + y) = 2(x + y). \tag{0.9}$$

Since, by (0.6), $x + y = 10$, we may invoke I_2 to substitute 10 for an occurrence of $x + y$ in (0.9) to obtain

$$2(x + y) = 2(10). \tag{0.10}$$

Using the Distributive Law, arithmetic, and I_2, we obtain

$$2x + 2y = 20. \tag{0.11}$$

Now, addition is a function allowing real-valued inputs, so if r_1 and r_2 are expressions denoting particular real numbers, then $r_1 + r_2$ denotes the same real number regardless of the symbolic forms of r_1 and r_2. Because of this, the number returned by adding the left sides of (0.7) and (0.11) must be the same number that results from adding the right sides of (0.7) and (0.11). Thus, we may perform the familiar operation of adding two equations, in this case (0.7) and (0.11). By doing this, and performing further simplifications according to the laws of arithmetic, we can obtain the new equation,

$$3x = 24. \tag{0.12}$$

Since division by a nonzero number is also a function, we can divide both sides of (0.12) by 3 to get $x = 8$, one of the values we were seeking. Again using I_2, we can substitute 8 for x in (0.7), to get

$$8 - 2y = 4. \tag{0.13}$$

Using reasoning and operations similar to those just described, we can prove that $y = 2$, and this finishes the problem.

In this example, we applied functions to equations. This type of operation can be illustrated in more general form. Suppose that we have a function, f, with one input argument, and suppose that we have an equation $\alpha = \beta$, in which α and β are terms denoting allowable input values to the function. Since f is a function, for any input, say α, f returns a unique value denoted by $f(\alpha)$. Because $f(\alpha)$ is a well-defined, unique object, by I_1, we have $f(\alpha) = f(\alpha)$. Now, since $\alpha = \beta$, we can use I_2 to obtain $f(\alpha) = f(\beta)$. In effect, we may apply a function to both sides of an equation in this manner. If the function has more than one argument, an analogous, but more general, result is justified in a similar way. As is common in mathematics, we will often use the laws of identity without explicitly mentioning them. This completes the review of background material. Any other background information that may be used is more specialized and will be presented as needed.

0.4 The Presentation

If I say "Bob is six feet tall," I am *using* the word 'Bob' to denote, or refer to, a particular person. If I say, " 'Bob' has three letters," I am not using 'Bob' to refer to anyone or anything. I am *mentioning* the word 'Bob,' i.e., 'Bob' is the subject of the discussion rather than a person Bob. When I mention a word or symbol, I will often enclose it inside single quotation marks to emphasize the fact that it is being mentioned rather than used. If it is set off in a special display, or the context makes it clear that it is being mentioned, then these single quotation marks may be omitted. Here are some examples of these conventions: 'Red' has three letters. 'Red' refers to a color. The word

red

has three letters. The string of symbols

$$\text{arxtbztt}$$

contains eight characters. The numeral

$$302356$$

has six digits. The expression

$$2x + (3 - x^y) = xz$$

has three occurrences of x.

The last example mentions both '$2x + (3 - x^y) = xz$' and 'x'. Single quotes were not used in that example, because it is clear that both the expression and the symbol are being mentioned, not used. In the study of logic, it is often necessary to mention a symbol, word, phrase, or sentence. The conventions of enclosing these mentioned items in single quotes, or setting them off in displays, is widely used. I will follow these conventions when it appears that confusion might otherwise result. I will occasionally use double quotation marks. This will be done for standard direct quotations and to indicate irony or a popular usage.

Woefully, this book contains a large number of definitions. This seems unavoidable given the number of new concepts it introduces. I have tried to introduce only terminology that is essential for developing the material in the text; in a few places, alternative terms are also mentioned because they are in wide use. All of the important terms are introduced in definitions set off from the main text. This should ease the task of referring back to definitions when needed. A few, less vital terms are defined more casually in the main text. In either case, the term or phrase that is being defined is italicized.

The text includes a large number of proofs of theorems. Most of these proofs will serve as useful examples of proof techniques and styles of presentation. In traditional editions of *Euclid's Elements*, see Heath (1956), the end of a proof is marked by Q.E.D., which abbreviates *quod erat demonstrandum*. This Latin phrase is translated as *which was to be proved*. I follow this tradition, and write **_Q.E.D._** at the end of proofs. In addition to serving as examples of techniques and style, these proofs should also function as models for many of the proofs that are in the exercises. However, the exercises are not mere variations of material in the main text; some of them require a fair amount of ingenuity. Definitions of new terms and applications of theorems are usually illustrated by examples. When such examples easily fit into the discussions in the main text, they are presented in this text. If an example requires some digression or is intended as a model for the exercises, then it is specially marked and set off from the main text. When this is done, a black square, ■, is used to mark the end of the example.

A list of selected references is included near the end of the book. Most of the items on this list are advanced articles or texts and are suggested for further reading. A number of the items listed are cited in the text; most of these citations are suggestions for more advanced, follow-up study, but a few could be used for collateral reading. The character and context of the citations should distinguish these types of references. A few of the citations are to historical sources that I found especially interesting, but I have made no attempt to mention historical background systematically or completely.

In addition to the references, the end sections include a detailed index and a glossary of symbols. This glossary lists the pages on which special symbols are defined or otherwise characterized. The literature on mathematical logic has traditionally made use of Greek letters, and several are used in this text. For reference, a list of the Greek alphabet is included in the end sections.

A final remark before we begin in earnest: A number of proper names of individual people are used in various examples given in this text. Except for a few famous historical characters, these names are purely fictitious.

Chapter 1
Sentential Calculus

Logic is concerned with methods and patterns of reasoning, in particular, reasoning that is presented in the form of an argument. For our purposes, an *argument* consists of a set of declarative sentences, the *premises*, from which another sentence, the *conclusion*, is inferred. There are several forms of arguments; two important types are *inductive* and *deductive* arguments. The former are used in everyday reasoning and in the empirical sciences when one infers that a conclusion is probably true on the basis of evidential data. For example, suppose that one identifies crows in terms of their size, shape, cawing sounds, eating habits, etc., except that color is not presupposed. Then one observes thousands of crows in North America and finds that every one that is observed is black. One might infer from this data that it is likely that the sentence 'All crows are black' is true. This inference is an example of an *inductive* inference. This book will not analyze such inferences.

Notice that, in spite of all the observed black crows, the sentence 'All crows are black' may still be false. There may be green crows in South America or mutant pink crows somewhere else. Just because they have not been observed does not guarantee that they do not exist. The supporting evidence used in an inductive inference should make its conclusion probable, but it does not guarantee the truth of this conclusion. On the other hand, a *sound* deductive argument is an argument such that, *if* its premises are all true, *then* its conclusion must also be true. In order to make this idea precise and useful, we must clarify what is meant by terms such as *sentence* and *truth*. It is also useful to study ways to construct sound arguments and (when possible) to test for soundness. To handle complex arguments, it would be helpful if we could program computers to perform these constructions and tests for us. To achieve the desired precision and to develop such computer programs, we need to use *formal languages* in which deductive arguments can be stated. One of the simplest of such languages is the *sentential calculus*. We will examine it first. The last chapter of this book describes a much more powerful formal language, *first-order predicate calculus*, which includes sentential calculus as a sublanguage.

1.1   Syntax of the Sentential Calculus

Suppose that we are given a set of symbols such as 'A', 'b', '*', '+', '34', '$\rightarrow$'. Let us say that a *string* is one of these symbols or else two or more of them written in a particular order. Thus, '+' and 'b' are strings. Also, 'bA' and ' $+ \rightarrow AAA$' are strings. Chapter 3 includes a systematic theory of strings, but for our present purposes, it is adequate to rely on this informal understanding of what counts as a string of characters. In addition, we will consider only finitely long strings, so every string contains a definite number, n, of characters, for some positive integer n.

The *syntax* of a language specifies the symbols it uses and states formation rules that determine how these symbols can be combined into legitimate expressions of the language. In general, an expression is a string of one or more symbols, but only special forms of such strings will be considered legitimate or *well-formed*. Syntax rules alone do not specify any meanings for the expressions of the language; the *semantics* of the language does this. In order to talk or write about a formal language, we use another language, called a *metalanguage*, which is already assumed to be known by the persons talking about the formal language. Our metalanguage will be ordinary English supplemented with some mathematical concepts and symbols. The syntax of the sentential calculus is simple: There are two principal types of symbols, *sentential connectives* and *sentential letters*; in addition, parentheses are used for grouping certain expressions together. The legitimate (or well-formed) expressions are *sentences*, of which the *atomic sentences* are a special kind.

Definition 1-1 The syntax of the sentential calculus ($\mathcal{SC}$) consists of *symbols* and *sentences* as follows:

> *Symbols*
>
>> *parentheses*: (,)
>>
>> *sentential connectives*: $\neg$, $\vee$, $\wedge$, $\rightarrow$, $\leftrightarrow$
>>
>> *sentential letters*: A, B, C, through X, Y, Z, and any of these letters with a subscript that is a positive Arabic numeral.
>
> *Sentences*
>
>> Any sentential letter is an $\mathcal{SC}$ sentence and is called an *atomic sentence*.

Also, any expression (finitely long string of symbols) that is obtainable by use of the following sentential calculus construction rule (SCCR) is a sentence.

SCCR. Let ϕ and ψ stand for $\mathcal{SC}$ sentences. Then expressions of the following forms are $\mathcal{SC}$ sentences:

$$\neg\phi, \ (\phi \vee \psi), \ (\phi \wedge \psi), \ (\phi \rightarrow \psi), \ (\phi \leftrightarrow \psi)$$

Any $\mathcal{SC}$ sentence is either atomic or obtainable by use of this construction rule. A sentence that is not an atomic sentence is a *compound sentence.*

Notice that Definition 1-1 actually introduces several items, including *sentential connective, sentential letter, atomic sentence, sentence,* and *compound sentence.* The commas used in this definition are not part of the sentential calculus language; they are just used for punctuation in the metalanguage to aid in reading expressions. The parentheses are used only to group certain types of expressions, as indicated in the construction rule, SCCR. The definition itself is stated in the metalanguage, and the Greek letters are used as metalinguistic variables that stand for various items in the formal language. The sentential calculus ($\mathcal{SC}$) is often called the *propositional calculus* and is sometimes called the *statement calculus.* Other terminology varies accordingly; for example, sentential letters are often called *propositional letters,* and the sentential connectives may be called *propositional connectives.* Some writers use different notation for the connectives, e.g., $\supset$ is often used in place of $\rightarrow$, and $\equiv$ in place of $\leftrightarrow$. It is usually easy to convert mentally from one set of symbols to another. *For the rest of the chapter, unless explicitly stated otherwise, 'sentence' will mean $\mathcal{SC}$ sentence.* Here are some examples of sentences in this language:

$$K$$

$$\neg K$$

K is a sentence because it is an atomic sentence, and it is an atomic sentence because it is a sentential letter. Let ϕ stand for K. Then $\neg K$ is a sentence because the construction rule states that any expression of the form $\neg\phi$ is a sentence if ϕ is. The definition of *sentence* is an example of a *recursive definition.* Definitions of this kind will be examined in detail in Chapter 3. For present purposes, I will use the term *recursive definition* in the following general way: It is a definition that first defines one or more *base,* or starting, cases (or instances) of some concept; it then defines additional instances of the concept by stating how these additional instances can be constructed or obtained from previously defined instances. In the definition of *sentence,* the base case states that any sentential letter is a sentence. These are called *atomic sentences.* Having defined these as sentences, by the construction rule, SCCR, we obtain more complex sentences, such as $\neg K$. Once $\neg K$ is defined, SCCR can be applied to it to obtain

$$\neg\neg K.$$

A and $\neg A$ are also sentences, so from the construction rule,

$$(\neg A \vee \neg K)$$

is a sentence. It is now easy to verify that the following are sentences:

$$\neg(B \leftrightarrow \neg(K \wedge \neg A)),$$

$$((\neg A \vee \neg K) \to \neg(B \leftrightarrow \neg(K \wedge \neg A))).$$

A blank space, such as the space between 'A' and 'B' in the expression '$A \quad B$', is a character just like A and B. It is made by pushing the spacebar on the keyboard. Because it is a character, a space could be considered to be a symbol of a language. However, a space is *not* a symbol of sentential calculus since it is not included in the specification of the syntax. I will use spaces in writing sentences just to make them easier to read, but according to the syntax rules, these spaces should be ignored.

When reading and speaking, it is convenient to have names for the sentential connectives. For this purpose, one may use the following: $\neg$ is *hoe*, $\wedge$ is *angle-cap*, $\vee$ is *angle-cup*, $\to$ is *right-arrow*, and $\leftrightarrow$ is *doublearrow*. Later I will introduce more meaningful names for these symbols. The sentential connectives are so named because they connect two sentences together to form expressions such as $(A \wedge A)$, $(A \wedge B)$, and $(A \to B)$. An exception is $\neg$, since it operates on one sentence at a time, to form expressions such as $\neg S$, $\neg\neg S$, and $\neg(A \wedge B)$. Although $\neg$ does not connect two sentences together, by convention it is also considered to be a sentential connective. Notice that when connecting two sentences with $\wedge$, $\vee$, $\to$, or $\leftrightarrow$, the result is enclosed in parentheses. If this were not done, the construction rule would not enable us to distinguish between sentences such as $(A \to (B \to C))$ and $((A \to B) \to C)$. Adding $\neg$ cannot lead to such ambiguities, so we do not close up with parentheses when $\neg$ is used. Thus, from A we obtain $\neg A$. Of course, $\neg$ can be used in front of a compound sentence that is already enclosed within parentheses, e.g., from

$$(A \to \neg B),$$

we can obtain

$$\neg(A \to \neg B).$$

Every sentence is either atomic or compound. If the former, it is a single letter, so it contains no (sentential) connectives. If it is compound, then it

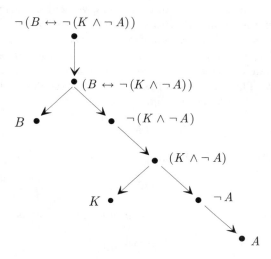

Figure 1.1. *Tree Structure of a Sentence*

contains one or more connectives and is built up according to the recursive construction rule, SCCR. When it is built up, there is always one last construction operation that is performed. For example, in building $\neg A$, the last operation was adding $\neg$ to the A. In building $(S \leftrightarrow T)$, the last operation was putting the $\leftrightarrow$ between S and T and closing up with parentheses. The connective added in this last operation is called the *main connective* of the sentence. When reading a sentence, it is usually convenient to locate the main connective, and then mentally decompose it by reversing the steps of the recursive operations that were used in constructing it. This requires that one repeatedly look for main connectives as one is decomposing the sentence until no more connectives remain. This process is illustrated in Figure 1.1. The main connective of $\neg(B \leftrightarrow \neg(K \wedge \neg A))$ is $\neg$. If we strip that off, we obtain $(B \leftrightarrow \neg(K \wedge \neg A))$ with the main connective $\leftrightarrow$. If we decompose this sentence at this main connective, we obtain the subsentences B and $\neg(K \wedge \neg A)$. B is atomic, so it cannot be decomposed. $\neg(K \wedge \neg A)$ is compound and is decomposed further as shown in Figure 1.1. Eventually, this process of decomposition halts when only atomic sentences remain.

The diagram in Figure 1.1 has the shape of an upside-down tree. The small black circles are called *nodes*. In this figure, the nodes represent sentences. The node at the top is called the *root node*. The terminal nodes (corresponding to atomic sentences) are sometimes called *leaves* or *leaf nodes*. Each node, except for the leaves, branches into one or two lower nodes; the leaves do not have branches. Diagrams in which nodes are connected by lines are called *graphs*; a tree is a special kind of graph. In later chapters, we will use graphs for several purposes. When first learning $\mathcal{SC}$ syntax, it is helpful to draw the tree structures of several sentences. *If an expression does not have a tree structure that can be built up according to the construction rule, then it is not a sentence.* If ϕ is an $\mathcal{SC}$ sentence, then any sentence represented by a node on its tree diagram is a *subsentence* of ϕ. This includes ϕ itself. A subsentence of ϕ that is different from ϕ is a *proper subsentence*. In Figure 1.1,

$$\neg(B \leftrightarrow \neg(K \wedge \neg A))$$

is a subsentence of itself. In addition,

$$(B \leftrightarrow \neg(K \wedge \neg A)),$$

is a proper subsentence. B is also a proper subsentence, as is

$$(K \wedge \neg A),$$

and so on, down all branches of the tree. Although it is not proved here, it should be obvious that any $\mathcal{SC}$ sentence corresponds to a unique (only one) tree. Hence, two different sentences, e.g.,

$$(A \to (B \to C)) \text{ and } ((A \to B) \to C),$$

will be decomposed in two different ways. This results from the use of parentheses for grouping subsentences, and this uniqueness is sometimes called *unique readability*. If the parentheses were omitted in these sentences, we would just have $A \to B \to C$, and we would not know how to *read* this expression. In fact, this expression is not even a sentence because it cannot be obtained by using the SCCR.

Exercises

E 1-1 According to precise and strict application of the syntax rules, which of the following expressions are $\mathcal{SC}$ sentences, and which are not?

1. $((P \vee \neg Q) \to S)$

2. $\neg\neg\neg A$

3. $(\neg A)$

4. $\neg(\neg A \wedge B)$

5. $((X \vee Y) \leftrightarrow \neg(A \leftrightarrow A) \wedge B))$

6. $(\neg A \leftrightarrow \neg(A \wedge B))$

7. $(\neg(P\neg Q))$

8. $(\neg(\neg P \leftrightarrow (A \vee B)) \rightarrow \neg(A \vee B))$

9. $(S_1 \rightarrow R_{13})$

10. $(\phi \rightarrow \psi)$

11. $((Q \rightarrow R) \rightarrow P \vee Q)$

12. $(G_2 \leftrightarrow (\neg G_2 \vee G_2))$

E 1-2 Sketch the tree structure of each of the expressions in the previous exercise that *is* a sentence. Also, for each of these sentences, list all of its proper subsentences.

1.2   Correspondence to Natural Languages

For present purposes, consider a natural language to be one that people use for everyday communication, such as English, Spanish, Arabic, Hindi, Chinese, Navajo, Swahili, American Sign Language, etc. Natural languages permit the formulation of atomic sentences, for instance, the English sentence, 'Bob is tall.' When considering the correspondence between natural language and sentential calculus, I will say that a *natural language atomic sentence* is a natural language sentence that cannot be decomposed into proper subsentences combined only by the use of natural language sentential connectives. (This characterization can be replaced by a more precise and positive one by reference to the predicate calculus, the subject of Chapter 4.) The sentence 'John is tall and Mary smiles' is a compound (nonatomic) sentence because it has the two subsentences 'John is tall' and 'Mary smiles' combined by 'and'. The sentential calculus is much simpler than any natural language, for it enables us to represent the structure of only certain types of compound sentences in terms of atomic sentences. It does not provide any way to represent the internal structure (e.g., grammatical subject and predicate) or specific meaning of atomic sentences. Atomic sentences in a natural language correspond to single $\mathcal{SC}$ letters. Section 1.3 presents the semantics of the sentential calculus; it will be seen that $\mathcal{SC}$ sentences can have the values of true and false. To help motivate this semantics, it is useful to consider a rough correspondence between the sentential connectives and some aspects of ordinary language, specifically English.

There is no unique and precise correspondence between the $\mathcal{SC}$ connectives and any English words. The connectives do correspond to some English words, when these words are used to relate *complete sentences* in certain ways. This emphasis on complete sentences is important. Consider these two examples using the English word 'and': (1) 'The day is warm and the bugs are buzzing.'

(2) 'Wilma's zebra is black and white.' Sentence (1) is a compound English sentence with the two complete subsentences 'The day is warm' and 'The bugs are buzzing.' Sentence (2) is not a compound sentence, nor does it mean the same as one; in particular, it does not mean the same as 'Wilma's zebra is black and Wilma's zebra is white.' Thus, at least until further notice, we will only consider some correspondences between the SC connectives and a few English words, when those words are used to connect English sentences. The correspondences that we will use are more likely to hold when the English sentences under consideration are used in mathematical or certain other technical contexts; and these are the types of uses that are of primary concern in this book. I call these correspondences *Conventional Sentential Calculus Translations* because they show how we will translate certain English words into sentential calculus connectives, and vice versa. These translations are given in Table 1.1, in which A and B are assumed to denote declarative sentences in English.

English	Sentential Calculus
not A	$\neg A$
it is not the case that A	$\neg A$
A and B	$(A \wedge B)$
A & B	$(A \wedge B)$
A, but B	$(A \wedge B)$
although A, B	$(A \wedge B)$
A or B	$(A \vee B)$
either A or B	$(A \vee B)$
A and/or B	$(A \vee B)$
if A, then B	$(A \rightarrow B)$
A only if B	$(A \rightarrow B)$
B is necessary for A	$(A \rightarrow B)$
A is sufficient for B	$(A \rightarrow B)$
A if B	$(B \rightarrow A)$
A if and only if B	$(A \leftrightarrow B)$
A iff B	$(A \leftrightarrow B)$
A is necessary and sufficient for B	$(A \leftrightarrow B)$

Table 1.1. *Conventional SC Translations*

Suppose A denotes the English sentence, 'The book is red.' This sentence can be negated or denied in at least two ways: 'The book is not red' and 'It is not the case that the book is red.' In the sentential calculus, both English sentences would correspond to $\neg A$. Note that $\neg A$ is true just in case A is not true (false).

Now let B denote 'The sky is blue.' We might then write in English, 'The book is red and the sky is blue.' This corresponds to $(A \wedge B)$ in the $\mathcal{SC}$ language. If the English word 'and' connects two English sentences, then this use of 'and' normally corresponds to the symbol $\wedge$. Other English sentences sometimes correspond to the $\wedge$. For example, an English sentence of the form, 'ϕ, but χ' often behaves like 'ϕ and χ' in the sense that both of these compound sentences are true just in case ϕ and χ are both true. Using 'but' instead of 'and' is sometimes useful to emphasize a contrast between the two subsentences. For example, 'She fell off the Eiffel Tower, but she lived.' In this case the contrast is surprise, since we do not normally expect one to survive a fall from the Eiffel Tower. The word 'although' can be used in a similar manner, as in 'Although she fell off the Eiffel Tower, she lived.' and 'She lived, although she fell off the Eiffel Tower.' Table 1.1 is not intended to be an exhaustive list of correlations between natural language and the sentential calculus.

Now consider 'The book is red or the sky is blue.'; this English sentence corresponds to $(A \vee B)$. It is true in exactly those situations in which one or the other of A, B is true, or in which both A and B are true. Thus, it is false only when both A and B are false. This is sometimes expressed as 'and/or' and is called *inclusive or* to emphasize that it is true when both A and B are true, as well as when only one of them is. By almost universal convention, logicians interpret 'either A or B' to mean exactly the same as 'A or B', in which 'or' is inclusive. The table shows that we also follow this convention. *In this book, both 'or' and 'either-or' mean inclusive 'or'.*

A so-called *exclusive or*, sometimes written as *xor*, has the following property: 'A xor B' is true if A is true or if B is true, but not when both A and B are true and not when both of them are false. If we want to express an 'xor', we will always do so explicitly by using 'and', 'not', and 'or' together. Thus, we will write 'A xor B' as $((A \vee B) \wedge \neg(A \wedge B))$. This can be expressed in English as 'A or B, and it is not the case that A and B'. A simpler statement is 'A or B, but not both A and B'. In this case the word 'but' again functions like 'and'. The sentence just considered is often stated even more briefly as 'A or B, but not both'. This example demonstrates that "translating" from natural language into $\mathcal{SC}$ requires a careful interpretation of exactly what the natural language sentence means, for it will often be expressed in an abbreviated or colloquial style.

Suppose that a large company maintains a computer database of information on its employees, including their salaries, work performance, etc. The company has several different rules about giving certain employees bonuses at the end of the year. Jane is a plant manager, and there are two ways a plant manager can earn a bonus: Cut costs while maintaining plant output, or increase plant output while maintaining costs. Let C denote 'Jane cuts costs (while maintaining output)' and B denote 'Jane gets a bonus.' Then the following will be true:

$$\text{If Jane cuts costs, then Jane gets a bonus.} \qquad (1.1)$$

This can be translated or symbolized as

$$(C \to B). \qquad (1.2)$$

Now consider the following possibilities. Suppose that Jane cuts costs, so C is true. Then she gets a bonus, so B is true. In other words, if the if-part of a true if-then sentence is true, then the then-part is also true. The if-part is called the *antecedent* and the then-part the *consequent*. As Table 1.1 shows, the right-arrow connective between an antecedent and a consequent corresponds to an if-then ordinary language sentence.

Now consider

$$\text{Jane cuts costs only if Jane gets a bonus.} \qquad (1.3)$$

This is a peculiar-sounding sentence in English. Compare it with 'The paper will burn only if it is dry.' In this case, background information suggests that what is meant is that being dry is a causally necessary condition for burning. In other words, in order for some possible cause (say, a spark) to cause the paper to burn, certain other conditions must also be true, for instance, that the paper is dry, that there is adequate oxygen, etc. In English, 'only if' is often used to express such causally necessary conditions. One reason why (1.3) sounds peculiar is because getting a bonus will occur after Jane cuts costs, so it is not a causally necessary condition. Yet, it still is a *necessary condition* in terms of the truth values of the component sentences. Using the designated letters, sentence (1.3) means this: In order for C to be true, it is necessary that B be true. We ignore the temporal relations between C and B and consider only their truth values. Thus, C is true *only if* B is also true, which is exactly what (1.3) states. But notice that *if* C is true, *then* B is also true; (1.3) states the same truth relationship between C and B that is stated by (1.1). Thus, at least for our purposes, both (1.1) and (1.3) have the same meaning and are symbolized by (1.2). This sameness of meaning is indicated in Table 1.1, where several forms of English sentences are related to $(A \to B)$.

In English, another way of stating that B is true when C is true is to say that B is necessary for C. This is also indicated in the table, which shows yet another way of saying the same thing: C is sufficient for B. This means that the truth of C is enough to assure the truth of B. In other words, if it is true that Jane cuts costs, then she gets a bonus. Slightly different locutions are often used, especially by mathematicians. When $(C \to B)$ is true, one often says that C is a sufficient condition for B. One can also say that B is a necessary condition for C. The choice among all of these different expressions of the same fact often depends on which feature one wants to emphasize. This should become clear from the use of these expressions later in this book.

In English, instead of saying 'If Jane cuts costs, then Jane gets a bonus', we may also say, 'Jane gets a bonus if Jane cuts costs.' The form 'B if C' means the same as 'If C, then B.' An if-then sentence is called a *conditional*, so we can state conditionals in English by using the form 'If Q, then P' or the form 'P if Q'. The latter form is often used for stating conditionals in programming languages that are used for computer construction of logical arguments. A widely used language of this kind is Prolog, which is used for a kind of "programming in logic."

Unfortunately, people often confuse the form 'P if Q' with the form 'P only if Q'. The former states that Q is sufficient for P; the latter states that Q is a necessary condition for P. These are different relationships. In our example, cutting costs is a sufficient condition for Jane to get a bonus. Yet, cutting costs is not a necessary condition for her to get a bonus. She could also get a bonus if she increases plant output while maintaining costs. If A is some statement, there may be several, or even many, different sufficient conditions for A. Also, A may have many different necessary conditions. Try to think of all the necessary conditions that must be satisfied in order for an automobile or a computer or a human body to operate in a normal manner.

Although sufficient conditions are usually not necessary, and vice versa, in special cases a condition may be both. Suppose that B is both necessary and sufficient for A. Since B is necessary, 'If A, then B' is true; since it is sufficient, 'If B, then A' is true. The latter can be rewritten as 'A if B' and the former as 'A only if B.' The two together amount to 'A if and only if B.' An example from plane geometry is this: Triangle ABC is equilateral if and only if it is equiangular. In English, I will usually use 'iff' in place of 'if and only if'; this is a standard abbreviation and is listed in Table 1.1. In the sentential calculus, these expressions correspond to the connective $\leftrightarrow$.

Table 1.1 is not an exhaustive list of English connectives, nor is the preceding discussion a thorough treatment of the ones in the table. Some logic texts devote much space to techniques and problems of translating ordinary language sentences into formal languages. The interested reader may consult Mates (1972), Kalish, Montague, and Mar (1980), Bonevac (1987), and Sainsbury (1991). Because we are mainly concerned with technical applications of logic, most of the English connectives that we will need to translate (or symbolize) are given in Table 1.1. Some additional ones will be briefly discussed later when they are needed. The next section presents the semantics of sentential calculus; it will help to clarify the relations between ordinary language and this formal language.

Previously, the connectives were given descriptive names like 'hoe', 'angle-cap', 'right-arrow', etc. Instead of these descriptive names, many people name the connectives by their conventional English correlates. When reading them aloud, one often uses the following: 'not' or 'it is not the case that' for ¬, 'and' for ∧, 'or' for ∨, 'if ⋯ , then —' or ' ⋯ only if —' for →, and 'if and only if' for ↔. This is a casual way of speaking. It is all right, provided that one remembers that the 𝒮𝒞 connectives do not literally mean the same as these English correlates. We will continue to use the English words in English text with their usual meanings. One should not confuse these usual meanings in informal English with the meanings of the connectives in a formal language. The formal meanings of the connectives are defined in the next section. These meanings are exactly what they are defined to be, no more and no less.

Exercises

Exercises for which answers are provided at the back of the book are marked with "Answer Provided."

E 1-3 Each of the following problems is presented as a simple argument in English. The conclusion of each argument is indicated by 'Therefore'. The arguments may be sound or unsound; that is not of concern here. The present task is to symbolize (translate) each argument into sentential calculus. Use 𝒮𝒞 letters for each atomic sentence in English, using the same 𝒮𝒞 letter *wherever* that English atomic sentence occurs in one of the *arguments*, and different 𝒮𝒞 letters for different English atomic sentences. For example, in the first argument, you could use M for 'the machine runs' and then use $¬M$ for 'the machine does not run' (i.e., 'it is not the case that the machine runs'). Use a different letter, say, S_1 for 'Switch1 is on'. For each 𝒮𝒞 letter you use, indicate the English sentence which it denotes. Write the 𝒮𝒞 symbolizations on separate, numbered lines that correspond to the numbered English lines. Some of the English atomic sentences may appear in different formats in various compound sentences because of stylistic or grammatical reasons. Be careful to parse the English sentences in the most plausible way, given the punctuation used and commonsense interpretations of the meanings. Use the Conventional 𝒮𝒞 Translations in Table 1.1 to help you interpret the English connectives.

1. (Answer Provided)

    (a) The machine runs only if both Switch1 is on and Switch2 is on.

    (b) For the machine to run, it is necessary that the power cord be plugged in.

    (c) If a fuse is blown, then the machine does not run.

    (d) Therefore: If the machine does not run and the power cord is plugged in and Switch1 is on and Switch2 is on, then the fuse is replaced.

2. (a) Part A has failed or part B has failed.

(b) If the gadget passes test number one, then it is not the case that part B has failed.

(c) If the battery is dead, then the green indicator light is not on.

(d) Therefore: The green indicator light is on only if the gadget does not pass test number one.

3. (a) Bob gets a raise or Bob gets a bonus or Bob gets nothing new.

(b) If Bob gets a raise, then it is not the case either that Bob gets a bonus or that he gets nothing new.

(c) Bob gets nothing new if and only if profits do not rise.

(d) Therefore: If the cost of widget production remains constant, then: If the demand for widgets rises, then profits rise and Bob gets a raise.

4. (a) In order for the patient to live, it is necessary that the doctor perform the surgery.

(b) The patient lives iff he does not die.

(c) The doctor performs the surgery iff there is not a power failure.

(d) There is a power failure if the nuclear plant melts down.

(e) Therefore: The nuclear plant melts down only if the patient dies.

5. (a) The red light was on or the green light was on.

(b) If the green light was on, he took path A.

(c) If the red light was on, he took path B.

(d) If he took path A, he encountered a dragon.

(e) If he took path B, he encountered two doors.

(f) If he encountered two doors, then he opened the left door or he opened the right door.

(g) If he encountered a dragon, then he lost the game.

(h) If he opened the left door, then he lost the game.

(i) He did not lose the game.

(j) Therefore: The red light was on and he opened the right door.

E 1-4 Make a list of all English atomic sentences in the following sentences, and denote each English atomic sentence by an $\mathcal{SC}$ letter. Use different $\mathcal{SC}$ letters for different English atomic sentences. For example, you could use O for 'the ferry operates', and then use $\neg O$ for 'the ferry does not operate' (i.e., 'it is not the case that the ferry operates'). Use a different letter, say, R for 'there is rain'. Symbolize each sentence in $\mathcal{SC}$, and write your symbolizations on separate, numbered lines that correspond to the numbered English lines. Some of the English atomic sentences may appear in different formats in various compound sentences because of stylistic or grammatical reasons. Be careful to parse the English sentences in the most plausible way, given the punctuation used and commonsense interpretations of the meanings. Use the Conventional *SC* Translations in Table 1.1 to help you interpret the English sentential connectives. Do not be concerned about the truth values of the sentences, or whether they are coherent together.

1. The ferry operates only if there is not rain.

2. She crosses the river on the bridge if the ferry does not operate and there is no lightning.

3. If there is lightning, then the ferry does not operate, and she does not cross the river on the bridge.

4. If she crosses the river on the bridge, it is necessary that there is no lightning.

5. If the ferry does not operate, then she does not cross the river on the bridge only if she goes through the tunnel.

6. She goes through the tunnel only if she is not claustrophobic.

7. Although the ferry operates and there is no lightning, she goes through the tunnel.

8. It is not the case that she crosses the river on the bridge only if the ferry does not operate.

9. She goes through the tunnel, or she crosses the river on the bridge if and only if it is not the case that both there is rain and she is not claustrophobic.

1.3   Semantics of the Sentential Calculus

This section will show how to give a precise interpretation to sentential calculus and how the truth value of a sentence depends on such an interpretation. First, however, it will be useful to define some additional syntactical terminology.

Definition 1-2 Let ϕ and ψ be any *SC* sentences. Then:

A sentence of the form $\neg\phi$ is a *negation*.

A sentence of the form $(\phi \lor \psi)$ is a *disjunction*, and ϕ and ψ are its *disjuncts*.

A sentence of the form $(\phi \wedge \psi)$ is a *conjunction*, and ϕ and ψ are its *conjuncts*.

A sentence of the form $(\phi \rightarrow \psi)$ is a *conditional*, with *antecedent* ϕ and *consequent* ψ.

A sentence of the form $(\phi \leftrightarrow \psi)$ is a *biconditional*, with ϕ and ψ as *components*.

The *converse* of the conditional $(\phi \rightarrow \psi)$ is the conditional $(\psi \rightarrow \phi)$.

The *contrapositive* of the conditional $(\phi \rightarrow \psi)$ is the conditional

$$(\neg\psi \rightarrow \neg\phi).$$

This definition describes five forms of compound sentences, and each of these forms is determined by the main connective of the sentence. By comparing Definition 1-2 with the definition of *sentence* in Definition 1-1, one can see that any sentence is either atomic, or a negation, or a disjunction, or a conjunction, or a conditional, or a biconditional. According to the construction rule, SCCR, these are all of the possible forms of sentences in our formulation of the sentential calculus. We will occasionally need to consider each of these forms in a systematic manner.

In order to define the semantical concepts, it is necessary to use sets and functions. In this chapter, we will work with an intuitive understanding of *set*, *set union* ($\cup$), and *set intersection* ($\cap$) as was discussed in the Introduction. Also, recall the discussion of *functions* in the Introduction. It was pointed out that a function may apply to objects other than numbers. This was illustrated by a function that maps a set of people into a set of two boldface characters. We will now consider functions that map sentential letters into *truth values*.

Definition 1-3 An *interpretation* of the sentential calculus is a function **I** that assigns to each sentential letter the value **T** (for *true*) or **F** (for *false*). (Sometimes other binary values such as 1 and 0 are used.)

Here are the first few values of two different interpretations, $\mathbf{I}_1$ and $\mathbf{I}_2$:

$$\mathbf{I_1}(A) = \mathbf{T}, \ \ \mathbf{I_1}(B) = \mathbf{T}, \ \ \mathbf{I_1}(C) = \mathbf{F}, \ \ \cdots$$

$$\mathbf{I_2}(A) = \mathbf{T}, \ \ \mathbf{I_2}(B) = \mathbf{F}, \ \ \mathbf{I_2}(C) = \mathbf{F}, \ \ \cdots$$

Of course, $\mathbf{I}_1$ has a **T** or **F** value for each sentential letter, so a complete specification of $\mathbf{I}_1$ would require consideration of an infinite number of sentential letters. The same holds for $\mathbf{I}_2$ and any other interpretation. However, we can see that $\mathbf{I}_1 \neq \mathbf{I}_2$ since they specify different values for B. If **I** and **I'** are any two interpretations, they are identical if and only if they assign exactly the same truth value to every sentential letter. This is a special case of a more general principle: If f and g are two functions defined on a set Γ, then $f = g$ iff, for every element x in Γ, $f(x) = g(x)$. This is a familiar principle in the case of typical numerical functions, but it also applies to any kind of function, including interpretations.

There is an infinite set of different sentential calculus interpretations determined by the infinite number of ways of assigning **T** or **F** to the infinite set of sentential letters. When referring to all possible interpretations (which we will often do), we will be referring to this infinite set of functions. Completely describing any particular interpretation requires specifying values for the infinite set of letters. In practice, we usually will work only with a finite subset of these letters, and it will be adequate to specify the values of these alone, since the values of letters that are not used in an application problem will not affect the work on this problem.

Suppose that one thinks of a sentential letter as denoting a particular declarative sentence, such as 'Jane cuts costs.' Then this sentence is either true or false in some set of circumstances. The sentential calculus is not a sufficiently powerful language for representing the meaning of an atomic sentence like 'Jane cuts costs' but it can represent the truth value of such a sentence. That is exactly what a sentential calculus interpretation does—an interpretation specifies a truth value for each sentential letter. Thus, we can symbolize 'Jane cuts costs' with a sentential letter and give this letter the same truth value as the original English sentence. The atomic sentences are given truth values (**T** or **F**) by an interpretation. We now want a way to determine the truth values of the compound sentences, so let ϕ be any $\mathcal{SC}$ sentence, and **I** be any $\mathcal{SC}$ interpretation. We now define a new *valuation function*, $\mathbf{V_I}(\phi)$, which returns the truth value of ϕ, given the interpretation **I**. $\mathbf{V_I}$ maps each $\mathcal{SC}$ sentence into either **T** or **F**. If ϕ is atomic, then its truth value will be that given directly by **I**. If ϕ is compound, we want some way to compute its truth value in terms of the truth values of the atomic sentences in it. The computation will proceed by breaking the sentence into smaller and smaller parts according to its tree structure. Because each sentence corresponds to a unique tree structure, i.e., we have unique readability, there will be a unique way of computing the truth value of any given sentence. Because there are different forms of sentences, we must consider all of these forms. The definition has six parts that correspond to the six forms of sentences, including the atomic sentences.

Definition 1-4 Let ψ, χ be any $\mathcal{SC}$ sentences and **I** be any $\mathcal{SC}$ interpretation. Then the *valuation function*, $\mathbf{V_I}$, is specified by the following:

  1. $\mathbf{V_I}(\psi) = \mathbf{I}(\psi)$ if ψ is atomic (an $\mathcal{SC}$ letter).

2. $\mathbf{V_I}(\neg\psi) = \begin{cases} \mathbf{F} & \text{if } \mathbf{V_I}(\psi) = \mathbf{T}, \\ \mathbf{T} & \text{if } \mathbf{V_I}(\psi) = \mathbf{F}. \end{cases}$

3. $\mathbf{V_I}(\psi \vee \chi) = \begin{cases} \mathbf{F} & \text{if } \mathbf{V_I}(\psi) = \mathbf{V_I}(\chi) = \mathbf{F}, \\ \mathbf{T} & \text{otherwise.} \end{cases}$

4. $\mathbf{V_I}(\psi \wedge \chi) = \begin{cases} \mathbf{T} & \text{if } \mathbf{V_I}(\psi) = \mathbf{V_I}(\chi) = \mathbf{T}, \\ \mathbf{F} & \text{otherwise.} \end{cases}$

5. $\mathbf{V_I}(\psi \rightarrow \chi) = \begin{cases} \mathbf{F} & \text{if } \mathbf{V_I}(\psi) = \mathbf{T} \;\&\; \mathbf{V_I}(\chi) = \mathbf{F}, \\ \mathbf{T} & \text{otherwise.} \end{cases}$

6. $\mathbf{V_I}(\psi \leftrightarrow \chi) = \begin{cases} \mathbf{T} & \text{if } \mathbf{V_I}(\psi) = \mathbf{V_I}(\chi), \\ \mathbf{F} & \text{otherwise.} \end{cases}$

The valuation function, $\mathbf{V_I}$, gives a truth value for each sentence, relative to a given interpretation $\mathbf{I}$. This is another recursive definition, in the general sense previously described: It has base cases, and it states how new, nonbase cases are determined by previously defined cases. Of course, the exact form of this definition is rather different from that of Definition 1-1. That definition recursively constructed an infinite set of $\mathcal{SC}$ sentences, building ever larger sentences by using the SCCR. The definition of $\mathbf{V_I}$ specifies a function that is defined on this infinite set of sentences. The base case of this definition covers the atomic sentences. For any compound sentence, the value of $\mathbf{V_I}$ is determined by the value or values of the one or two subsentences that result when it is decomposed at its main connective. This determination is given by the parts of the definition numbered two through six. Reading all of these details is tedious. It is customary to summarize the information contained in parts two through six in the form of *truth tables* for negation, disjunction, etc., presented here. The first of these, for negation, Table 1.2, requires no discussion.

The truth table for disjunction, Table 1.3, is also obvious when one recalls that $\vee$ corresponds intuitively to inclusive 'or'. A disjunction of ψ with χ is true in any of these three cases: ψ is true, or χ is true, or both ψ is true and χ is true; otherwise, it is false, which occurs only if ψ is false and χ is false. The truth table lists all four possible combinations of truth values that ψ and χ could have and exhibits the truth value of the disjunction for each of these combinations. The table summarizes the same information that is given in part

ψ	$\neg\psi$
T	**F**
F	**T**

Table 1.2. *Negation*

ψ	χ	$(\psi \lor \chi)$
T	**T**	**T**
T	**F**	**T**
F	**T**	**T**
F	**F**	**F**

Table 1.3. *Disjunction*

three of the preceding definition of $\mathbf{V_I}$. Because we have a recursive definition, it is assumed that, whatever ψ and χ may be, their truth values are determined before that of their disjunction. These prior values are then used to compute the truth value of the disjunction.

A conjunction is specified to be true if both conjuncts are true, and false otherwise; see Table 1.4. This corresponds closely to the ordinary meaning of 'and'.

ψ	χ	$(\psi \land \chi)$
T	**T**	**T**
T	**F**	**F**
F	**T**	**F**
F	**F**	**F**

Table 1.4. *Conjunction*

Notice Table 1.5 for the conditional; it clearly exhibits the fact that $\mathbf{V_I}$ of a conditional is **T** when the antecedent is **F**. This troubles some people. I believe this is largely because they expect to find a strong correlation between $\rightarrow$ and the use of 'if-then' in natural language. It has already been pointed out that 'if-then' in natural language is often used with a more complex meaning than can be expressed in sentential calculus, so we should not be surprised to find that a strong correlation does not exist. The conventional translations are useful conventions for some purposes, but they have definite limitations. Yet

ψ	χ	$(\psi \to \chi)$
T	**T**	**T**
T	**F**	**F**
F	**T**	**T**
F	**F**	**T**

Table 1.5. *Conditional*

the truth values for $\to$ are not as strange as they may at first appear. Suppose that a sentence of the form $(\psi \to \chi)$ is true. As long as we consider only truth values and not extraneous factors such as causal connections, the truth values for this sentence correlate well with 'ψ only if χ', 'ψ is sufficient for χ', and 'χ is necessary for ψ'. Intuitively, we want χ to be true in situations where ψ is true. This is what the first row of the table for the conditional expresses. On the other hand, if the antecedent is true and the consequent false, we want the conditional to be false. This is what the second row says. Finally, suppose again that the conditional is true, and we also have a false antecedent. Because ψ is false, nothing is determined about the consequent χ; it could be either **T** or **F**. The last two rows of the table represent these situations. Although these considerations should give the truth conditions for conditionals more plausibility, the real test of the adequacy of the definition of **V$_I$** is in its use. The six-part definition for truth valuations works well, but further developments are required to demonstrate that this is true.

Table 1.6 is for the biconditional. One way to understand this table is by treating $(\psi \leftrightarrow \chi)$ as having the same truth values as $((\psi \to \chi) \land (\chi \to \psi))$. The latter is true when both conjuncts are true. This happens if both ψ and χ are true and also if both are false. On the other hand, if one of ψ or χ is true and the other false, then one of the conjuncts is false, so the entire conjunction is false.

ψ	χ	$(\psi \leftrightarrow \chi)$	$((\psi \to \chi) \land (\chi \to \psi))$
T	**T**	**T**	**T**
T	**F**	**F**	**F**
F	**T**	**F**	**F**
F	**F**	**T**	**T**

Table 1.6. *Biconditional*

The discussion of $((\psi \to \chi) \wedge (\chi \to \psi))$ is an example of computing a truth table for a compound sentence. Given the definition of the valuation function, one can compute the truth value that any compound sentence has for any interpretation. When this is done for *all* interpretations, it is customary to put the results in tabular form. This is called *making a truth table for the sentence*. An example, using the sentence in Figure 1.1, is presented next.

Example.   The truth table for $\neg(B \leftrightarrow \neg(K \wedge \neg A))$ is Table 1.7.

A	B	K	$\neg(B \leftrightarrow \neg(K \wedge \neg A))$
T	**T**	**T**	**F**
T	**T**	**F**	**F**
T	**F**	**T**	**T**
T	**F**	**F**	**T**
F	**T**	**T**	**T**
F	**T**	**F**	**F**
F	**F**	**T**	**F**
F	**F**	**F**	**T**

Table 1.7.

The table does not show the intermediate calculations, but it is easy to obtain the results shown in this table. For instance, by substituting the values A: **T**, B: **T**, and K: **T**, the rightmost value in the first row is computed as follows:

$$
\begin{aligned}
\mathbf{V_I}(\neg(B \leftrightarrow \neg(K \wedge \neg A))) &= \neg\,(\mathbf{T} \leftrightarrow \neg(\mathbf{T} \wedge \neg \mathbf{T})) \\
&= \neg\,(\mathbf{T} \leftrightarrow \neg(\mathbf{T} \wedge \mathbf{F})) \\
&= \neg\,(\mathbf{T} \leftrightarrow \neg \mathbf{F}) \\
&= \neg(\mathbf{T} \leftrightarrow \mathbf{T}) \\
&= \neg\,\mathbf{T} \\
&= \mathbf{F}. \quad ■
\end{aligned}
$$

We have not defined expressions like $(\mathbf{T} \wedge \mathbf{F})$, but they can be considered abbreviations for $\mathbf{V_I}(\mathbf{T} \wedge \mathbf{F})$, and $\mathbf{V_I}(\mathbf{T} \wedge \mathbf{F})$ can be considered the same as $\mathbf{V_I}(A \wedge B)$ when A is true and B false. Notice that each row of the truth table corresponds to an infinite set of $\mathcal{SC}$ interpretations that result from various possible assignments of **T** or **F** to each of the remaining sentential letters not in the table. Considering only the three letters (A, B, and K) in the sentence, we can assign A either **T** or **F**. With each of these two assignments, we can assign B either **T** or **F**, giving four possible joint assignments to A and B. With each of these four combinations, we can assign K either **T** or **F**, yielding a total of eight joint assignments to the three letters. Thus, the truth table has eight rows. These eight distinct assignments can be easily constructed using the systematic arrangement of **T**s and **F**s shown in the table. For a sentence with exactly n letters, the complete truth table has 2^n rows. Computing all rows of a large truth table requires an enormous number of operations.

Given any interpretation **I** and any sentence ϕ, Definition 1-4 yields a unique value for $\mathbf{V_I}(\phi)$. If this were not the case, we would not be able to assign a truth value to ϕ, and $\mathbf{V_I}$ would not be a function. I will not prove that it is a function, but the following considerations indicate why it is. If ϕ is atomic, then $\mathbf{V_I}(\phi) = \mathbf{I}(\phi)$, and the latter has a unique **T** or **F** value from Definition 1-3. If ϕ is compound, then it has a tree structure with leaves that are atomic sentences. The values of these atomic sentences are determined by **I**. Using these values and the truth tables (which summarize parts of Definition 1-4), one can successively compute the values of each sentence in the tree, ending with ϕ. Since each sentence ϕ has a unique tree structure, there will be one and only one value computed for $\mathbf{V_I}(\phi)$. The preceding example illustrates how this is done. Moreover, *any* function $\mathbf{V_I}$ satisfying Definition 1-4 will have its values computed in the same way, so there is a unique $\mathbf{V_I}$ for any given **I**. Using the kind of mathematical methods introduced in Chapter 3, it is possible to refine these sketchy considerations into a rigorous proof that $\mathbf{V_I}$ is indeed a unique function.

Now that we can compute truth tables, let us return briefly to the truth table, Table 1.5, for the conditional. The definition of $\mathbf{V_I}$ for each form of compound sentence depends only on the truth values of its subsentences. The valuation function does not recognize causal connections or any other special connection. Thus, $\mathbf{V_I}$ of a conditional must be defined in such a way that it returns **T** or **F** for each possible assignment of truth values to the antecedent and the consequent of the conditional. Let $(\psi \rightarrow \chi)$ be an arbitrary conditional. How should one define (the values of) $\mathbf{V_I}(\ (\psi \rightarrow \chi)\ )$ for each of the possible truth value assignments to ψ and to χ? Again think of $(\psi \rightarrow \chi)$ as corresponding to 'ψ only if χ', where we consider only the truth values of the subsentences. If 'ψ only if χ' is true, then the truth of ψ guarantees the truth of χ. Thus, it should not be the case that ψ is true and χ is false. This last condition corresponds in the sentential calculus to $\neg(\psi \wedge \neg\chi)$, and we can compute the truth table of this sentence using just $\neg$ and $\wedge$, as shown in Table 1.8. This truth table has exactly the same rows as the one that was used previously to define $\mathbf{V_I}$ for the conditional. Thus, what can be expressed by $\rightarrow$ can also be expressed in terms of $\wedge$ and $\neg$. It can be seen that $\rightarrow$ has a special meaning that is different from the meanings of many natural language conditionals. Yet

ψ	χ	$\neg(\psi \wedge \neg\chi)$
T	**T**	**T**
T	**F**	**F**
F	**T**	**T**
F	**F**	**T**

Table 1.8. *Truth Table for* $\neg(\psi \wedge \neg\chi)$

it is a *clear* meaning; an $\mathcal{SC}$ conditional is false when it has a true antecedent and a false consequent, otherwise, it is true.

Before moving on to the next section, we should reflect on the kinds of definitions that are being introduced. Notice the difference between the recursive definition of an $\mathcal{SC}$ sentence and the recursive definition of $\mathbf{V_I}$. The former defines an infinite set of expressions (the set of all $\mathcal{SC}$ sentences) that are considered to be well-formed. Given an interpretation $\mathbf{I}$, the latter defines a function that maps the set of sentences to the set { **T**, **F** }. One might wonder what counts as a good definition and ask whether these particular definitions are good.

Very generally, a definition is used to introduce (or make more precise) some term or phrase. The term or phrase that is being defined is often used as a substitute for some longer expression that can already be formulated in the language. For instance, in English one might define the term *x is a bachelor* as a substitute for the longer expression, *x is an adult male who has never been married*, where it is assumed that the meanings of the words in the second expression are already known. Definition 1-1 can be viewed this way: The term *sentence* is a substitute for *an expression that is either an $\mathcal{SC}$ letter or else can be built up by using the construction rule.* It is not so clear that the definition of $\mathbf{V_I}$ can be viewed in this fashion, but it can. Assuming that there is a unique function that satisfies conditions 1 through 6 in Definition 1-4, one can consider $\mathbf{V_I}$ as a short name for this function (that otherwise has a complex description). But this begs another question: How can we be sure that such a function exists?

It is not simple to say what 'f exists' means when f is an object like a set, function, or other kind of abstract mathematical entity. Philosophers and mathematicians have been concerned with this issue for a long time, and some of this concern has led to the development of improved theories of sets, natural numbers, and real numbers, as well as a better understanding of the nature of computable functions. Most of this is beyond the scope of this book, although later chapters will define many different types of sets, functions, and other mathematical entities; and these definitions will be justified as rigorously as is practical in an introductory book on logic and set theory.

Later chapters will introduce specific adequacy requirements that should be satisfied by certain kinds of definitions. These requirements will eventually cover most of the types of definitions that are commonly used in mathematics and the theory of computing. In the meantime, one can be guided by the following: First, if a definition leads to a contradiction, i.e., the definition enables one to deduce some statement of the form (ϕ and not ϕ) for some sentence ϕ, and this contradiction is not deducible without the definition, then this is a bad definition. Unfortunately, in general it is difficult to decide whether such a situation exists, although in some special cases it clearly does. The study of logic helps to clarify these ideas.

Second, when defining a function, one must check that the definition guarantees that the function returns a unique value for the function for any specific value(s) (input(s)) of the argument(s) of this function. Many of the sets and functions that we will use will be specified by procedures that permit calculating their elements or values, respectively. In such cases, one should check that the procedures always terminate instead of getting caught in an eternal loop or some other endless process. If the function is specified independently of some procedure given for it, then one should check that the procedure not only terminates, but also returns the correct values according to the independent specification.

The earlier discussion of syntax showed that one can determine, in a finite number of steps, whether or not a given string is or is not a sentence. Given a sentence and an interpretation **I**, we can compute the truth value of this sentence with respect to **I** in a finite number of steps. I have not proved these facts but hope that the reader is persuaded of their truth. Chapter 3 introduces techniques that enable one to construct such proofs.

Exercises

E 1-5 For each of the following *forms* of sentences, write a specific $\mathcal{SC}$ sentence that is an instance (example) of that form.

1. a negation of an atomic sentence

2. a negation of a conjunction

3. a disjunction of two conjunctions

4. a conditional with an antecedent that is a negation of a disjunction

5. a biconditional with a left side that is also a biconditional and with a right side that is the disjunction of two atomic sentences

6. a conditional that has a consequent that is the negation of another conditional

E 1-6 Write the converse and the contrapositive of each of the following:

1. $(\neg(A \vee B) \rightarrow (B \leftrightarrow (C \wedge B)))$
2. $((A \wedge \neg B) \rightarrow \neg\neg(A \vee A))$
3. $(((C \leftrightarrow \neg B) \rightarrow H) \rightarrow \neg T)$

E 1-7 (Answers provided for 1 and 2) Make a truth table for each of the following $\mathcal{SC}$ sentences:

1. $\neg((P \rightarrow Q) \rightarrow Q)$
2. $((A \rightarrow (B \rightarrow C)) \leftrightarrow (A \wedge (B \wedge C)))$
3. $(((A \vee \neg B) \wedge ((B \vee C) \wedge (\neg A \vee C))) \rightarrow C)$
4. $\neg(((P \rightarrow Q) \rightarrow P) \rightarrow P)$
5. $((A \rightarrow (B \rightarrow C)) \leftrightarrow ((A \wedge B) \rightarrow C))$
6. $((D \leftrightarrow E) \leftrightarrow ((D \wedge E) \vee (D \vee E)))$

E 1-8 The sentence $(P \wedge Q)$ is true (under an interpretation) if both P and Q are true (under this interpretation); it is false otherwise. The sentence $(P \wedge \neg Q)$ is true if P is true and Q is false; it is false otherwise. Corresponding remarks apply to other conjunctions of P or $\neg P$ with Q or $\neg Q$. Also, notice that $((P \wedge Q) \vee (P \wedge \neg Q))$ is true iff either both P and Q are true, or P is true and Q is false. These remarks can be generalized to sentences of three or more sentential letters. Use these observations to write an $\mathcal{SC}$ sentence ϕ composed of the three letters P, Q, R that has the following properties: (1) It is a disjunction. (2) Each disjunct of ϕ is a conjunction of P or $\neg P$ with a conjunction of Q or $\neg Q$ with R or $\neg R$. (3) The sentence ϕ is true if P, Q, R are all true or P, Q, R are all false; otherwise, it is false. Make a truth table for ϕ.

E 1-9 Use the information given in the previous exercise to write an $\mathcal{SC}$ sentence ψ that is true if one and only one of P, Q, R is true and is false otherwise.

1.4   Some Metatheoretical Concepts

We use the metalanguage (English with some mathematical concepts) to characterize and discuss features of the sentential calculus. Most of the terms that have been defined, e.g., *sentence, conditional, interpretation,* $\mathbf{V_I}$, are in this metalanguage. I will now define some additional metalinguistic terms that denote important logical concepts as they apply to the sentential calculus. Most of these logical concepts have generalizations to languages that are more powerful than sentential calculus; we will consider some of these generalizations in the last chapter of this book. To simplify expressions, I will sometimes use the following convention.

Notational Abbreviation: The outermost parentheses of an $\mathcal{SC}$
sentence may be omitted when convenient.

Thus, instead of $(P \to Q)$, one may write $P \to Q$. This form of abbreviation
applies only to parentheses that enclose everything else in the sentence. If
one later uses this sentence as a part of a larger sentence, then these omitted
parentheses must be restored. So, if we wish to negate $R \to S$, we must write
$\neg(R \to S)$.

Definition 1-5 Let ϕ be an $\mathcal{SC}$ sentence and **I** an $\mathcal{SC}$ interpretation. We
say that ϕ *is true under* **I** iff $\mathbf{V_I}(\phi) = \mathbf{T}$; if ϕ is not true under **I**, then it
is false under **I**. We also say that **I** *satisfies* ϕ iff $\mathbf{V_I}(\phi) = \mathbf{T}$. A sentence
ϕ is *satisfiable* iff there exists at least one interpretation that satisfies it. A
tautologous $\mathcal{SC}$ sentence (a *tautology*) ϕ is an $\mathcal{SC}$ sentence that is true under
every interpretation of the sentential calculus. An $\mathcal{SC}$ sentence is *contingent* iff
it is satisfiable and not tautologous.

   In general, the truth value of a sentence varies with the interpretation one
is using. To show that a sentence is tautologous, one must show that it is *true
under every interpretation*. To show that it is not a tautology, one need only
exhibit *one* interpretation under which it is false. If a sentence is contingent, it
is true under some interpretation(s). Since it is not tautologous, it is also false
under some interpretation(s). The only parts of an interpretation that affect
the truth value of a sentence are the assignments of **T** or **F** to the letters in
that sentence. Thus, the truth values of a sentence are completely listed for
all interpretations by all rows of a truth table for that sentence. A sentence is
tautologous iff the calculations of its truth values yield **T** in every row of its
truth table.

Examples.

   The following are tautologies:

$$P \vee \neg P, \quad P \to (P \vee Q), \quad \neg(P \wedge \neg P).$$

It is easy to check that each of these has a truth table that yields (returns) the
value **T** in each row.

   The following are not tautologous:

$$P, \quad P \vee Q, \quad P \to Q, \quad (P \wedge \neg P).$$

No atomic sentence (such as P) can be tautologous, since some interpretations
give it the value **T** and others the value **F**. Also, $P \vee Q$ is false under (i.e., gets
the value **F** from) any interpretation that assigns **F** to both P and Q. $P \to Q$

is false when P is true and Q is false, and the fourth sentence is false under every interpretation. It is not satisfiable. The sentences P, $P \vee Q$, and $P \rightarrow Q$ are contingent. ∎

The preceding definition of *satisfies* is now extended to sets of sentences.

Definition 1-6 Let Γ be a set of $\mathcal{SC}$ sentences and **I** be an $\mathcal{SC}$ interpretation. Then **I** *satisfies* Γ iff every sentence in Γ is true under **I**. A set Γ of $\mathcal{SC}$ sentences is *satisfiable* iff there is some $\mathcal{SC}$ interpretation that satisfies Γ (i.e., some interpretation under which each sentence in Γ is **T**). A set of $\mathcal{SC}$ sentences Γ is *unsatisfiable* iff Γ is not satisfiable.

From Definitions 1-5 and 1-6, a single sentence ϕ is *satisfiable* iff $\{\phi\}$ is satisfiable. Also, an $\mathcal{SC}$ sentence ϕ is *unsatisfiable* iff $\{\phi\}$ is unsatisfiable.

Examples.

1. $\{\, P, (P \vee Q), (R \rightarrow \neg P) \,\}$ is satisfiable.

To show that it is satisfiable, consider any interpretation that assigns the values: P: **T**, Q: **T**, R: **F**. Then we have

$$\mathbf{V_I}(P) = \mathbf{T},$$

$$\mathbf{V_I}(P \vee Q) = (\mathbf{T} \vee \mathbf{T}) = \mathbf{T},$$

$$\mathbf{V_I}(R \rightarrow \neg P) = (\mathbf{F} \rightarrow \neg\mathbf{T}) = \mathbf{T}.$$

We also could have used the interpretation with those values of P and R, along with Q: **F**.

2. $P \wedge \neg P$ is unsatisfiable.

To show that it is not satisfiable, one could construct its entire truth table. This table has only two rows; the sentence is false in each row. The unsatisfiability can also be seen directly by seeking a truth value for P under which the sentence is true. P is either true or false. If we try $\mathbf{I}(P) = \mathbf{T}$, then $\neg P$ is false, so the conjunction is false. If we try $\mathbf{I}(P) = \mathbf{F}$, the conjunction is again false. ∎

By definition, a set of sentences Γ is satisfiable iff there is some $\mathcal{SC}$ interpretation under which all sentences in Γ are true. Thus, Γ is unsatisfiable iff there does not exist such an interpretation. So Γ is unsatisfiable iff for any interpretation **I**, there is at least one sentence in Γ that is false under **I**.

Example.

Problem Statement. Give a brief argument to show that the following set of sentences is unsatisfiable: $\{\ P \rightarrow Q,\ \ (Q \wedge R) \rightarrow \neg P,\ \ P \wedge R\ \}$.

Discussion. We could make a truth table for this one. We would need to show that, for any combination of input truth values for P, Q, and R, at least one of the three sentences in the set is false. But making the whole truth table is unnecessary for this problem. One can immediately see that both P and R need to be true in order for the third sentence to be true. But since P must be true, then in order for the first sentence to be true, we must also have Q: **T**. It is now easy to write up the desired brief argument.

Written Solution. Let $\Gamma = \{\ P \rightarrow Q,\ \ (Q \wedge R) \rightarrow \neg P,\ \ P \wedge R\ \}$, and suppose that **I** is an $\mathcal{SC}$ interpretation that satisfies Γ, i.e., all three sentences in Γ are true under **I**. In order for $\mathbf{V_I}(P \wedge R) = \mathbf{T}$, we must have P: **T** and R: **T**. Since P must be true, then it is necessary to have Q: **T** to make the first sentence true. But if P, Q, and R are all true, then

$$\mathbf{V_I}((Q \wedge R) \rightarrow \neg P) = ((\mathbf{T} \wedge \mathbf{T}) \rightarrow \neg\mathbf{T}) = (\mathbf{T} \rightarrow \mathbf{F}) = \mathbf{F}.$$

Thus, the only interpretation that satisfies the first and last sentence makes the middle sentence false, so no interpretation can satisfy Γ.

Remark. In general the question of satisfiability of $\mathcal{SC}$ sentences is computationally very complex. Yet, with practice, one can gain the skill to solve simple problems, like the preceding one, in "one's head." Writing the solution is then straightforward. ■

I should mention a common misconception here. Suppose that Γ and Δ are each sets of sentences. A sentence ϕ is in the union $\Gamma \cup \Delta$ iff ϕ is in Γ or in Δ. Suppose that the interpretation **I** satisfies $\Gamma \cup \Delta$. By associating 'or' with union, it is sometimes erroneously believed that **I** must satisfy Γ or satisfy Δ. Actually, if **I** satisfies $\Gamma \cup \Delta$, then every sentence in $\Gamma \cup \Delta$ is true under **I**. So, if ϕ is in Γ, then ϕ is true under **I**, or if ϕ is in Δ, then ϕ is true under **I**. Hence, **I** satisfies Γ *and* **I** satisfies Δ.

I will now prove a simple, but important, metatheorem. A metatheorem is a theorem *about* a formal language; it is proved in a metalanguage that is used to describe this formal language. In this metatheorem, we use one of the standard notations, $\varnothing$, to denote the empty set, i.e., the set with no elements in it. In the next chapter, it will be proved that there is only one empty set; for now we will just assume this obvious fact. The empty set is a set of $\mathcal{SC}$ sentences, namely, the set that has no sentences in it. Is $\varnothing$ satisfiable? Yes, indeed—every interpretation satisfies it.

Metatheorem 1-7   The empty set, ∅, of *SC* sentences is satisfied by every *SC* interpretation.

Proof. Let **I** be any *SC* interpretation. We want to show that **I** satisfies ∅. **I** satisfies ∅ iff every sentence in ∅ is true under **I**. Thus, **I** satisfies ∅ iff no sentence in ∅ is false under **I**. In other words, **I** satisfies ∅ iff there does not exist a sentence in ∅ that is false under **I**. The latter is true because there does not exist any sentence that is in ∅, since it is empty. Therefore, **I** satisfies ∅. ***Q.E.D.***

Notice that the proof used various logical inferences in the metalanguage. This is unavoidable; we must rely on logical intuitions that are used in the metalanguage. The proof is given in standard "mathematical English." One of the aims of this book is to introduce several different kinds of theorems and proof styles, as well as various methods for proving theorems and metatheorems. One way to learn how to state and prove theorems is by carefully studying many examples, but passive reading is usually not sufficient for gaining a working understanding. The theorems that are stated and proved in this book are not only useful for their content; they are also intended to be models of logical and mathematical reasoning and writing style. After reading a theorem, one should try to reproduce the proof. It will be good practice to do this with all the theorems in this book. If this is done before attempting the Exercises, many of these problems should be easier to solve. Eventually, one should be able to prove most of the theorems that are in this text.

The next definition is one of the most important in this chapter. It was stated near the beginning of this chapter that a *sound* deductive argument is an argument such that, *if* its premises are all true, *then* its conclusion must also be true. This is the basic, intuitive idea of what a sound deductive argument is. One of the main tasks of formal logic is to define this idea precisely. We are now in a position to do this for arguments in the sentential calculus.

Definition 1-8 Let Γ be a set of *SC* sentences and ϕ be an *SC* sentence. Then ϕ is a *tautological consequence* of Γ (denoted by $\Gamma \vDash_T \phi$) iff there is no interpretation that satisfies Γ and does not satisfy ϕ. When $\Gamma \vDash_T \phi$, we may also say that Γ *tautologically implies* ϕ. If Γ is a set containing only one sentence, say, $\Gamma = \{\rho\}$, then all of this terminology can be applied to ρ just as it applies to $\Gamma = \{\rho\}$.

So, for the sentential calculus, the idea of a sound argument is explicated as an argument in which the conclusion, ϕ, is a tautological consequence of its set of premises Γ. In other words, the premises tautologically imply the conclusion. If we have such an argument, and all of its premises are true under an *SC* interpretation **I**, then its conclusion must also be true under **I**. The truth of all of its premises under an interpretation guarantees the truth of its conclusion under that interpretation. It is very important to remember that

$\mathcal{SC}$ is a weak language; not all intuitively sound arguments can be adequately represented in the sentential calculus. Chapter 4 is about a much more powerful formal language, predicate calculus, of which $\mathcal{SC}$ is just a small sublanguage. Yet the basic idea of Definition 1-8 has a natural generalization for the predicate calculus. Whatever language we use, and however interpretation and truth are defined for this language, a deductively sound argument is one in which there is no interpretation that satisfies the premises and does not satisfy the conclusion. The preceding definition is just a special case of this very general idea.

In the following example, and others like it, we temporarily assign specific values to metalinguistic variables such as Γ and ϕ. In effect, for the sake of the example, we treat these metalinguistic variables as metalinguistic constants that *denote* certain specific objects. One could say, for instance, "Let ϕ denote ' $\neg R$ '." in which the single quotes indicate that we are *referring to* (mentioning) the symbol string composed of $\neg$ and R. In most contexts, there will be little chance for confusion about what is meant. Instead of using such tedious expressions, I will follow common practice and say, "Let $\phi = \neg R$." Sometimes the 'let' will not even be written, it will be implied by the context. It must be remembered that this use of the identity sign (=) in the metalanguage does not mean that the symbol 'ϕ' is literally identical with the symbol ' $\neg R$ '; it means only that we are temporarily using ϕ to stand for $\neg R$.

Example.

Problem Statement. Let

$$\Gamma = \{ \, P, \, (S \vee \neg S), \, (\neg P \vee Q), \, (\neg Q \leftrightarrow R) \, \},$$

and $\phi = \neg R$. Prove that $\Gamma \vDash_T \phi$.

Discussion. By Definition 1-8, $\Gamma \vDash_T \phi$ means that any interpretation that satisfies Γ must also satisfy ϕ. One way to show this is to calculate a truth table with rows for every possible combination of truth values of P, Q, R, and S. Each of these rows would have a place for the corresponding truth values of P, $(S \vee \neg S)$, $(\neg P \vee Q)$, $(\neg Q \leftrightarrow R)$, and $\neg R$. If every row that has *each* of P, $(S \vee \neg S)$, $(\neg P \vee Q)$, $(\neg Q \leftrightarrow R)$ true also has $\neg R$ true, then every interpretation that satisfies Γ also satisfies ϕ. Unfortunately, this method of proof requires considerable calculation of truth values. Fortunately, with this kind of problem, one can often shorten the work by examining the particular sentences involved in more detail, thereby reducing the possible truth values to consider. Here is an example of how to do this by writing a metalinguistic proof.

Written Solution. Let **I** be *any* interpretation that satisfies Γ, so all four sentences in Γ are true under **I**. Thus, it must be the case that $\mathbf{I}(P) = \mathbf{T}$. But then Q must also be true under **I**, for if it were false, then $(\neg P \vee Q)$ would be false. Thus, so far, **I** *must* make P true and Q true. But since Q is true under **I**, $\neg Q$ is false, so R must be false in order for the biconditional to be true. But since R is false under **I**, the conclusion $\neg R$ must be true under **I**. This shows that any interpretation that satisfies Γ must also satisfy ϕ. **Q.E.D.**

Remark. The assertion that "*any* interpretation that satisfies Γ must also satisfy ϕ" is true iff "there is *no* interpretation that satisfies Γ and does not satisfy ϕ." Also, notice that this proof did not require determining a truth value for S, since S occurs only in $(S \vee \neg S)$, and it is a tautology, so is true regardless of the truth value of S. ▪

Although we avoided calculating a large truth table, using this method to prove that $\Gamma \vDash_T \phi$ is usually tedious because one must prove that *any* interpretation that satisfies Γ must also satisfy ϕ. On the other hand, to prove that Γ does not tautologically imply ϕ, it is only necessary to prove that there is *one* interpretation satisfying Γ that does not satisfy ϕ. To do this, one must construct the interpretation (usually by using some trial-and-error guesses), and then prove that it has the required features. It is convenient to introduce a new term to describe such interpretations.

Definition 1-9 Let Γ be a set of *SC* sentences and ϕ be an *SC* sentence. An *SC* interpretation **I** is a *counterexample to the claim that* Γ *tautologically implies* ϕ iff **I** satisfies Γ, but ϕ is false under **I**.

Example.

Problem Statement. Show that S is not a tautological consequence of $\Gamma = \{ (Q \rightarrow P),\ (R \vee S),\ (R \rightarrow P) \}$.

Discussion. To show that S is not a tautological consequence of Γ we need a counterexample interpretation. Mathematicians and logicians often do not show how they find a particular example (or counterexample). For many purposes, it suffices to describe the example and prove that it has some specified features, and this is usually the most concise form in which to communicate the result. Yet if there are techniques for constructing certain kinds of examples, it is often useful to know what they are.

Examples sometimes seem merely "to occur" to someone; in other cases, one might find an example by a somewhat random trial-and-error process. Fortunately, with small problems, one can usually find a counterexample by fairly direct reasoning. Here is how we can construct a counterexample interpretation for this problem. We must have $\mathbf{I}(S) = \mathbf{F}$ in order to have a counterexample. We also want the interpretation to satisfy Γ. In as much as S is false under **I**, in order to have $(R \vee S)$ true, we need R true. Then to have $(R \rightarrow P)$ true, we need P true. Because P is true, $(Q \to P)$ is true regardless of the value of Q. In order to have a particular counterexample interpretation, let's choose Q false. After going through such a construction process, one should state the interpretation clearly and show that it is a counterexample.

Written Solution. Let **I** be the interpretation with the following truth value assignments, P: **T**, Q: **F**, R: **T**, S: **F**. Under **I**, the conclusion S is false. Whereas Q is false, $(Q \to P)$ is true. Since R is true, the disjunction is true. Also, $\mathbf{V_I}(R \to P) = (\mathbf{T} \to \mathbf{T}) = \mathbf{T}$. Hence, **I** satisfies Γ and makes S false, so **I** is a counterexample to the argument, i.e., S is not a tautological consequence of Γ. ∎

The preceding solution is a little misleading in referring to *the* interpretation **I**, since the solution really describes only a small part of an infinite set of interpretations with different values for all of the $\mathcal{SC}$ letters except those used in the example. But because we are interested in only those four letters, use of 'the' refers to the particular set of truth value assignments to these four letters.

Example.

$\neg S$ is not a tautological consequence of

$$\Gamma = \{\ (Q \to P),\ (R \vee S),\ (R \to P)\}.$$

This is very easy. For practice, one should construct a suitable interpretation and show that it is a counterexample. The interpretation must satisfy Γ and make the conclusion false. ∎

The next example is more complex.

Example.

Problem Statement. Show that $X \to (Q \to A)$ is not a tautological consequence of

$$\Gamma = \{\ \neg P \to (Q \vee X),\ (H \vee Q) \wedge \neg H,\ X \to P,\ P \leftrightarrow (A \to Q)\ \}.$$

Discussion. Again, we need a counterexample interpretation, i.e., we must exhibit a particular interpretation, **I**, that satisfies Γ and does not satisfy ϕ, i.e., ϕ is false under **I**. How do we find one?

One way to search for such an interpretation is to compute the truth values of all of the sentences

$$\neg P \to (Q \vee X),\ (H \vee Q) \wedge \neg H,\ X \to P,\ P \leftrightarrow (A \to Q),\ X \to (Q \to A),$$

for all possible **T/F** values of the $\mathcal{SC}$ letters A, H, P, Q, X. If we did this completely, it would amount to making a truth table with 32 rows and computing the value of each of the five sentences in each row. This is considerable work. However, we need only one counterexample interpretation. As soon as we find a row (set of input values of A, H, P, Q, X) that satisfies the first four sentences, the elements of Γ, and makes the last one false, then we will have a counterexample interpretation. Yet, we might need to compute all 32 rows to find one. (If no counterexample exists, then ϕ is a tautological consequence of Γ.) Using a truth table to solve our problem could be called the "brute force" method.

Fortunately, when faced with a problem of this type, one can usually see ways of reducing the set of possible input values to consider. This is done by making use of special features of the particular sentences under consideration. Consider the five sentences:

$$\neg P \to (Q \lor X), \quad (H \lor Q) \land \neg H, \quad X \to P, \quad P \leftrightarrow (A \to Q), \quad X \to (Q \to A).$$

We want the first four true and the fifth false. Look over the sentences and try to find ones that require fairly specific input values in order to yield the desired output value. The only values that make the fifth false are X: **T**, Q: **T**, A: **F**. Now consider the second sentence. In order for it to be true, we need H: **F**. Also, in order for the third sentence to be true, we need P: **T**, since X is already **T**. In that case, the values of the five $\mathcal{SC}$ letters have already been determined. It is easy to check that the remaining sentences are true, as desired. We can now write up our solution.

Written Solution. Let

$$\Gamma = \{\, \neg P \to (Q \lor X), \; (H \lor Q) \land \neg H, \; X \to P, \; P \leftrightarrow (A \to Q) \,\},$$

and let ϕ denote the sentence $X \to (Q \to A)$. Let **I** be an interpretation with the following truth-value assignments:

$$A: \mathbf{F}, \; H: \mathbf{F}, \; P: \mathbf{T}, \; Q: \mathbf{T}, \; X: \mathbf{T}.$$

Under **I**, the sentences have the following truth values:

$$\mathbf{V_I}(\neg P \to (Q \lor X)) = \neg\mathbf{T} \to (\mathbf{T} \lor \mathbf{T}) = \mathbf{F} \to \mathbf{T} = \mathbf{T},$$

$$\mathbf{V_I}((H \lor Q) \land \neg H) = (\mathbf{F} \lor \mathbf{T}) \land \neg\mathbf{F} = \mathbf{T} \land \mathbf{T} = \mathbf{T},$$

$$\mathbf{V_I}(X \to P) = \mathbf{T} \to \mathbf{T} = \mathbf{T},$$

$$\mathbf{V_I}(P \leftrightarrow (A \to Q)) = \mathbf{T} \leftrightarrow (\mathbf{F} \to \mathbf{T}) = \mathbf{T} \leftrightarrow \mathbf{T} = \mathbf{T},$$

$$\mathbf{V_I}(X \to (Q \to A)) = \mathbf{T} \to (\mathbf{T} \to \mathbf{F}) = \mathbf{T} \to \mathbf{F} = \mathbf{F}.$$

Therefore, $X \to (Q \to A)$ is not a tautological consequence of Γ, since $\mathbf{I}$ satisfies Γ but not $X \to (Q \to A)$.

Remark. It is not always necessary to show the truth-value calculations in as much detail as is shown here, but one should show enough work to make it obvious what values the sentences have under $\mathbf{I}$. ■

Definition 1-10 Let Γ be a set of $\mathcal{SC}$ sentences, and let ϕ be an $\mathcal{SC}$ sentence. ϕ is *tautologically consistent* with Γ iff $\Gamma \cup \{\, \phi \,\}$ is satisfiable. ϕ is *tautologically independent* of Γ iff ϕ is not a tautological consequence of Γ.

Suppose that $\neg\phi$ is not a tautological consequence of Γ, so there is an interpretation $\mathbf{I}$ that satisfies Γ, such that $\neg\phi$ is false under $\mathbf{I}$. So, ϕ is true under $\mathbf{I}$, so $\Gamma \cup \{\, \phi \,\}$ is satisfiable, so ϕ is tautologically consistent with Γ. There is an earlier example to show that $\neg S$ is not a tautological consequence of $\Gamma = \{\, (Q \to P),\ (R \lor S),\ (R \to P)\}$. This shows that $\neg S$ is tautologically independent of Γ, and that S is tautologically consistent with Γ. There was an even earlier example that showed that S is not a tautological consequence of Γ, so this example also shows that $\neg S$ is tautologically consistent with Γ. Thus, we see that both S and $\neg S$ can be tautologically consistent with a set of sentences, and that S can be both tautologically independent of, and consistent with, a set of sentences.

Now consider any two $\mathcal{SC}$ sentences, ϕ and ψ. It is possible that neither tautologically implies the other; for example, this is true for any two distinct sentential letters. It is possible that one tautologically implies the other, but not conversely (i.e., it is not the case that the implication holds in the opposite direction). For example, $(P \land Q) \vDash_T P$, but P does not tautologically imply the conjunction. Finally, it is possible that ϕ tautologically implies ψ and conversely, i.e., ψ also tautologically implies ϕ, as is shown by the next example.

Example.

$$(P \land Q) \vDash_T (Q \land P) \text{ and } (Q \land P) \vDash_T (P \land Q).$$

This is very easy to prove. Let $\mathbf{I}$ be any interpretation that satisfies $(P \land Q)$. Then, by Definition 1-4, both P and Q must be true under $\mathbf{I}$. But then $\mathbf{I}$ also satisfies $(Q \land P)$. This establishes the first of the two tautological implications. The second implication is proved similarly by the same kind of argument applied in the opposite direction. ■

Definition 1-11 Let ϕ and ψ be $\mathcal{SC}$ sentences; ϕ and ψ are *tautologically equivalent* iff $\phi \vDash_T \psi$ and $\psi \vDash_T \phi$.

The concepts of tautological implication and tautological equivalence will play a fundamental role in the next section. For now, notice that the preceding example can easily be generalized into a metatheorem about any pair of $\mathcal{SC}$ sentences, not just P and Q. Whenever one finds an interesting example, one should try to generalize it into an even more interesting theorem.

Metatheorem 1-12   Let ϕ and ψ be $\mathcal{SC}$ sentences. Then $(\phi \wedge \psi)$ and $(\psi \wedge \phi)$ are tautologically equivalent.

Proof. We must prove that each one tautologically implies the other.

Let **I** be any interpretation that satisfies $(\phi \wedge \psi)$. Then (by Definition 1-4) both ϕ and ψ must be true under **I**. But then **I** also satisfies $(\psi \wedge \phi)$. This establishes that $(\phi \wedge \psi)$ tautologically implies $(\psi \wedge \phi)$. The converse implication (in the opposite direction) is proved similarly. **Q.E.D.**

Here are a few more simple metatheorems. They state useful facts, and their proofs are good examples of how to prove such facts. In order to make the proofs relatively easy to understand, I present them in more detail than is customary. Throughout this book, when a new kind of proof is first encountered, I will present it in great detail, and often include explanatory remarks that are not required as part of the proof. When similar proofs occur later in the book, they will be written in briefer, more standard mathematical style.

Metatheorem 1-13   Let ϕ and ψ be $\mathcal{SC}$ sentences. Then $\phi \vDash_T \psi$ iff $\phi \to \psi$ is tautologous.

Discussion of Proof. We need to prove an 'iff' statement. Since 'iff' abbreviates 'if and only if', we must prove both the 'if' part and the 'only if' part. It is often convenient, though not always necessary, to break the proof into two parts corresponding to the 'if' part and to the 'only if' part. I will do that here.

Proof.

Part 1. I will first prove: $\phi \vDash_T \psi$ only if $\phi \to \psi$ is tautologous. This means the same as: If $\phi \vDash_T \psi$, then $\phi \to \psi$ is tautologous. In other words, $\phi \vDash_T \psi$ is *sufficient* for $\phi \to \psi$ to be tautologous.

Suppose that $\phi \vDash_T \psi$. I claim that this assumption is *sufficient* for $\phi \to \psi$ to be tautologous. Let **I** be any $\mathcal{SC}$ interpretation. Then $\mathbf{V_I}(\phi) = \mathbf{T}$ or $\mathbf{V_I}(\phi) = \mathbf{F}$, so we have two cases to consider.

If $\mathbf{V_I}(\phi) = \mathbf{F}$, then $\phi \to \psi$ has a false antecedent. By the truth table for $\to$, $\phi \to \psi$ is true under **I**.

If $\mathbf{V_I}(\phi) = \mathbf{T}$, then, since $\phi \vDash_T \psi$, we have ψ true under $\mathbf{I}$. In this case, $\phi \rightarrow \psi$ is true, since it has a true consequent.

In both cases, i.e., regardless of whether ϕ is true or false under $\mathbf{I}$, it turns out that $\phi \rightarrow \psi$ is true. Because $\mathbf{I}$ is any interpretation, $\phi \rightarrow \psi$ is a tautology. Hence, the assumption that $\phi \vDash_T \psi$ is sufficient for $\phi \rightarrow \psi$ to be a tautology. In other words: If $\phi \vDash_T \psi$, then $\phi \rightarrow \psi$ is tautologous. Thus, $\phi \vDash_T \psi$ only if $\phi \rightarrow \psi$ is tautologous.

Part 2. We now need to prove: $\phi \vDash_T \psi$ if $\phi \rightarrow \psi$ is tautologous. In other words, $\phi \rightarrow \psi$ being tautologous is sufficient for $\phi \vDash_T \psi$. So suppose that $\phi \rightarrow \psi$ is a tautology. Let $\mathbf{I}$ be any $\mathcal{SC}$ interpretation that satisfies ϕ. Since $\phi \rightarrow \psi$ is tautologous, it is true under $\mathbf{I}$. Since ϕ is also true, ψ must be true under $\mathbf{I}$. Therefore, any interpretation that satisfies ϕ also satisfies ψ, so $\phi \vDash_T \psi$. Hence, if $\phi \rightarrow \psi$ is tautologous, then $\phi \vDash_T \psi$. **_Q.E.D._**

Metatheorem 1-13 helps to illuminate the difference between a sentential calculus conditional sentence, $\phi \rightarrow \psi$, and a tautological implication, $\phi \vDash_T \psi$. In general, the truth value of a conditional sentence varies from one interpretation to another. Even if its antecedent is true under an interpretation $\mathbf{I}$, its consequent may not be true under $\mathbf{I}$, in which case the entire conditional is false under $\mathbf{I}$. If ϕ is true under $\mathbf{I}$ and ψ false under $\mathbf{I}$, then ϕ does not tautologically imply ψ. However, _if the conditional is a tautology_, then it is true under _every_ interpretation. In that case, the truth of ϕ under $\mathbf{I}$ _does_ guarantee the truth of ψ. In other words, ϕ tautologically implies ψ, which is what Metatheorem 1-13 says. The converse is also true: If ϕ tautologically implies ψ, then $\phi \rightarrow \psi$ is a tautology. The next metatheorem relates conditionals to biconditionals. It codifies a fact that was already pointed out in the earlier discussion of the truth table for the biconditional.

Metatheorem 1-14 Let $\mathbf{I}$ be any $\mathcal{SC}$ interpretation, and ϕ, ψ be any $\mathcal{SC}$ sentences. Then $\phi \leftrightarrow \psi$ has the same truth value under $\mathbf{I}$ as $(\phi \rightarrow \psi) \wedge (\psi \rightarrow \phi)$.

Proof. Let $\mathbf{I}$ be any $\mathcal{SC}$ interpretation. Under $\mathbf{I}$, ϕ is either true or false, and ψ is either true or false. Table 1.9 shows the possible values for the conjunction.

By comparison with the truth table for $\leftrightarrow$, it can be seen that this conjunction has the same truth values as the biconditional. **_Q.E.D._**

By Metatheorem 1-14, if $\phi \leftrightarrow \psi$ is true under an interpretation $\mathbf{I}$, then $(\phi \rightarrow \psi) \wedge (\psi \rightarrow \phi)$ is true under $\mathbf{I}$. Conversely, if $(\phi \rightarrow \psi) \wedge (\psi \rightarrow \phi)$ is true, then so is $\phi \leftrightarrow \psi$. Thus, by Definition 1-11, they are tautologically equivalent. In English, we use 'iff' to abbreviate 'if and only if', and 'ϕ iff ψ' means 'if ϕ,

ϕ	ψ	$(\phi \to \psi) \wedge (\psi \to \phi)$
T	T	T
T	F	F
F	T	F
F	F	T

Table 1.9.

then ψ, and, if ψ, then ϕ'. Therefore, as expected, and indicated in Table 1.1, 'iff' in English behaves in a manner analogous to $\leftrightarrow$ in sentential calculus.

Metatheorem 1-15 Let ϕ and ψ be $\mathcal{SC}$ sentences. Then ϕ and ψ are tautologically equivalent iff $(\phi \leftrightarrow \psi)$ is a tautology.

Proof.

Part 1. Suppose that ϕ and ψ are tautologically equivalent. Then, from Definition 1-11, we have $\phi \vDash_T \psi$ and $\psi \vDash_T \phi$. So, from Metatheorem 1-13, $(\phi \to \psi)$ and $(\psi \to \phi)$ are tautologous. Hence, $(\phi \to \psi) \wedge (\psi \to \phi)$ is tautologous. From the previous metatheorem, under any interpretation, $\phi \leftrightarrow \psi$ has the same truth value as the conjunction. Hence, $\phi \leftrightarrow \psi$ is true under every interpretation, so it is a tautology.

Part 2. Conversely, suppose that $\phi \leftrightarrow \psi$ is a tautology. From the previous metatheorem, $(\phi \to \psi) \wedge (\psi \to \phi)$ is a tautology. Hence, $(\phi \to \psi)$ and $(\psi \to \phi)$ are tautologous. From Metatheorem 1-13, $\phi \vDash_T \psi$ and $\psi \vDash_T \phi$, so ϕ and ψ are tautologically equivalent. **Q.E.D.**

Metatheorem 1-16 Let ϕ and ψ be $\mathcal{SC}$ sentences. Then ϕ and ψ are tautologically equivalent iff for any $\mathcal{SC}$ interpretation, they have the same truth value under that interpretation.

Proof. Suppose that ϕ and ψ are tautologically equivalent. From the previous metatheorem, $\phi \leftrightarrow \psi$ is tautologous, so it is true under every interpretation. Thus, under any interpretation, ϕ and ψ have the same truth value, according to Part 6 of Definition 1-4.

Conversely, suppose that, under any interpretation **I**, ϕ and ψ have the same truth value. Then $\phi \leftrightarrow \psi$ is true under any interpretation, so it is tautologous. From the previous metatheorem, ϕ and ψ are tautologically equivalent.

Q.E.D.

This is a very important result. It says that, with respect to their truth values under an interpretation, tautologically equivalent sentences are "the same." Of course, they may be different sentences. For example, it was previously proved that $\phi \wedge \psi$ and $\psi \wedge \phi$ are tautologically equivalent. We now know that they must have the same truth values under any interpretation. By the Commutative Law for addition, $x+y = y+x$ for any real numbers x, y. The number $x + y$ is the same as the number $y + x$. Metatheorem 1-12 shows that $\wedge$ is also commutative. Sentential calculus is not concerned with real number values, but rather with truth values. To say that $\wedge$ is commutative amounts to saying that $\phi \wedge \psi$ and $\psi \wedge \phi$ have the same truth value (under any interpretation).

Example.

Problem Statement. Metatheorem. If Γ is a satisfiable set of $\mathcal{SC}$ sentences and ρ is any $\mathcal{SC}$ sentence, then either $\Gamma \cup \{ \rho \}$ or $\Gamma \cup \{ \neg\rho \}$ is satisfiable.

Discussion. This problem simply states a metatheorem. We will follow the convention that if a problem merely states a theorem or metatheorem, then the task is to prove that theorem or metatheorem, so we must try to write a proof. The antecedent of this theorem gives little information—we know only that Γ is satisfiable. Our proof will use this information, together with definitions and theorems stated earlier in the book.

Notice that the consequent of the theorem is a disjunction. Stated precisely it says: Either $\Gamma \cup \{ \rho \}$ is satisfiable or $\Gamma \cup \{ \neg\rho \}$ is satisfiable. Thus, in the metalanguage, it has the general form, 'Either $\mathcal{M}$ or $\mathcal{N}$', and this form of statement is true iff at least one of $\mathcal{M}$ or $\mathcal{N}$ is true. But if we apply the truth table for conditionals to 'If not $\mathcal{M}$, then $\mathcal{N}$', it too is true iff at least one of $\mathcal{M}$ or $\mathcal{N}$ is true. In other words, in the metalanguage, a statement of the form 'Either $\mathcal{M}$ or $\mathcal{N}$' has the same truth value as 'If not $\mathcal{M}$, then $\mathcal{N}$', under any combination of truth value assignments to $\mathcal{M}$ and to $\mathcal{N}$. Therefore, these two forms of statements are tautologically equivalent in the metalanguage, and proving one is equivalent to proving the other. This provides a strategy for proving 'Either $\mathcal{M}$ or $\mathcal{N}$': Instead of proving this directly, try to prove 'If not $\mathcal{M}$, then $\mathcal{N}$'. We can attempt this by assuming not $\mathcal{M}$, and trying to use this assumption to prove $\mathcal{N}$. This strategy is often very helpful. Of course, 'If not $\mathcal{N}$, then $\mathcal{M}$' is also equivalent, and we could try the same strategy with it. In effect, we can assume that one of the disjuncts is false, and use this assumption to try to prove that the other disjunct must then be true. If this is done, it shows that at least one of the disjuncts must be true.

In the current problem, not $\mathcal{M}$ is the statement that $\Gamma \cup \{ \rho \}$ is not satisfiable, so we want to prove from this that $\Gamma \cup \{ \neg\rho \}$ is satisfiable. We are given that Γ is satisfiable, so there exists at least one interpretation **I** that satisfies Γ. Hence, **I** does not satisfy ρ, since otherwise $\Gamma \cup \{ \rho \}$ would be satisfiable.

We now use **I** to complete the proof, which is very simple.

Written Solution. Metatheorem. If Γ is a satisfiable set of $\mathcal{SC}$ sentences and ρ is any $\mathcal{SC}$ sentence, then either $\Gamma \cup \{\rho\}$ or $\Gamma \cup \{\neg\rho\}$ is satisfiable.

Proof. Assume that Γ is satisfiable and that $\Gamma \cup \{\rho\}$ is not satisfiable. Since Γ is satisfiable, there is some interpretation **I** that satisfies Γ. Then **I** does not satisfy ρ, since otherwise $\Gamma \cup \{\rho\}$ would be satisfiable. Thus, ρ is false under **I**, so $\neg\rho$ is true under **I**. Hence, **I** satisfies both Γ and $\{\neg\rho\}$, so **I** satisfies $\Gamma \cup \{\neg\rho\}$. **Q.E.D.** ■

We have now proved several metatheorems regarding interpretations, tautological implications, and related concepts. We still do not have efficient *rules* for the construction of arguments; these are introduced in the next section. It is worth noting at this point that the concept of satisfiability has both practical and theoretical significance in the science of computation. Given an $\mathcal{SC}$ sentence, ϕ (which could be the conjunction of all sentences in a finite set, Γ, of sentences), one may wish to know whether or not ϕ is satisfiable. This question can be answered by constructing the truth table for ϕ. If ϕ has n $\mathcal{SC}$ letters, the complete table will have 2^n rows. If one is lucky, a suitable interpretation might be found on the first row. If ϕ is unsatisfiable, one might have to compute all 2^n rows to prove this. Suppose the sentence contains 500 $\mathcal{SC}$ letters. The number of rows is 2^{500}, and

$$\log_{10}(2^{500}) = 500(\log_{10}2) = 500(.30103) = 150.515.$$

There are about 10^{150} rows. Assume that we have a computing machine that takes an average of one second to calculate the truth value of a row. There are $3.1536(10^7)$ seconds in a 365-day year. It would take about $3.2(10^{142})$ years to compute the truth table. This is a long time. The age of the earth is reckoned to be of the order of 10^9 years.

These time estimates may seem silly. Yet there are practical problems, such as problems about scheduling and route-finding, that are connected with the problem of determining whether certain $\mathcal{SC}$ sentences are satisfiable. The $\mathcal{SC}$ sentences related to these problems often have hundreds of letters. There is consequently both practical concern and theoretical interest in developing procedures (algorithms) for determining the satisfiability of $\mathcal{SC}$ sentences. These problems have been investigated for several years, and this research continues. These subjects are beyond the scope of this book. For later reading, a nice introduction is Chapter 7 of Harel (1992); a more technical treatment is in Chapter 11 of Horowitz and Sahni (1978). A good basic understanding of logic should be acquired before attempting serious study of these difficult problems.

It should also be mentioned here that experienced human beings generally solve complex scheduling problems by trial-and-error methods that are guided by heuristic rules (i.e., rules of thumb that are not guaranteed to find a solution, but often help one to do so). Artificial intelligence research has contributed to the analysis and implementation of such heuristic methods. A good basic understanding of logic is also a prerequisite for approaching these problems. Some useful texts are listed in the References.

Exercises

General Instructions

Throughout this book, if an exercise is simply the statement of a theorem or a metatheorem, then the task is to prove this theorem or metatheorem. In your proof, you may use the definitions given in the text. Unless instructed to the contrary, you may also use theorems stated in the text.

Exercises for which answers are provided at the back of the book are marked with "Answer Provided."

Finally, please note that exercises that introduce important principles or useful applications, are especially difficult, or are special in some way, are indicated by **bold** print. In some cases, parenthetical comments describe the special features of the problem. If such comments are missing, the exercise is nonetheless of special value in introducing a new concept, principle, or technique.

E 1-10 Construct three $\mathcal{SC}$ sentences: One should be a tautology, one an unsatisfiable sentence, and the third a contingent sentence. Make up sentences that are different from the examples in the text. Construct the truth table for each sentence.

E 1-11 Show that $\neg Q \rightarrow (P \rightarrow (Q \rightarrow R))$ is tautologous by writing a complete truth table for this sentence.

E 1-12 Without computing the complete truth table, write a brief argument to show that $A \rightarrow (B \rightarrow (C \rightarrow A))$ is tautologous.

E 1-13 Show that $(((A \rightarrow B) \wedge \neg B) \wedge (C \vee A)) \rightarrow (C \vee D)$ is tautologous and that its converse is not.

E 1-14 Write a brief argument to show that $\neg(X \leftrightarrow Y)$ and $(\neg X \leftrightarrow Y)$ are tautologically equivalent.

E 1-15 Most English dictionaries give 'if not' as one standard meaning of 'unless'. Consider the example, 'The patient will die unless the doctor operates', which means 'The patient will die if the doctor does not operate'. Thus, 'A

unless B' means the same as 'A if not B', which means the same as 'If not B, then A'. Let us extend Table 1.1 by adding $(\neg B \to A)$ as the Conventional $\mathcal{SC}$ Translation of 'A unless B'.

1. Symbolize the following as $\mathcal{SC}$ sentences: 'A unless B', 'If not A, then B', 'A or B', and 'B unless A'.

2. Prove that all of the resulting $\mathcal{SC}$ sentences are pairwise tautologically equivalent to each other, i.e., any two of them are equivalent.

3. 'Unless' is often used with a negative part as in 'You will not get the job unless you pass the test', and 'The patient will not live unless the doctor operates'. Symbolize these two sentences and informally explain how they express necessary conditions.

E 1-16 (Answer Provided) Give an interpretation that satisfies the following set of sentences, and briefly show the calculations of the truth values of each sentence.

$$\{\, P \to \neg Q, \;\; Q \wedge R, \;\; \neg R \vee Q \,\}$$

E 1-17 Give an interpretation that satisfies the following set of sentences, and briefly show the calculations of the truth values of each sentence.

$$\{\, A \vee (B \to C), \;\; B \wedge \neg D, \;\; A \to D, \;\; \neg(E \to (C \wedge D)) \,\}$$

E 1-18 Give an interpretation that satisfies the following set of sentences, and briefly show the calculations of the truth values of each sentence.

$$\{\, A \wedge (D \vee \neg E), \;\; B \to \neg D, \;\; \neg(C \wedge A), \;\; C \vee (\neg A \vee B) \,\}$$

E 1-19 Show that the following set of sentences is satisfiable.

$$\{\, A \vee (B \vee \neg C), \;\; \neg A \wedge P, \;\; P \to \neg B, \;\; C \leftrightarrow A \,\}$$

E 1-20 Show that the conclusion of the following argument is not a tautological consequence of the (three) premises:

1. $A \to B$

2. $C \to B$

3. $D \to B$

4. Therefore: $(A \wedge C) \to D$

E 1-21 Show that the conclusion of the following argument is a tautological consequence of the (three) premises. (Hint. This can be done tediously by using truth tables, but for a short argument, consider what is required in order for the conclusion to be false.)

1. $A \rightarrow B$

2. $C \rightarrow B$

3. $B \rightarrow D$

4. Therefore: $(A \wedge C) \rightarrow D$

E 1-22 Show that $\neg(A \vee S) \rightarrow B$ is not a tautological consequence of $\Gamma = ((A \vee B) \leftrightarrow R), \ \neg(A \vee R))$.

E 1-23

1. Show that $\{ Q, (Q \rightarrow H) \leftrightarrow R, (R \rightarrow S) \}$ does not tautologically imply $(Q \wedge \neg S) \rightarrow H$.

2. Show that $\{ (P \vee D) \vee Q, D \rightarrow B, \neg D \rightarrow (Q \vee P) \}$ does not tautologically imply $P \vee B$.

3. Show that $\{ \neg(A \rightarrow B), \neg B, (A \rightarrow C) \vee E \}$ does not tautologically imply $C \leftrightarrow B$.

E 1-24 In a murder trial, an expert medical witness testifies, "The physical evidence is consistent with the proposition that the victim was hit on the head with a hammer." Although it is an oversimplification, for the sake of the example, interpret the use of 'consistent' here as tautologically consistent. Explain why the expert's statement *does not* mean or imply: "The physical evidence (tautologically) implies the proposition that the victim was hit on the head with a hammer."

E 1-25 Let $\Gamma = \{ S, P \leftrightarrow Q, Q \vee R \}$ and ϕ be $S \rightarrow \neg R$. (1). Show that ϕ is not a tautological consequence of Γ. (2). Show that $\Gamma \cup \{ \phi \}$ is satisfiable.

E 1-26 Construct a set of $\mathcal{SC}$ sentences with the following properties: It contains three different sentences and it is unsatisfiable. Show that it is unsatisfiable.

E 1-27 Let $\Gamma = \{ (P \vee Q) \rightarrow R, Q \rightarrow (P \vee R), R \rightarrow Q \}$. Show that Q is tautologically consistent with Γ and that Q is tautologically independent of Γ.

E 1-28 For each of the following sets of sentences, either show that it is satisfiable with a suitable $\mathcal{SC}$ interpretation, or prove that it is not satisfiable by writing an argument about possible interpretations.

1. $\{\, X \leftrightarrow \neg Y,\ (W \vee Z) \rightarrow X,\ Y \vee \neg X,\ \neg W \leftrightarrow Z \,\}$

2. $\{\, K \rightarrow (L \wedge M),\ \neg K \rightarrow (L \wedge \neg L) \,\}$

3. $\{\, A \rightarrow (Q \vee C),\ T \rightarrow A,\ \neg C \wedge T,\ T \leftrightarrow \neg Q \,\}$

4. $\{\, Y \vee (Z \rightarrow X),\ X \rightarrow Y,\ (Y \vee \neg Z) \leftrightarrow X,\ \neg X \,\}$

5. $\{\, R \rightarrow (S \vee T),\ S,\ R \vee S,\ (R \vee T) \leftrightarrow \neg S \,\}$

6. $\{\, A \vee B,\ A \vee \neg B,\ \neg A \vee B,\ \neg A \vee \neg B \,\}$

E 1-29 Use truth tables to show that $\neg(P \wedge Q)$ is tautologically equivalent to $\neg P \vee \neg Q$.

E 1-30 Use truth tables to show that $P \wedge (Q \vee R)$ is tautologically equivalent to $(P \wedge Q) \vee (P \wedge R)$.

E 1-31 Use truth tables to show that $\neg(A \rightarrow B)$ is tautologically equivalent to $A \wedge \neg B$.

E 1-32 Let Γ and Δ be sets of $\mathcal{SC}$ sentences. Either give a proof or give a counterexample for each of the following statements.

1. If Γ is satisfiable and Δ is satisfiable, then $\Gamma \cup \Delta$ is satisfiable.

2. If Γ is satisfiable and Δ is satisfiable, then $\Gamma \cap \Delta$ is satisfiable.

3. If $\Gamma \cup \Delta$ is satisfiable, then Γ is satisfiable.

E 1-33 (Answer Provided) *Metatheorem.* Let τ be a tautology and Γ be any set of $\mathcal{SC}$ sentences. Then $\Gamma \vDash_T \tau$. (See the general instructions at the start of this exercise set.)

E 1-34 (Answer Provided) *Metatheorem.* Let ρ be an $\mathcal{SC}$ sentence. Then $\varnothing \vDash_T \rho$ iff ρ is tautologous.

E 1-35 *Metatheorem.* Let ρ be any $\mathcal{SC}$ sentence and let τ be a tautology. Then $\tau \vDash_T \rho$ iff ρ is tautologous.

E 1-36 *Metatheorem.* Let ϕ, ψ, χ be $\mathcal{SC}$ sentences. If $(\phi \rightarrow \psi)$, $(\psi \rightarrow \chi)$, $(\chi \rightarrow \phi)$ are each tautologous, then each of ϕ, ψ, χ is tautologically equivalent to any other one. In other words, ϕ, ψ, χ are pairwise tautologically equivalent.

E 1-37 *Metatheorem.* Let Γ, Δ be sets of $\mathcal{SC}$ sentences, and let ϕ, ψ, χ be $\mathcal{SC}$ sentences. Suppose that $\Gamma \vDash_T \phi$ and $\Delta \vDash_T \psi$. Then

Assume $A \not\perp B$.

$$\text{if } \{\, \phi,\, \psi \,\} \vDash_T \chi, \text{ then } \Gamma \cup \Delta \vDash_T \chi.$$

E 1-38 *Metatheorem.* Let Γ be a set of $\mathcal{SC}$ sentences and ϕ be an $\mathcal{SC}$ sentence.

1. If Γ is unsatisfiable, then for any ϕ, $(\phi \wedge \neg\phi)$ is a tautological consequence of Γ.

2. If there is some ϕ such that $(\phi \wedge \neg\phi)$ is a tautological consequence of Γ, then Γ is unsatisfiable.

E 1-39 *Metatheorem.* Let ϕ, ψ be any $\mathcal{SC}$ sentences. Then $(\phi \wedge \neg\phi) \vDash_T \psi$.

E 1-40 *Metatheorem.* Let Γ be a set of $\mathcal{SC}$ sentences and ϕ be an $\mathcal{SC}$ sentence, then $\Gamma \vDash_T \phi$ iff $\Gamma \cup \{\, \neg\phi \,\}$ is unsatisfiable.

E 1-41 *Metatheorem.* Let ϕ_1, ϕ_2, ψ be $\mathcal{SC}$ sentences. Then

1. If $\{\, \phi_1,\, \phi_2 \,\} \vDash_T \psi$, then $\phi_1 \to (\phi_2 \to \psi)$ is tautologous.

2. If $\phi_1 \to (\phi_2 \to \psi)$ is tautologous, then $(\phi_1 \wedge \phi_2) \to \psi$ is tautologous.

3. If $(\phi_1 \wedge \phi_2) \to \psi$ is tautologous, then $\{\, \phi_1,\, \phi_2 \,\} \vDash_T \psi$.

E 1-42 *Metatheorem.* Let Γ be a set of $\mathcal{SC}$ sentences, and ϕ be an $\mathcal{SC}$ sentence. If $\Gamma \vDash_T \phi$, then for any $\mathcal{SC}$ sentence ψ, $\Gamma \cup \{\, \neg\phi \,\} \vDash_T \psi$.

E 1-43 *Metatheorem.* Let ϕ, ψ, δ, γ be $\mathcal{SC}$ sentences, and let Γ be a set of $\mathcal{SC}$ sentences. If $\Gamma \vDash_T (\phi \vee \psi)$, $\Gamma \vDash_T (\phi \to \delta)$, $\Gamma \vDash_T (\psi \to \gamma)$, then $\Gamma \vDash_T (\delta \vee \gamma)$.

E 1-44 *Metatheorem.* Let Γ be a set of $\mathcal{SC}$ sentences, and τ be a tautology. Then, Γ is unsatisfiable iff $\Gamma \vDash_T \neg\tau$. (Remark: Remember that "iff" means "if and only if." Since the conclusion of this theorem is an iff-statement, one must prove that the left side of the iff-statement implies the right side, and also prove that the right side implies the left side.)

E 1-45 Either prove that the following assertion is true, or prove that it is false by a counterexample: For any satisfiable set Γ of $\mathcal{SC}$ sentences, and any contingent $\mathcal{SC}$ sentence ψ, $\Gamma \cup \{\, \psi \,\}$ is satisfiable.

E 1-46 Repeat the previous exercise with the one change that ψ is a tautologous $\mathcal{SC}$ sentence.

1.5   Principles for Sentential Calculus Derivations

Definition 1-8 characterizes *tautological consequence* in terms of interpretations. Since interpretations are part of the semantics of sentential calculus, it is said that this definition provides a semantical concept of tautological consequence. This semantical concept is very important, since it can be generalized to other formal languages with other kinds of interpretations. It also provides the basis for counterexamples to unsound arguments. However, it does not yield a very efficient means for finding tautological consequences of premises or for proving that a given sentence is a tautological consequence of a set of premises. For these purposes, it is usually more convenient to construct proofs (or derivations) within the sentential calculus itself. The general investigation of such proofs is the concern of the *proof theory* of sentential calculus, as opposed to its *semantical theory*. The proof theory of a formal language begins with the characterization of the general form of a proof and the statement of rules (and sometimes axioms) for constructing proofs. We will usually refer to proofs as *derivations*, but both terms will be used. This section describes the general structure of SC derivations and introduces the rules for constructing them. Section 1.6 discusses the adequacy of the rules and provides guidance for using them in proofs and applications.

Definition 1-17 A sentential calculus (SC) *derivation* (or *proof*) is a finite sequence of lines, each with the following form:

$$\{\, Pr_{i_1},\; Pr_{i_2},\; \cdots ,\; Pr_{i_k} \,\} \quad (i) \quad \phi$$

which satisfies the following conditions:

The number i is a positive integer and is the *line number* of the line. For the first line, $i = 1$. If i is the line number of any line, the number of the next line (if any) is $i + 1$.

ϕ is an SC sentence called *the sentence on line i*.

For each j such that $1 \leqslant j \leqslant k$, Pr_{i_j} is a *premise name* and denotes the sentence on line i_j.

When nonempty, $\{\, Pr_{i_1},\; Pr_{i_2},\; \cdots ,\; Pr_{i_k} \,\}$ is the *set of premise names of line i*. The set of sentences on lines $i_1,\; i_2,\; \cdots ,\; i_k$, is the set of *premises* of line i.

Each line of the sequence can be obtained (or generated) by an application of one of the *sentential calculus derivation rules* presented in this section.

This is a complicated definition that requires some time to comprehend fully. To illustrate some of the ideas, a line of a derivation could look like this:

$$\{ Pr_1, Pr_3, Pr_4 \} \qquad (9) \qquad P \to \neg Q$$

In this example, the line number is 9, and the sentence on that line is $P \to \neg Q$. The set of premise names of this line is $\{Pr_1, Pr_3, Pr_4 \}$. Referring to the definition, we have $i = 9$, and $Pr_{i_1} = Pr_1$. Thus, $i_1 = 1$, $i_2 = 3$, $i_3 = 4$, and $k = 3$. Also, Pr_1 denotes the sentence on line (1), Pr_3 denotes the sentence on line (3), and Pr_4 denotes the sentence on line (4). These three sentences on lines (1), (3), and (4) are not shown in the example. Whatever they may be, they constitute the *premises* of line (9).

The last condition in this definition does not mean that a line *could not* be obtained by using a different rule, or by using several rules together. It means that the application of one rule is sufficient to generate the line. Many different sets of derivation rules can be used. Some logicians prefer rule sets with very few rules. This has some theoretical advantages, especially for proving certain metatheorems. Unfortunately, if one begins with very few rules, SC derivations are often difficult to construct and lack intuitive appeal. The set of rules that we will use is rather large, but after one becomes familiar with these rules and some strategies for proofs, most derivations will be easy.

Unless otherwise specified, for the remainder of this chapter, 'derivation' will mean the same as 'SC derivation'. Because every line of a derivation has both a sentence on that line as well as a set of premise names, the rules must specify how to obtain both the sentence and the set of premise names. The premise names are not SC sentences, but they serve as aliases for SC sentences. It will turn out that each premise name denotes exactly one SC sentence that is used as a *premise* (or assumption) somewhere in the proof. The set of premise names of a line will serve as an alias for the set of premises of that line, i.e., a set of sentences on earlier lines of the proof that are assumptions from which the current line has been derived by using rules. It will be seen that, in special situations, the set of premise names may be empty. We want to develop a system of rules for derivations that has *at least* the following property:

> If ϕ is the sentence on any line of any derivation constructed in accordance with the rules, then ϕ is a tautological consequence of the set of premises of that line.

Using sets of premise names is just an abbreviated way of writing sets of premises; we could do without the names and just write the sets of premises. This discussion should alleviate the surprise that the first rule often evokes. Each rule justifies a sentence being placed on a new line of the proof and also specifies the set of premise names of this new line. A rule is correctly used only if both of these items agree with the rule. Here is the first rule:

RULE P (Introduction of premises). Any $\mathcal{SC}$ sentence may be placed on a new line (n) of a derivation. The set of premise names is $\{ Pr_n \}$.

In other words, we may write:

$$\{ Pr_n \} \qquad (n) \qquad \phi$$

where ϕ is the $\mathcal{SC}$ sentence on the line. By the definition of $\mathcal{SC}$ derivation, n is either 1 (if this is the first line of a proof) or else is the next integer greater than the line number of the previous line of the proof. Referring back to the definition of a derivation, we see that the 'n' in Rule P corresponds to i in the definition. Also, $k = 1$ and $i_1 = n$; the only premise name required by Rule P is Pr_n. *It is very important to notice that the n in Pr_n is the same number as the line number (n)*! The rule is stated in this way so that this subscript points directly back to the line on which ϕ is introduced as a premise.

To aid the reader of proofs, we add a fourth column to each line. This column contains a brief comment that documents which rule is being used and to which earlier line(s) (if any) it is applied. Doing this is similar to writing comments in computer programs; these comments are not part of the derivation, just as program comments are not part of the executable program code; but we will follow the good practice of documenting proofs with comments. Thus, instead of just writing the three items previously mentioned, we write:

$$\{ Pr_n \} \qquad (n) \qquad \phi \quad \text{P}$$

where the 'P' is a comment that this line is obtained by an application of Rule P.

Rule P often surprises people because, at first glance, it may appear hare-brained to allow *any* $\mathcal{SC}$ sentence ϕ to be placed on a line. But one need not worry. We are not claiming that ϕ is true (under any particular interpretation). The line of the proof claims only that ϕ is a tautological consequence of its set of premises. This set of premises is the set of sentences indicated by the set of premise names of this line, which is $\{ Pr_n \}$; but Pr_n denotes the sentence on line n, which is just ϕ. Thus, a line added by Rule P amounts to saying that ϕ tautologically implies ϕ, which is certainly true. The chart titled "Tautological Implication Rules" states several rules that permit one to infer a new line of a derivation on the basis of one or more previous lines. These rules are stated in very general form; their use is discussed and illustrated with examples on the following pages.

TAUTOLOGICAL IMPLICATION RULES. Each of these rules has the general form: From Γ, infer ρ, where Γ is a set of sentences of *specified forms* on existing lines of a derivation, and ρ is a sentence *of a specified form* to be placed on a new line of the derivation. For each of these rules, the set of premise names of the new line (with ρ) consists of the union of the sets of premise names of all the lines to which the rule is applied. The order in which the sentences in Γ occur on earlier lines of the proof does not matter. The specific rules are listed below. Some are presented in more than one form for convenience. Each rule has a traditional or conventional name together with an abbreviation of this name. The lowercase Greek letters stand for $\mathcal{SC}$ sentences.

Modus Ponens (MP)

From $\{\ \phi,\ \phi \rightarrow \psi\ \}$ infer ψ

Modus Tollens (MT)

From $\{\ \phi \rightarrow \psi,\ \neg\psi\ \}$ infer $\neg\phi$

From $\{\ \phi \rightarrow \neg\psi,\ \psi\ \}$ infer $\neg\phi$

Disjunctive Syllogism (DS)

From $\{\ \phi \lor \psi,\ \neg\phi\ \}$ infer ψ

From $\{\ \phi \lor \psi,\ \neg\psi\ \}$ infer ϕ

From $\{\ \neg\phi \lor \psi,\ \phi\ \}$ infer ψ

From $\{\ \phi \lor \neg\psi,\ \psi\ \}$ infer ϕ

Simplification (Simp)

From $\{\ \phi \land \psi\ \}$ infer ϕ

From $\{\ \phi \land \psi\ \}$ infer ψ

Conjunction (Conj)

From $\{\ \phi,\ \psi\ \}$ infer $\phi \land \psi$

Hypothetical Syllogism (HS)

From $\{\ \phi \rightarrow \psi,\ \psi \rightarrow \eta\ \}$ infer $\phi \rightarrow \eta$

Addition (Add)

From $\{\ \phi\ \}$ infer $\phi \lor \psi$ (ψ may be any sentence)

From $\{\ \phi\ \}$ infer $\psi \lor \phi$ (ψ may be any sentence)

Contradictory Premises (ContraPrm)

From $\{\ \phi, \neg\phi\ \}$ infer ψ (ψ may be any sentence)

Cut Rule (Cut)

From $\{\ \phi \vee \psi, \neg\phi \vee \eta\ \}$ infer $\psi \vee \eta$

From $\{\ \phi \vee \psi, \eta \vee \neg\phi\ \}$ infer $\psi \vee \eta$

From $\{\ \phi \vee \psi, \neg\psi \vee \eta\ \}$ infer $\phi \vee \eta$

From $\{\ \phi \vee \psi, \eta \vee \neg\psi\ \}$ infer $\phi \vee \eta$

Law of Clavius (Clav)

From $\{\ \neg\phi \to \phi\ \}$ infer ϕ

From $\{\ \phi \to \neg\phi\ \}$ infer $\neg\phi$

Constructive Dilemma (CD)

From $\{\ \phi \vee \psi, \phi \to \delta, \psi \to \gamma\ \}$ infer $\delta \vee \gamma$

In order to apply these rules, one must carefully *match* the general sentence *forms* in the rules to the particular sentences in one's proof. Here is a short derivation example using some of these rules.

$\{\ Pr_1\ \}$	(1)	$P \to (R \vee S)$	P
$\{\ Pr_2\ \}$	(2)	$\neg R$	P
$\{\ Pr_3\ \}$	(3)	P	P
$\{\ Pr_1, Pr_3\ \}$	(4)	$R \vee S$	MP (1), (3)
$\{\ Pr_1, Pr_2, Pr_3\ \}$	(5)	S	DS (2), (4)

Lines (1) – (3) are premises. Notice that the sentences on lines (1) and (3) match the patterns $\phi \to \psi$ and ϕ, so we can apply Modus Ponens to obtain line (4). Line (4) is documented with the comment in the rightmost column; this comment notes that Modus Ponens was applied to lines (1) and (3).

There is nothing mysterious about the pattern-matching process. MP says: From $\{\ \phi, \phi \to \psi\ \}$ infer ψ. Let $\phi = P$, $\psi = (R \vee S)$, and substitute these specific sentences into the statement of MP. The result is

From $\{\ P, P \to (R \vee S)\ \}$ infer $(R \vee S)$.

This statement is a specific instance of the general pattern of inference expressed by the MP rule. The set of sentences on the left is exactly the set of sentences on lines (1) and (3) (their order is irrelevant). So, we are entitled by MP to infer $(R \lor S)$ and place it on a new line of the derivation. This is done on line (4), where the outer parentheses were omitted by convention. The set of premise names of (4) is the union of the sets of premise names of (1) and (3).

Continuing the proof, lines (2) and (4) match one of the four patterns of the Disjunctive Syllogism Rule, so this rule is used to obtain (5) and its premise names. This derivation ends with (5); it shows that S can be derived from the set of premises named by { Pr_1, Pr_2, Pr_3 }, which are the sentences on lines (1), (2), and (3). Notice that the only thing we have done is to *derive* the conclusion by the following rules. It is not yet clear how such a derivation is related to tautological consequences. I stated earlier that we want the sentence on the line to be a tautological consequence of the premises denoted by the premise names of that line. This condition is true, but it has not been proved. The reason it is true is discussed in the next section.

We can illustrate this derivation with an argument in English that has the same form. Let the sentential letters stand for English sentences as follows. P: there is a short circuit, R: the fuse is burned out, and S: a wire is sparking. Using this representation, the $\mathcal{SC}$ derivation corresponds to the following argument:

1. If there is a short circuit, then either the fuse is burned out or a wire is sparking. (Premise)

2. The fuse is not burned out. (Premise)

3. There is a short circuit. (Premise)

4. Either the fuse is burned out or a wire is sparking. (From 1 and 3)

5. (Therefore) A wire is sparking. (From 2 and 4)

The forms of sentences specified in the tautological implication rules must match *entire lines*. One type of error often made by beginners is to violate this pattern-matching constraint. Here is an example.

Error Example.

| { Pr_1 } | (1) | $(P \land Q) \to R$ | P |
| { Pr_1 } | (2) | $Q \to R$ | Simp (1) **rule use error** |

Line (2) does *not* result from an application of the Simplification Rule! This rule can be applied only to an entire line that is a conjunction, but (1) is not a conjunction; it is a conditional. It is easy to prove by counterexample that

(2) is not a tautological consequence of (1). Use the interpretation given by $\mathbf{I}(P) = \mathbf{F}, \mathbf{I}(Q) = \mathbf{T}, \mathbf{I}(R) = \mathbf{F}$. Under $\mathbf{I}$, (1) is true, but (2) is false. This shows the power of the semantical theory developed in the previous two sections.

Using commonsense logical intuitions, one can see that this is not a sound *form* of argument. Consider this instance of the form:

If Bob is rich and Bob is famous, then Bob is happy.

Therefore: If Bob is famous, then Bob is happy. **error**

Most people see right away that this is a bad argument. Being rich *and* being famous are sufficient for Bob to be happy. To be sure, Bob has rather questionable sufficient conditions for happiness, but they do not imply that the even more limited single condition, being famous, is sufficient. In general, people have correct intuitions about simple, concrete arguments, but they sometimes produce horrible fallacies when learning to construct formal derivations. One moral is this: When constructing a formal derivation, do not just manipulate the symbols, always try to think about what they *could* mean, i.e., think about the possible interpretations that can be applied to them. ∎

In addition to applying to entire lines, there is another feature common to all of the tautological implication rules: Each of them permits us to infer a sentence on a new line from a set of sentences on previous lines that tautologically implies the sentence on the new line. This has the effect that they are all *one-way rules*, i.e., they permit us to move from a set of sentences on previous lines to a sentence on a new line. Yet these rules do not enable us to make inferences in the opposite direction, from the inferred sentence to the sentences that were on the previous lines. On the other hand, the previous section showed that some pairs of sentences are *tautologically equivalent* in the sense that each implies the other. This suggests that there might also be some useful rules that are two-way rules and can be applied in either direction. This is true, and the rules we will use are stated in the chart titled, "Tautological Equivalence Rules." When studying and using these rules, keep in mind that sentences ϕ and ψ are tautologically equivalent iff $(\phi \leftrightarrow \psi)$ is tautologous, as was proved in Metatheorem 1-15. Also remember that a *subsentence* of a sentence ϕ is either ϕ itself or else a string of symbols within ϕ that is also an $\mathcal{SC}$ sentence. The previous error example showed that the (Tautological) Implication Rules must be applied to entire lines. Because the Equivalence Rules permit substitutions (replacements) for subsentences, that restriction does not apply to the Equivalence Rules.

TAUTOLOGICAL EQUIVALENCE RULES. Let $\theta \leftrightarrow \rho$ be one of the tautologous biconditional forms listed below. Let $\phi(\theta)$ be a sentence containing θ as a subsentence (so it is possible that θ is all of ϕ). Let $\phi(\theta/\rho)$ be the result of substituting one occurrence of θ in ϕ with one occurrence of ρ. If $\phi(\theta)$ is on an existing line of a derivation, then $\phi(\theta/\rho)$ may be placed on a new line. Similarly, if $\phi(\rho)$ is on an existing line of a derivation, then $\phi(\rho/\theta)$ may be placed on a new line, where $\phi(\rho/\theta)$ results from replacing an occurrence of ρ in ϕ by an occurrence of θ. In either case, the set of premise names of the new line is the same as the set of premise names of the existing line. The tautologous biconditionals are listed here with traditional or conventional names. Some of the rules have more than one form. The lowercase Greek letters stand for $\mathcal{SC}$ sentences.

Double Negation (DN)

$$\phi \leftrightarrow \neg\neg\phi$$

De Morgan's Theorem (DeM)

$$\neg(\phi \wedge \psi) \leftrightarrow (\neg\phi \vee \neg\psi)$$
$$\neg(\phi \vee \psi) \leftrightarrow (\neg\phi \wedge \neg\psi)$$

Commutation (Comm)

$$(\phi \vee \psi) \leftrightarrow (\psi \vee \phi)$$
$$(\phi \wedge \psi) \leftrightarrow (\psi \wedge \phi)$$

Association (Assoc)

$$(\phi \vee (\psi \vee \chi)) \leftrightarrow ((\phi \vee \psi) \vee \chi)$$
$$(\phi \wedge (\psi \wedge \chi)) \leftrightarrow ((\phi \wedge \psi) \wedge \chi)$$

Distribution (Dist)

$$(\phi \wedge (\psi \vee \chi)) \leftrightarrow ((\phi \wedge \psi) \vee (\phi \wedge \chi))$$
$$(\phi \vee (\psi \wedge \chi)) \leftrightarrow ((\phi \vee \psi) \wedge (\phi \vee \chi))$$

Contraposition (ContraPos)

$$(\phi \rightarrow \psi) \leftrightarrow (\neg\psi \rightarrow \neg\phi)$$

Conditional–Disjunction (CDis)

$$(\phi \rightarrow \psi) \leftrightarrow (\neg\phi \vee \psi)$$

Negation of Conditional (NC)

$$\neg(\phi \rightarrow \psi) \leftrightarrow (\phi \wedge \neg\psi)$$

"Tautology" Laws (Taut)
$$\phi \leftrightarrow (\phi \wedge \phi)$$
$$\phi \leftrightarrow (\phi \vee \phi)$$

Equivalence (Equiv)
$$(\phi \leftrightarrow \psi) \leftrightarrow ((\phi \rightarrow \psi) \wedge (\psi \rightarrow \phi))$$
$$(\phi \leftrightarrow \psi) \leftrightarrow ((\phi \wedge \psi) \vee (\neg\psi \wedge \neg\phi))$$

SC Derivation Example.

$\{ Pr_1 \}$	(1)	$P \rightarrow \neg(P \wedge Q)$	P
$\{ Pr_1 \}$	(2)	$P \rightarrow (\neg P \vee \neg Q)$	DeM (1)
$\{ Pr_1 \}$	(3)	$\neg P \vee (\neg P \vee \neg Q)$	CDis (2)
$\{ Pr_1 \}$	(4)	$(\neg P \vee \neg P) \vee \neg Q$	Assoc
$\{ Pr_1 \}$	(5)	$\neg P \vee \neg Q$	Taut (4)
$\{ Pr_1 \}$	(6)	$P \rightarrow \neg Q$	CDis (5)

Notice that DeM was applied to a subsentence of line (1), and line (5) results from applying Taut to part of (4). This example illustrates how the tautological equivalence rules can be usefully applied in transforming a sentence into a tautologically equivalent sentence in a different form. In this case, the sentence on line (1) is transformed into the one on (6). It is instructive to write another derivation beginning with $P \rightarrow \neg Q$ on line (1) and ending with $P \rightarrow \neg(P \wedge Q)$ on line (6). This will illustrate how the tautological equivalence rules can be used in either direction. Of course, it has not yet been *justified* that (1) and (6) are tautologically equivalent. The general issue of justification of the rules is discussed in the next section.

Here is an argument in English that is an instance of this sentential calculus derivation:

If she runs for President (P), then it is not the case that she both runs for president (P) and runs for the Senate (Q).

If she runs for President, then either she does not run for President or she does not run for the Senate.

She does not run for President, or either she does not run for President or she does not run for the Senate.

Either she does not run for President or she does not run for President, or she does not run for the Senate.

She does not run for President or she does not run for the Senate.

If she runs for President, then she does not run for the Senate. ■

There are some interesting facts about conditionals. Suppose that we have derived a sentence ψ on a line with premise names Γ. By the Addition Rule, we can obtain $\neg \phi \vee \psi$, where ϕ is any sentence. Then, by CDis, we obtain $\phi \rightarrow \psi$, with premise names Γ. And so, if we have derived ψ, we can use any sentence ϕ as the antecedent of a conditional, $\phi \rightarrow \psi$, and derive this conditional on a line with the same premise names as those of ψ. Also, if we have derived $\neg\phi$ with premise names Γ, then by Add, we obtain $\neg\phi \vee \psi$, where ψ is any sentence. Thus, we can also obtain $\phi \rightarrow \psi$, with the same premise names as those of $\neg\phi$. These are procedures that are sometimes useful in derivations.

Conditionals are often proved by arguments that are more subtle and complex than those described in the preceding paragraph. This is especially true in mathematics, for most theorems are conditionals. To take a familiar example, recall that, in plane Euclidean geometry, an isosceles triangle is one that has two sides of equal length. One of the basic theorems of geometry states that the angles opposite the equal sides of an isosceles triangle are equal. To prove this, one begins with an assumption or premise something like: 'Let ABC be a triangle with side AB equal to side AC.' One then proceeds with a sequence of inferences ending with 'Angle ABC equals angle ACB.' Assuming that the sequence of inferences is correct, this shows that 'ABC is a triangle with side AB equal to side AC' is *sufficient* for 'Angle ABC equals angle ACB.' Most proofs that I have seen stop at this point. The writers of these proofs assume that the person reading them understands that the proof implies that 'If ABC is a triangle with side AB equal to side AC, then angle ABC equals angle ACB', which is one way to state the theorem. In other words, suppose that we assume ϕ as a premise and deduce ψ from it. This shows that ϕ is sufficient for ψ. We then infer the conditional, 'If ϕ, then ψ.' In the next section, I will justify this principle of inference. For now it is introduced as the last of the $\mathcal{SC}$ derivation rules, after some more terminology.

Definition 1-18 Any line in an $\mathcal{SC}$ derivation that has the general form

$$\{ \, Pr_k \, \} \quad (k) \quad \phi$$

is called a *premise line*.

A premise line is a line obtained by use of Rule P. In practice, the line is documented with P. However, even if P is absent, one can still recognize a premise line. A line is a premise line iff the number, k, of this line is the same as the subscript in the one premise name, Pr_k. The concept of a premise line is critical in the statement of Rule C. The rule is stated in two parts to emphasize two useful ways of obtaining conditionals. Actually, the first part is

not new, since it results from the combined use of Add and CDis to produce a
conditional in the way previously described.

RULE C (Conditionalization). Suppose that an $\mathcal{SC}$ derivation has a line of
the form

Γ $\qquad$ (n) $\quad$ ψ $\quad$ [comment]

Part 1. Let ϕ be any $\mathcal{SC}$ sentence, then $\phi \to \psi$ may be placed on a later line
with the same set of premise names Γ.

Part 2. If there is a premise line (k) (where possibly $k = n$) of the form

$\{\, Pr_k \,\}$ $\quad$ (k) $\quad$ ϕ $\qquad$ P

and Pr_k is in Γ, then $\phi \to \psi$ may be placed on a later line, using the premise
names Γ with Pr_k deleted.

There are two general ways of using Rule C. Schematically, the first way
looks like this:

$\cdots$ $\qquad\qquad$ $\cdots$ $\qquad\qquad\quad$ $\cdots$

Γ $\qquad\qquad$ (n) $\quad$ ψ $\qquad\qquad$ [comment]

$\cdots$ $\qquad\qquad$ $\cdots$ $\qquad\qquad\quad$ $\cdots$

Γ $\qquad\qquad$ (r) $\quad$ $\phi \to \psi$ $\quad$ C (n)

where ϕ is any sentence whatsoever. This schema uses Part 1 of Rule C, and
the set of premise names Γ is *not* changed from line (n) to line (r). The second
schema for using C is used much more often. Schematically,

$\cdots$ $\qquad\qquad$ $\cdots$ $\qquad\qquad\quad$ $\cdots$

$\{\, Pr_k \,\}$ $\qquad$ (k) $\quad$ ϕ $\qquad\qquad$ P (for CP)

$\cdots$ $\qquad\qquad$ $\cdots$ $\qquad\qquad\quad$ $\cdots$

$\Delta \cup \{\, Pr_k \,\}$ $\quad$ (n) $\quad$ ψ $\qquad\qquad$ [comment]

$\cdots$ $\qquad\qquad$ $\cdots$ $\qquad\qquad\quad$ $\cdots$

Δ $\qquad\qquad$ (r) $\quad$ $\phi \to \psi$ $\quad$ C (k), (n)

In this situation, $\Gamma = \Delta \cup \{\, Pr_k \,\}$, and Pr_k is not in Δ, so Δ refers to premises
other than ϕ. Thus, the set of premise names of line (r) is the same as that of
line (n) except that Pr_k has been deleted. This schema uses Part 2 of Rule C.

When using this second form of C, it is *absolutely essential* that the antecedent ϕ be an earlier premise, and that the premise name, Pr_k, that is

deleted refers to this premise ϕ. To help keep track of this premise to be deleted, we will normally comment this premise line with "P (for CP)", where "CP" stands for "Conditional Proof." This is explained in more detail in the next few paragraphs. The key idea is that we may delete the premise name of the antecedent, when this antecedent ϕ is an earlier premise in the proof. *Deleting any premise name other than the name of the antecedent ϕ is a misuse of Rule C and invalidates the derivation from that point on.*

It is important to notice that the first schema can always be used. It is a consequence of Add and CDis, as was previously mentioned. So, even if the situation fits the second schema, one can use the first form of the C Rule, with the result that no premise name is deleted. One is not forced to delete Pr_k. The most striking feature of this rule is the fact that it permits one to delete a premise name in certain situations. None of the previous rules allows this. Of course, the sentence ϕ in line (k) of the second schema does not totally disappear. Although its premise name is removed from line (r), ϕ is made the antecedent of the conditional on that line. The rule is quite natural. Suppose that, given the other premises Δ, ϕ is sufficient for ψ. This supposition corresponds to the situation in line (n) of the second schema. To say that 'ϕ is sufficient for ψ' is, intuitively, equivalent in meaning to saying '$\phi \rightarrow \psi$'. Thus, if given Δ, we have that ϕ is sufficient for ψ, then given Δ, we have '$\phi \rightarrow \psi$'. This corresponds to the situation in line (r). This pattern of inference will be proved to be sound in the next section using the semantical theory of the sentential calculus.

The capability of deleting premise names makes the $\mathcal{SC}$ derivation system more flexible. It even permits one to derive sentences that are on lines that have the empty set of premise names, for example,

| $\{\ Pr_1\ \}$ | (1) | P | P (for CP) |
| $\varnothing$ | (2) | $P \rightarrow P$ | C (1), (1) |

Rule C permits k and n to be the same number. Notice that (2) follows from (1) by using C with $\phi = P$ and $\psi = P$. Therefore, lines (k) and (n) in the second schema are both line (1) in this short derivation. Hence, line (2) is documented by C (1), (1).

Because Rule C always yields a conditional, it is used to infer conditional sentences, as in the previous derivation; and in so far as many mathematical theorems have the form of conditionals, Rule C is very useful. This rule leads to a natural strategy for proving conditionals.

> ***Conditional Proof* (CP) *Strategy*:** Suppose that we want to derive a sentence of the form $\phi \rightarrow \psi$. Assume ϕ as a new (additional) premise, and try to derive ψ with the help of this new premise. If this can be done, then by (the second part of) Rule C, one can infer

the desired goal sentence, $\phi \to \psi$, while at the same time deleting the new, assumed premise.

This strategy was used in the previous two-line proof, in which the "additional premise" was the only premise used. When using the Conditional Proof strategy, it is helpful to comment premises used for CP with the comment, "P (for CP)." The strategy is also illustrated in the following examples:

Examples.

Problem Statement. Given the one premise $A \to B$, the goal is to prove

$$(C \lor A) \to (C \lor B).$$

Discussion. The goal sentence to be proved is a conditional. This suggests trying the Conditional Proof strategy. In particular, add $C \lor A$ as a new premise and try to derive $C \lor B$. If this can be done, then invoke Rule C (the second part) to obtain $(C \lor A) \to (C \lor B)$ while deleting the new premise. Of course, additional work must be done between the line containing the new premise and the line that invokes Rule C, but this proof is quite short and simple.

Written Solution.

$\{\, Pr_1 \,\}$	(1)	$A \to B$	P
$\{\, Pr_2 \,\}$	(2)	$C \lor A$	P (for CP)
$\{\, Pr_1 \,\}$	(3)	$\neg A \lor B$	CDis (1)
$\{\, Pr_1, Pr_2 \,\}$	(4)	$C \lor B$	Cut (2), (3)
$\{\, Pr_1 \,\}$	(5)	$(C \lor A) \to (C \lor B)$	C (2), (4)

Further Discussion. Line (2) was added as an additional premise using Rule P. Again, the extra comment, '(for CP)', indicates that this line is intended to be used as a premise in a *Conditional Proof strategy* for deriving a desired conditional sentence. When this strategy is successful, the new premise (in this example, $C \lor A$) eventually becomes the antecedent of the desired conditional sentence (5), and the conclusion that is derived ($C \lor B$, on line (4)) becomes its consequent. Here is a similar derivation of a related result.

$\{\, Pr_1 \,\}$	(1)	$A \to B$	P
$\{\, Pr_2 \,\}$	(2)	$C \to A$	P (for CP)
$\{\, Pr_1, Pr_2 \,\}$	(3)	$C \to B$	HS (1), (2)
$\{\, Pr_1 \,\}$	(4)	$(C \to A) \to (C \to B)$	C (2), (3)

Both derivations use the CP (Conditional Proof) strategy, although they differ in the internal details of deriving the desired consequent with the help of the added premise. Incidentally, these derivations provide useful logical inferences. For example, suppose that

If it rains, then the streets are wet.

By the second derivation, it follows that

If it is the case that if there is a hurricane, then it rains; then, if there is a hurricane, then the streets are wet.

This conclusion can also be more simply stated as

If there is a hurricane only if it rains, then there is a hurricane only if the streets are wet. ∎

Additional proof strategies and examples will be presented later, but here is another example that illustrates some important features of proof construction. All of these discussions and proofs illustrate my way of thinking, but there are always alternative approaches to try. When studying any example, one should experiment with other avenues of proof construction.

Example.

Problem Statement. Write an $\mathcal{SC}$ derivation of $(P \rightarrow \neg Q) \rightarrow R$ from the set of premises $\Gamma = \{\ P \lor R,\ P \rightarrow (R \land Q),\ R \rightarrow (P \land Q)\ \}$.

Discussion. The final goal sentence to be proved is

$$(P \rightarrow \neg Q) \rightarrow R.$$

It is a conditional, i.e., has the form $\phi \rightarrow \psi$, in which the antecedent ϕ is itself a conditional, $P \rightarrow \neg Q$. This suggests using the Conditional Proof (CP) strategy: We use the antecedent $P \rightarrow \neg Q$ as an additional premise to help us derive the consequent R. If this succeeds, we then *conditionalize*: Use Rule C to obtain

$$(P \rightarrow \neg Q) \rightarrow R,$$

while simultaneously deleting the additional premise. Before starting this proof, we should consider whether using $P \rightarrow \neg Q$ as a new premise would be productive. One should always be on the lookout for significant patterns. The sentences in Γ fit the pattern of the Constructive Dilemma (CD) Rule, and if that rule is applied, we obtain

$$(R \land Q) \lor (P \land Q).$$

By two applications of commutation, this sentence is equivalent to

$$(Q \wedge R) \vee (Q \wedge P),$$

and the latter is equivalent to $Q \wedge (R \vee P)$, from which Q can be separated out by the Simp Rule. Applying MT to Q and the additional premise yields $\neg P$, which together with the very first premise yields R. Thus, the CP strategy will work here. The reasoning in this discussion should be simple and obvious to anyone sufficiently familiar with the rules. However, it takes time and practice to become familiar with all of these rules.

Written Solution.

$\{ Pr_1 \}$	(1)	$P \vee R$	P
$\{ Pr_2 \}$	(2)	$P \to (R \wedge Q)$	P
$\{ Pr_3 \}$	(3)	$R \to (P \wedge Q)$	P
$\{ Pr_4 \}$	(4)	$P \to \neg Q$	P (for CP)
$\{ Pr_1, Pr_2, Pr_3 \}$	(5)	$(R \wedge Q) \vee (P \wedge Q)$	CD (1), (2), (3)
$\{ Pr_1, Pr_2, Pr_3 \}$	(6)	$(Q \wedge R) \vee (P \wedge Q)$	Comm (5)
$\{ Pr_1, Pr_2, Pr_3 \}$	(7)	$(Q \wedge R) \vee (Q \wedge P)$	Comm (6)
$\{ Pr_1, Pr_2, Pr_3 \}$	(8)	$Q \wedge (R \vee P)$	Dist (7)
$\{ Pr_1, Pr_2, Pr_3 \}$	(9)	Q	Simp (8)
$\{ Pr_1, Pr_2, Pr_3, Pr_4 \}$	(10)	$\neg P$	MT (4), (9)
$\{ Pr_1, Pr_2, Pr_3, Pr_4 \}$	(11)	R	DS (1), (10)
$\{ Pr_1, Pr_2, Pr_3 \}$	(12)	$(P \to \neg Q) \to R$	C (4), (11) ■

Rule C is the last of the $\mathcal{SC}$ derivation rules we will use. The entire set of rules is summarized in Table 1.10. Additional derivation strategies and examples are in the next section, where we will also make use of Definition 1-19.

Premise Introduction	P
Tautological Implications	MP, MT, DS, Simp, Conj, HS, Add, ContraPrm, Cut, Clav, CD
Tautological Equivalences	DN, DeM, Comm, Assoc, Dist, ContraPos, CDis, NC, Taut, Equiv
Conditionalization	C

Table 1.10. *Summary of $\mathcal{SC}$ Derivation Rules*

Definition 1-19 Let Γ be a set of $\mathcal{SC}$ sentences and ϕ be an $\mathcal{SC}$ sentence. An $\mathcal{SC}$ derivation is *a derivation of ϕ from Γ* iff ϕ is on the last line of this derivation and all premises of that line are members of Γ. (Γ may have additional sentences that are not premises of the line.) ϕ is $\mathcal{SC}$ *derivable from* Γ (denoted by $\Gamma \vdash_T \phi$) iff there exists an $\mathcal{SC}$ derivation of ϕ from Γ. We write $\vdash_T \phi$, and say that ϕ is an $\mathcal{SC}$ *theorem*, iff $\varnothing \vdash_T \phi$.

$\Gamma \vdash_T \phi$ refers to *our* set of rules, although other sets of rules could have been used. We will use $\mathcal{SC}$ *proof* and $\mathcal{SC}$ *provable* as synonyms for $\mathcal{SC}$ *derivation* and $\mathcal{SC}$ *derivable*. The earlier two-line derivation of $P \rightarrow P$ from the empty set is a proof that $P \rightarrow P$ is an $\mathcal{SC}$ theorem. The relationship between the semantical concept of *tautological consequence* and the proof-theoretical concept of $\mathcal{SC}$ *derivable* is discussed in the next section.

Exercises

A General Instruction

Exercises for which answers are provided at the back of the book are marked with "Answer Provided."

E 1-47 (Answer Provided for 2) For each of the following arguments, give an $\mathcal{SC}$ derivation of the conclusion from the premise(s). The premises are given before the 'therefore' sign, $/ \therefore$, and the conclusion follows this sign. For instance, in the second problem the conclusion is C; and so, the task is to derive C from the premises, which are $(A \vee B) \rightarrow C$, $R \rightarrow A$, and R. The derivation (proof) must be a *formal derivation* using the format given in Definition 1-17 and using only the $\mathcal{SC}$ rules listed in Table 1.10.

1. $A \rightarrow B$, $C \vee A$ $/ \therefore (A \wedge B) \vee C$
2. $(A \vee B) \rightarrow C$, $R \rightarrow A$, R $/ \therefore C$
3. $A \vee \neg(B \wedge C)$ $/ \therefore B \rightarrow (C \rightarrow A)$
4. $P \vee Q$, $\neg(P \wedge A)$, $\neg Q$ $/ \therefore A \rightarrow B$
5. $P \rightarrow (Q \vee R)$, $R \rightarrow \neg P$, $\neg(P \wedge Q)$ $/ \therefore (P \rightarrow S)$
6. $L \rightarrow \neg A$, $\neg L \rightarrow (Z \vee Y)$, $\neg Y$ $/ \therefore (Y \vee A) \rightarrow Z$

E 1-48 Derive $A \rightarrow (A \vee B)$ from the empty set of premises. (Hint. Assume the antecedent, A, as a premise, derive $A \vee B$, then use Rule C.)

E 1-49 Derive the following sentence from the empty set of premises:

$$((P \rightarrow Q) \rightarrow P) \rightarrow P$$

E 1-50 The following is an **incorrect** "derivation" of $A \to C$ from $B \to C$.
This derivation contains a kind of mistake that is unfortunately rather common.
Identify the **error** in the alleged proof. Also, construct an interpretation that
is a counterexample to the claim made by line (6).

$\{\ Pr_1\ \}$	(1)	$B \to C$	P
$\{\ Pr_2\ \}$	(2)	$A \wedge B$	P
$\{\ Pr_2\ \}$	(3)	B	Simp (2)
$\{\ Pr_1, Pr_2\ \}$	(4)	C	MP (3), (1)
$\{\ Pr_2\ \}$	(5)	A	Simp (2)
$\{\ Pr_1\ \}$	(6)	$A \to C$	C (5), (4)

E 1-51 Derive K from $\Gamma = \{\ A \to (B \vee C),\ \neg B \wedge A,\ A \to \neg C\ \}$. (Hint. Γ is
unsatisfiable. If you can derive a contradiction from Γ, e.g., C on a line and
$\neg C$ on a line, then any sentence follows by ContraPrm.)

E 1-52 Give an $\mathcal{SC}$ derivation of $B \to T$ from the premises:

$$\{\ M \vee (B \wedge X),\ (M \leftrightarrow \neg B),\ \neg X\ \}.$$

(Hint. Notice that T is not in any premise. Which rule(s) does this suggest
might be useful?)

1.6   Adequacy and Use of Sentential Calculus Derivation Rules

1.6.1   Soundness and Completeness

It has been previously stated that rules for $\mathcal{SC}$ derivations should satisfy certain
adequacy requirements. The most important, indeed essential, requirement is
that the rules are guaranteed to produce tautological consequences.

Definition 1-20 A system of rules for $\mathcal{SC}$ derivations is said to be *sound* iff
for any sentence ϕ and any set of sentences Γ, if ϕ is $\mathcal{SC}$ derivable from Γ by
these rules, then ϕ is a tautological consequence of Γ (symbolically, if $\Gamma \vdash_T \phi$,
then $\Gamma \vDash_T \phi$).

We certainly want our set of derivation rules to be sound, for an unsound
set of rules would permit the inference of a sentence that is not a tautological
consequence of its premises. This would be a fallacious $\mathcal{SC}$ argument; one could
construct a counterexample to it. Fortunately, our rules are sound. This will
not be proved here, but let us consider what would be involved in such a proof.

First of all, consider Rule P. Because of the way it is stated, it does not
even appear to be an inference rule; yet it is. Consider any premise line:

{ Pr_k } (k) ϕ P

Like all lines in a derivation, this line has a set of premises denoted by the premise names of the line. In this case, there is only one premise, ϕ itself, so P states that we may infer ϕ from ϕ. But ϕ is a tautological consequence of itself, $\phi \vDash_T \phi$. If we had a one-line proof, that line would have to be a premise line. It would be a proof of ϕ from ϕ, and soundness would hold for it. To prove that derivations longer than one line are sound, one needs to do a lot of work. The basic idea is this: we prove that for any line,

Γ (n) ϕ [comment]

ϕ is a tautological consequence of premises of that line. To prove this, it is necessary to prove that no matter how a line is obtained, it satisfies this condition. Whereas new lines can be obtained by using any one of the many rules in our system, we need a proof for each rule. For example, consider MP. It works like this:

                From:
Γ (i) ϕ [comment]
$\ldots$ $\ldots$ $\ldots$

                and

$\ldots$ $\ldots$ $\ldots$
Δ (j) $\phi \to \psi$ [comment]
$\ldots$ $\ldots$ $\ldots$

                infer

$\ldots$ $\ldots$ $\ldots$
$\Gamma \cup \Delta$ (k) ψ MP (i), (j)

Suppose that ϕ is a tautological consequence of the premises of its line, named by Γ, and suppose that $\phi \to \psi$ is similarly a consequence of the premises Δ. Then we want to prove that ψ is a consequence of the premises of its line. Consider this metatheorem:

Metatheorem 1-21   Let Γ, Δ be sets of $\mathcal{SC}$ sentences, and ϕ, ψ be $\mathcal{SC}$ sentences. Suppose that $\Gamma \vDash_T \phi$ and $\Delta \vDash_T \phi \to \psi$. Then $\Gamma \cup \Delta \vDash_T \psi$.

Proof. Assume the antecedent of the conditional in the metatheorem, and let **I** be any interpretation that satisfies $\Gamma \cup \Delta$. We need to show that **I** also satisfies ψ. Because **I** satisfies every sentence in the union, $\Gamma \cup \Delta$, it must satisfy every sentence in Γ and every sentence in Δ, so **I** satisfies Γ and it satisfies Δ. Thus, **I** satisfies ϕ and it satisfies $\phi \to \psi$. Hence, **I** satisfies ψ, so $\Gamma \cup \Delta \vDash_T \psi$. **Q.E.D.**

We already know that the sentence on a one-line proof is a tautological consequence of the premise of that line. This metatheorem shows that, if the sentences on lines (i) and (j) are consequences of their premises, then applying MP to them yields another line whose sentence is a consequence of its premises (the union of the premises of line (i) and line (j)). We should prove similar metatheorems for all of our rules. Doing this for the tautological implication rules is easy but tedious. The proofs are all short and simple, like the one for MP.

The tautological equivalence rules require a little more work. It is easy to prove that all of the biconditionals we use are tautologous forms; this has already been done for the commutativity of conjunction in Metatheorem 1-12. These proofs essentially amount to construction of truth tables for the forms. It is also necessary to prove that making the substitutions of these tautologically equivalent forms is sound. The key idea is this: If θ and ρ are tautologically equivalent, then under any interpretation they have exactly the same truth value. This was proved in Metatheorem 1-16. Thus, substituting one for the other within another sentence ϕ produces a sentence tautologically equivalent to ϕ. Proving this requires some work and will not be done here, but it should be fairly obvious that it is true.

Because Rule C may permit one to delete a premise name, it may appear to be the most questionable of the rules, but its use can be justified by the following metatheorem. This metatheorem pertains to the second schema for applying Rule C; the first case of the rule has already been seen to follow from use of Add and CDis.

Metatheorem 1-22   (**Conditionalization**). If Γ is a set of $\mathcal{SC}$ sentences, and ϕ, ψ are $\mathcal{SC}$ sentences, then: $\Gamma \cup \{\phi\} \vDash_T \psi$ iff $\Gamma \vDash_T (\phi \rightarrow \psi)$.

Proof.

(1) Suppose that $\Gamma \cup \{\phi\} \vDash_T \psi$. Let **I** be any interpretation that satisfies Γ. We need to show that it also satisfies $\phi \rightarrow \psi$. Now $\mathbf{V_I}(\phi) = \mathbf{T}$ or $\mathbf{V_I}(\phi) = \mathbf{F}$. If $\mathbf{V_I}(\phi) = \mathbf{F}$, then $\mathbf{V_I}(\phi \rightarrow \psi) = \mathbf{T}$. If $\mathbf{V_I}(\phi) = \mathbf{T}$, then **I** satisfies $\Gamma \cup \{\phi\}$, so by hypothesis (the supposition), it satisfies ψ. Hence, **I** satisfies $\phi \rightarrow \psi$. In both cases, $\phi \rightarrow \psi$ is true under **I**, so $\Gamma \vDash_T (\phi \rightarrow \psi)$.

(2) Conversely, suppose that $\Gamma \vDash_T (\phi \rightarrow \psi)$. Let **I** be any interpretation that satisfies $\Gamma \cup \{\phi\}$. Then **I** satisfies Γ and ϕ. But, since **I** satisfies Γ and $\Gamma \vDash_T (\phi \rightarrow \psi)$, $\phi \rightarrow \psi$ is also true under **I**. Thus, ϕ and $\phi \rightarrow \psi$ are both true under **I**. Therefore, ψ is true under **I**. Hence, $\Gamma \cup \{\phi\} \vDash_T \psi$. **Q.E.D.**

The 'only if' part of this metatheorem (corresponding to the first part of the proof) is what justifies use of the second form of Rule C. It is interesting to see that the 'iff' statement can also be proved. Now suppose that we prove that each of the individual rules works in a sound manner as discussed above.

How do we know that they do so when all are used together? The reasoning goes like this: We know that a one-line proof is sound. We *assume* that any proof of k lines is sound, i.e., that a sentence on any of its lines is a tautological consequence of the premises of that line. If we add an additional line to the derivation, with a sentence ρ on this new line, then ρ is obtained by use of one of the rules. But we have, by hypothesis, already proved that each of these rules works in a sound manner. Thus, ρ must also be a tautological consequence of the premise(s) of its line. These premises are ρ (if we used Rule P), the premises of all earlier lines from which ρ was inferred (if we used a Tautological Implication or Equivalence Rule), or the premises specified by Rule C (if it was used). It follows from this that no matter how many lines a proof has, it is sound. This last step of the argument uses a method of proof called *mathematical induction*, which is the main topic of Chapter 3.

This discussion is not a proof of soundness, but it is a good outline for such a proof. We have seen that Rule P is sound. Most of the work in proving soundness consists of proving that each of the other rules is sound, i.e., they yield tautological consequences of the sentences to which they are applied. Metatheorem 1-21 proves this for MP. Proofs for the other Tautological Implications are similar. We also saw the general justification for the Tautological Equivalence rules. Each of the biconditionals used with this rule must be proved to be tautologous; this was done for commutativity of conjunction in Metatheorem 1-12. The easiest way to justify most of the tautologous biconditionals is by constructing truth tables for them. Finally, the soundness of Rule C was proved in the previous metatheorem.

Metatheorem 1-23   (**Soundness Theorem for Sentential Calculus**). Let ϕ be an $\mathcal{SC}$ sentence and Γ be a set of SC sentences. If $\Gamma \vdash_T \phi$, then $\Gamma \vDash_T \phi$.

Proof. Omitted, but the proof is outlined in the preceding discussion. Soundness for predicate calculus, which includes sentential calculus as a sublanguage, is proved in Chapter 4.

The Soundness Theorem has an important corollary. Recall from Definition 1-19 that $\vdash_T \phi$ is an abbreviation for $\varnothing \vdash_T \phi$.

Metatheorem 1-24   Let ϕ be an $\mathcal{SC}$ sentence. If $\vdash_T \phi$, then ϕ is tautologous.

Proof.   Suppose that $\vdash_T \phi$, so ϕ is $\mathcal{SC}$ derivable from the empty set. Then, by the Soundness Theorem, $\varnothing \vDash_T \phi$. But Metatheorem 1-7 states that any interpretation satisfies $\varnothing$, so any interpretation must satisfy ϕ. Hence, ϕ is tautologous. ***Q.E.D.***

Let ϕ be an $\mathcal{SC}$ sentence and Γ a set of sentences. Suppose that one has a derivation of $\phi \wedge \neg\phi$ from Γ. By the Soundness Theorem, it follows that

$\Gamma \vDash_T \phi \wedge \neg\phi$. Now suppose that some interpretation satisfied Γ. Then this interpretation would also have to satisfy $\phi \wedge \neg\phi$. But $\phi \wedge \neg\phi$ is unsatisfiable, so no interpretation satisfies it, and thus no interpretation can satisfy Γ, so Γ is unsatisfiable. This is an important result: in order to show that a set of sentences Γ is unsatisfiable, we need only derive an explicit contradiction from Γ, i.e., if we can obtain some ϕ on one line of the proof, with $\neg\phi$ on another line, we are done (since they could always be combined into $\phi \wedge \neg\phi$ by the Conjunction Rule). Demonstrating that a set Γ is unsatisfiable by considering all possible interpretations is usually tedious and very time consuming. It is often easier to derive a contradiction from Γ using the rules of derivation. Now that we know that soundness holds for our rules, it is natural to wonder about the converse conditional.

Definition 1-25 A system of rules for $\mathcal{SC}$ derivations is said to be *complete* iff for any sentence ϕ and any set of sentences Γ, if ϕ is a tautological consequence of Γ, then ϕ is $\mathcal{SC}$ derivable from Γ using these rules (if $\Gamma \vDash_T \phi$, then $\Gamma \vdash_T \phi$).

The property of completeness is the converse of soundness. We will make limited use of the next theorem, but it and related theorems have many uses in advanced logic. One should understand what completeness is, and know that our system of rules is complete.

Metatheorem 1-26 (**Completeness Theorem for Sentential Calculus**). Let ϕ be an $\mathcal{SC}$ sentence and Γ be a set of $\mathcal{SC}$ sentences. If $\Gamma \vDash_T \phi$, then $\Gamma \vdash_T \phi$.

Proof. The proof is beyond the scope of this book. Proofs for predicate calculus, as well as for sentential calculus, can be found in Mates (1972), Enderton (2001), Mendelson (1987), and in most books on mathematical logic. These proofs depend on the particular set of rules in question, but essentially the same proof techniques can be adapted to any good system of rules.

The Completeness Theorem has a corollary that is the converse of the previous corollary of the Soundness Theorem.

Metatheorem 1-27 Let ϕ be an $\mathcal{SC}$ sentence. If ϕ is tautologous, then $\vdash_T \phi$.

Proof. Suppose that ϕ is a tautology. Then it is true under every interpretation, so it is a tautological consequence of any set of sentences. In particular, it is a consequence of $\varnothing$. Therefore, by the Completeness Theorem, ϕ is derivable from $\varnothing$. **Q.E.D.**

Suppose that a set of sentences Γ tautologically implies a sentence ϕ. Also suppose that Γ is a subset of a set of sentences Δ. If an interpretation $\mathbf{I}$ satisfies Δ, then it also satisfies Γ, because every sentence in Γ is in Δ. But then, because $\Gamma \vDash_T \phi$, $\mathbf{I}$ also satisfies ϕ. Thus, $\Delta \vDash_T \phi$. This shows that, if Γ tautologically implies ϕ and Γ is a subset of Δ, then Δ tautologically implies

ϕ. The analogous fact holds for derivability. If we can derive ϕ from Γ, then we can surely derive ϕ from Δ, when Γ is a subset of Δ. We can simply use the same derivation again and ignore the extra premises that are provided by Δ.

Here is a special case of the preceding observation: If ϕ is derivable from the empty set, then it is also derivable from any set of premises, so a tautology is derivable from any set of premises. Whereas the negation of a tautology is unsatisfiable, one might also expect to find some interesting facts about that kind of sentence and, more generally, about sets of unsatisfiable premises. The Rule of Contradictory Premises states: From $\{\phi, \neg\phi\}$ infer any sentence ψ. If we have a contradictory pair of sentences in a proof, we can derive any ψ. Notice that $\{\phi, \neg\phi\} \vDash_T \psi$, because there is no interpretation that satisfies $\{\phi, \neg\phi\}$ and does not satisfy ψ (since $\{\phi, \neg\phi\}$ is unsatisfiable). Thus, we can derive any ψ from a contradictory pair of sentences, and a contradictory pair is an unsatisfiable set. Can we derive any ψ from any unsatisfiable set of sentences? The answer is in the next metatheorem, which can be proved as another corollary of the Completeness Theorem.

Metatheorem 1-28   Let Γ be an unsatisfiable set of $\mathcal{SC}$ sentences, and let ψ be any $\mathcal{SC}$ sentence. Then $\Gamma \vdash_T \psi$.

Proof. Because Γ is unsatisfiable, there is no interpretation that satisfies it (and does not satisfy ψ). Hence, $\Gamma \vDash_T \psi$, so by the Completeness Theorem, $\Gamma \vdash_T \psi$.

$$\mathbf{Q.E.D.}$$

The previous two metatheorems are important. The former tells us that a tautology can be derived from the empty set (and hence any set) of premises. The latter tells us that any sentence can be derived from an unsatisfiable set. These results are elegant, but constructing derivations still requires work. The Completeness Theorem asserts that certain proofs *exist*, but it does not describe a *procedure* for finding the proofs. The derivation rules determine what counts as a proof, but they do not give a procedure for constructing a proof when one exists. The strategies in the next section are helpful in constructing derivations.

If one is developing *any* formal language for deductive reasoning, it is important to have a semantical theory about interpretations and truth for that language. In terms of these, one defines the concept of logical consequence for that language. If a system of rules for formal derivations is then developed, it is essential that these rules be *sound* with respect to the concept of consequence for that language. Soundness is the most important adequacy requirement for a system of deductive rules. It is also often desirable that the system of rules be complete, but this is not always possible to achieve. Suppose that a computer program is developed that uses the rules to construct derivations from premises. Soundness is required to guarantee that the program produces genuine consequences of given premises (assuming the program and machine operate correctly). Completeness is not so important in practice.

Even if the rules are complete, it is possible that the machine could run a very long time and still not find a proof. To repeat: Completeness tells us only that a proof *exists*; it does not provide a procedure for constructing proofs. Proof-construction procedures have been extensively investigated for various formal languages; this is a complex research area.

Much of the research in logic during the early part of the twentieth century was motivated by the search for *effective procedures* for solving certain types of problems. Roughly speaking, an *effective procedure* is an ordered set of deterministic operations such that the entire set of operations is specified by a finite list of precise instructions. An example could be a precise set of instructions for adding two five-digit decimal numbers. Of course, the instructions would be formulated in terms of some given, basic operations, such as addition tables for smaller numbers. It is difficult to give a precise characterization of *effective procedure*; this is another topic for more advanced study, but we will rely on the rough description just given. Throughout this book, unless otherwise stated, *procedure* and *algorithm* will mean *effective procedure*.

In mathematics, one typically wants to know whether or not a sentence follows logically from a set of premises. One might even desire a procedure for answering this type of question. In the case of sentential calculus, we can state the following special case of this type of situation. Suppose that we are given any finite set Γ of $\mathcal{SC}$ sentences and any sentence ϕ. We seek a *decision procedure* for the question whether or not $\Gamma \vDash_T \phi$. This is to be an effective procedure that guarantees to tell us a definite answer after a finite number of operations. Because Γ is finite, we can form a conjunction γ of all the sentences in it. Then $\Gamma \vDash_T \phi$ iff $\gamma \rightarrow \phi$ is tautologous. Thus, all we need for our problem is a decision procedure that answers whether or not an arbitrary $\mathcal{SC}$ sentence ρ is tautologous. The point will not be argued here, but it should be intuitively clear that constructing a truth table for ρ constitutes such a decision procedure. Suppose that ρ has n $\mathcal{SC}$ letters. Then if all 2^n rows of its truth table yield the value **T**, we know that ρ is tautologous; if at least one row yields **F**, it is not tautologous. Notice that ρ is not tautologous iff $\neg\rho$ is satisfiable, so the present decision problem is related to the satisfiability problem discussed at the end of Section 1.4. Although truth-table calculations decide whether or not a sentence is a tautology, this is a very inefficient procedure. But it is an effective procedure, and this is theoretically important. There are no effective general procedures for certain types of decision problems in some languages that are more complex than sentential calculus.

1.6.2   Strategies for Constructing Proofs

There are some heuristic strategies that are useful in building proofs. 'Heuristic' means that the strategies are rules of thumb that may be helpful in constructing a proof, but following the rules does not guarantee that a proof will be found. These strategies are not proof-construction *procedures*. Most of the strategies are also somewhat vague; they suggest general approaches, but often leave the details of the application of the strategies unstated. The following should be considered general hints of what to try if one has premises of a certain form or a desired conclusion (goal) of a certain form. The "premises" and "conclusions" mentioned here need not be those of the entire derivation; they can be components of an intermediate subproof within a larger proof. Also, notice that some of the strategies make use of ("call") other strategies. The first few strategies are really just techniques for making use of certain forms of sentences. These may be called *exploitation techniques*. Most of the later strategies are directed toward constructing proofs of certain types of goals, although the last one that will be discussed has very broad applicability.

Exploitation Techniques

Using a Negation. Suppose a premise is a negation, $\neg\phi$. It can be used (exploited) with rules involving negations, such as MT, DS, and ContraPrm.

Using a Conjunction. If a premise has the form $\phi \wedge \psi$, use Simp to obtain ϕ and ψ on separate lines.

Using a Conditional. Suppose a premise is $\phi \rightarrow \psi$. It can be exploited in several ways. If ϕ can be obtained, use MP. If $\neg\psi$ can be obtained, use MT. If another suitable conditional can be found, HS may be useful. Sometimes it helps to convert the conditional into $\neg\phi \vee \psi$ by CDis, or into $\neg\psi \rightarrow \neg\phi$ by Contraposition.

Using a Negated Conditional. Sentences of the form $\neg(\phi \rightarrow \psi)$ occur rather frequently. The Rule of Negation of Conditional (NC) conveniently converts such a sentence into $\phi \wedge \neg\psi$, which can be exploited as a conjunction.

Using a Biconditional. A biconditional, $\phi \leftrightarrow \psi$, can be converted by the Equivalence Rule into $(\phi \rightarrow \psi) \wedge (\psi \rightarrow \phi)$, which can be exploited as a conjunction of two conditionals. It can also be converted into $(\phi \wedge \psi) \vee (\neg\phi \wedge \neg\psi)$. This is occasionally helpful when used together with other rules pertaining to conjunctions and disjunctions. I have listed a few sentential calculus theorems at the end of Section 1.6.3. Some of these theorems state useful properties of biconditionals. Familiarity with these theorems and their proofs can be helpful when working with biconditionals. However, our set of derivation rules does not provide a way to use these $\mathcal{SC}$ theorems. These theorems should be used with our rules only to suggest approaches to derivations.

In addition to using biconditional sentences, one can often make good use of the tautological equivalence rules. These rules allow one to substitute equivalent sentences for one another, thus transforming one sentence into another, tautologically equivalent one. We have seen an example of such a derivation earlier, and Section 1.7 describes an important application of this technique.

Using a Disjunction. Exploiting disjunctions can sometimes be troublesome. One way is to use Disjunctive Syllogism; this requires obtaining the negation of one of the disjuncts. A disjunction can always be converted by CDis into a tautologically equivalent conditional. If this conditional can be usefully exploited, such a conversion may help.

A common method of using a disjunction is *argument by cases*, which we have already used informally in the proofs of some metatheorems. Suppose that $\phi \vee \psi$ is on a line, and we have a disjunctive goal, $\delta \vee \gamma$. We try two cases: Assume ϕ and derive δ, and assume ψ and derive γ. These are separate subproofs. Suppose they succeed. By two uses of Rule C, we obtain $\phi \to \delta$ and $\psi \to \gamma$. Now apply the Constructive Dilemma Rule to obtain the disjunctive goal $\delta \vee \gamma$. CD is included in our set of rules specifically to facilitate argument by cases.

If the goal is not disjunctive, we can still use argument by cases. Suppose the goal is just δ, and $\phi \vee \psi$ is on a line. Again consider two cases and try to use CP to establish $\phi \to \delta$ and $\psi \to \delta$. If this is accomplished, by CD we obtain $\delta \vee \delta$, which simplifies to δ by use of one of the Tautology Laws.

Goal-Directed Strategies

Goal Reduction. A very general problem-solving strategy is this: If we want to achieve goal G, try to decompose G into subgoals, $G_1, G_2, \cdots, G_k$, such that each G_i can be attained relatively easily, and such that, if every G_i is achieved, then G can be easily achieved. This is sometimes called the *problem-reduction* or *divide-and-conquer* strategy. It applies to many kinds of problems, not just construction of proofs, and is a strategy that is often used informally by people and also implemented in computer programs. In the construction of proofs, it suggests looking for subgoals, which is achieved by subproofs, such that the main goal results from the subgoals.

Goal Is a Conditional. Use the method of Conditional Proof (CP) that was described in the previous section. Let $\phi \to \psi$ be the goal. Assume ϕ as a new premise by Rule P. Derive ψ (if possible), and then use Rule C to get the conditional goal. This is probably the most frequently used strategy; the real work is in proving that ϕ (perhaps together with other premises) is sufficient for ψ. If there is difficulty in developing this proof of sufficiency, try the same strategy on $\neg\psi \to \neg\phi$. If it is obtained, use Contraposition to get the original goal, $\phi \to \psi$.

It is very important to understand that CP is a strategy for proving a *goal* sentence that is a conditional. Beginners sometimes confuse this situation with one in which there is a conditional *premise*, say, $\eta \rightarrow \rho$. They add η as a new premise, then use MP to obtain ρ, and go on from there. Although any sentence may be entered into a derivation as a premise, adding the antecedent of a conditional premise is not the beginning of a CP. In my experience, it is usually not helpful to add the antecedent of a conditional premise. One may have the impression that the proof is progressing, but in the end, it is usually impossible to delete the premise name of the assumed antecedent. Sometimes a rule is misused in an erroneous attempt to delete this premise name. Suggestions for using a conditional premise have already been given; they are quite different from CP.

Goal Is a Conjunction. Suppose the goal is $\phi \wedge \psi$. Use goal reduction: Construct separate subproofs of ϕ and of ψ. Combine these results into $\phi \wedge \psi$ by the Conjunction Rule. If ϕ or ψ is a conjunction, first apply this strategy to them.

Goal Is a Biconditional. In special cases, one biconditional can be directly derived from another by repeated use of Tautological Equivalence Rules. This can be tried if there is a premise that is a biconditional. If this is successful, and uses only the Equivalence Rules, it has the added advantage that the proof is reversible, i.e., one has proved that the conclusion is tautologically equivalent to the premise. An important application of this technique is described in Section 1.7.

The technique described in the previous paragraph is often not applicable. Another standard strategy treats the biconditional goal as the conjunction of two conditionals. Let the goal be $\phi \leftrightarrow \psi$. By the Equivalence Biconditional, this is tautologically equivalent to $(\phi \rightarrow \psi) \wedge (\psi \rightarrow \phi)$. Applying the previous strategy, we reduce the proof to separate subproofs of $\phi \rightarrow \psi$ and $\psi \rightarrow \phi$. Try to make separate CP subproofs of these two conditionals. This is the same approach that we used informally in the metalanguage to prove some of the earlier "iff" metatheorems.

Goal Is a Disjunction. Let $\phi \vee \psi$ be the goal. Several approaches exist. If one can prove ϕ or prove ψ, then the disjunction can be obtained by using the Addition Rule.

Another approach is to try CP to get $\neg\phi \rightarrow \psi$ or to get $\neg\psi \rightarrow \phi$. Suppose that the former is obtained. Use CDis to get $\neg\neg\phi \vee \psi$. Then from DN one obtains $\phi \vee \psi$. If the latter is obtained, use the same transformation, plus commutation, on it. Notice that this approach is not just an artifact of the formal language; it is very intuitive. We want to prove $\phi \vee \psi$. We assume $\neg\phi$, and then prove ψ. When this is done, it establishes the disjunction, because it shows that if ϕ is not true, then the other disjunct ψ must be, so at least one of the disjuncts must be true.

A further approach is to assume $\neg(\phi \vee \psi)$ and use RAA, which will be explained shortly.

Possible Tautological Goal. As shown in the previous section, it follows from the Completeness Theorem that a tautology is derivable from any set of premises. If one has a given set of premises together with a goal sentence that is a tautology, the given set of premises is redundant. If it is suspected that the goal might be a tautology, try to derive it from $\varnothing$. Using the premises may just complicate things unnecessarily.

Goals Containing an $\mathcal{SC}$ Letter Not in the Premises. Occasionally a goal sentence may contain an $\mathcal{SC}$ letter, say Q, that is not in any of the premises. This suggests that the Add or ContraPrm rule may be required in the proof. Even when there is not a new $\mathcal{SC}$ letter in the conclusion, one should always consider the possibility that the original premises of the argument may be unsatisfiable. If they are, we know from the previous section that any sentence is derivable from them. In particular, for any sentence ϕ, it is possible to find a derivation of ϕ on one line and $\neg\phi$ on another line. If this is done, then *any conclusion* follows by the Rule of Contradictory Premises.

Reductio Ad Absurdum

This is a general strategy that can be applied to any kind of goal sentence, although it applies best to certain types of goal sentences. Some systems of logic introduce a special rule, *RAA*, or *Indirect Proof*, for Reductio Ad Absurdum, but it is not necessary to do so. I will just call it the *RAA strategy*.

Let the desired conclusion be ψ and the original premises be Γ. Use Rule P to add $\neg\psi$ as a new premise. Now use this expanded set of premises to derive (if possible) *any contradiction* (a sentence and its negation) from $\Gamma \cup \{\, \neg\psi \,\}$.

Suppose that this is done, with θ on a line of the proof and $\neg\theta$ on another line. By the Contradictory Premises Rule, any sentence can now be placed on a new line, so put the goal ψ on a new line. It sometimes happens that ψ itself is derived from $\Gamma \cup \{\, \neg\psi \,\}$, so the contradiction is between $\neg\psi$ and ψ. In this case, the step using the Contradictory Premises Rule is not needed. In either case, we have this situation:

$$\Gamma_1 \cup \{\, \neg\psi \,\} \vdash_T \psi$$

where the elements of Γ_1 are also elements of Γ. (In many cases, $\Gamma_1 = \Gamma$.) In terms of the derivation, this means that ψ is on a line with premise names corresponding to the premises $\Gamma_1 \cup \{\, \neg\psi \,\}$.

Applying Rule C, one now obtains this situation,

$$\Gamma_1 \vdash_T \neg\psi \to \psi$$

which means that $\neg\psi \to \psi$ is on a line of the proof. Applying the Law of Clavius to this line, one finally obtains

$$\Gamma_1 \vdash_T \psi$$

which corresponds to a derivation of the goal ψ from the original premises Γ. It is important to understand that RAA does not require any new derivation rules. In fact, RAA just uses the Conditional Proof (CP) strategy to derive a conditional of the form,

$$\neg\psi \to \psi.$$

Of course, the final goal ψ is then obtained by using Clav.

Here is a schematic example of the use of RAA. Suppose that we want to prove that $\Gamma \vdash_T \psi$, where $\Gamma = \{\,\phi_1,\ \phi_2,\ \phi_3\,\}$. The assumption here of three original premises is merely illustrative. There could be any finite number of original premises, including none, in a derivation. We begin the schematic derivation this way:

$$
\begin{array}{llll}
\{\,Pr_1\,\} & (1) & \phi_1 & \text{P} \\
\{\,Pr_2\,\} & (2) & \phi_2 & \text{P} \\
\{\,Pr_3\,\} & (3) & \phi_3 & \text{P} \\
\{\,Pr_4\,\} & (4) & \neg\psi & \text{P (for RAA)}
\end{array}
$$

Lines (1)–(4) of the derivation contain the three original premises and the RAA premise, $\neg\psi$. This is a CP premise, but it is useful to document it with RAA to help the reader understand the overall proof stategy being used. Using the three original premises, plus the RAA premise, we try to derive any contradiction. This is represented by lines (i) and (j) in the next part of the derivation. Of course, there will almost certainly be some intermediate lines of work, which are indicated by the ellipses.

$$
\begin{array}{llll}
\cdots & \cdots & \cdots & \text{[intermediate work]} \\
\{\,Pr_1,\ Pr_2,\ Pr_4\,\} & (i) & \theta & \text{[comment]} \\
\cdots & \cdots & \cdots & \text{[intermediate work]} \\
\{\,Pr_1,\ Pr_3,\ Pr_4\,\} & (j) & \neg\theta & \text{[comment]}
\end{array}
$$

After a contradiction is obtained, we use ContraPrm to infer the desired conclusion sentence, ψ. But we still need to remove the extra, RAA, premise. There is only one way to delete a premise, by Rule C, which is used on line $(j+2)$.

The derivation is then finished using the Law of Clavius.

$\{\, Pr_1,\ Pr_2,\ Pr_3,\ Pr_4\,\}$	$(j+1)$	ψ	ContraPrm (i), (j)
$\{\, Pr_1,\ Pr_2,\ Pr_3\,\}$	$(j+2)$	$\neg\psi \to \psi$	C (4), (j+1)
$\{\, Pr_1,\ Pr_2,\ Pr_3\,\}$	$(j+3)$	ψ	Clav (j+2)

In many derivations of this type, the set of premise names of each of lines (i) and (j) will refer to all initial premises and thus be $\{\, Pr_1,\ Pr_2,\ Pr_3,\ Pr_4\,\}$. More generally, the set of premise names of line $(j+1)$ will include Pr_4, and some (usually all) of the initial premises.

By ContraPrm, a contradiction tautologically implies any sentence. Therefore, if we have any contradiction, such as lines (i) and (j), this contradiction implies any sentence and also the negation of this sentence. For example, from (i) and (j), we can infer A and $\neg A$, and infer $(A \vee \neg C)$ and $\neg(A \vee \neg C)$, etc. In particular, whatever ψ may be, we can infer both ψ and $\neg\psi$. Of course, $\neg\psi$ is already on line (4), so if we can obtain ψ on a line, then we will already have a contradiction. This often happens in RAA proofs, resulting in a schema like this one:

$\{\, Pr_1\,\}$	(1)	ϕ_1	P
$\{\, Pr_2\,\}$	(2)	ϕ_2	P
$\{\, Pr_3\,\}$	(3)	ϕ_3	P
$\{\, Pr_4\,\}$	(4)	$\neg\psi$	P (for RAA)
$\cdots$	$\cdots$	$\cdots$	[intermediate work]
$\{\, Pr_1,\ Pr_2,\ Pr_3,\ Pr_4\,\}$	(i)	ψ	[comment]
$\{\, Pr_1,\ Pr_2,\ Pr_3\,\}$	$(i+1)$	$\neg\psi \to \psi$	C (4), (i)
$\{\, Pr_1,\ Pr_2,\ Pr_3\,\}$	$(i+2)$	ψ	Clav (i+1)

Although this form of RAA derivation does not require the ContraPrm step, it uses the same RAA starting move: Assume the negation of the desired goal sentence and try to derive a contradiction.

The RAA strategy is useful, especially when the goal sentence is small, e.g., a single $\mathcal{SC}$ letter or perhaps a sentence with only one connective. If the goal sentence is long and complicated, then its negation is likely to be long and complicated. In such cases, RAA often does not help much. RAA is occasionally helpful in proving a conditional goal, $\phi \to \psi$. Using the CP strategy, one assumes ϕ, derives ψ, then uses C; or else one does the same with the contraposition, $\neg\psi \to \neg\phi$. If neither of these leads to progress, try RAA. Just assume $\neg(\phi \to \psi)$, which yields $\phi \wedge \neg\psi$ by the Negation of Conditional

Rule (NC). Try to derive any contradiction from the latter. Often one just derives $\neg\phi$, in which case CP probably would work with the contrapositive. Occasionally a contradiction other than $\phi \wedge \neg\phi$ is obtained more easily, so this approach can be helpful.

RAA often takes a special format when the goal is a conditional. By NC, $\neg(\phi \rightarrow \psi)$ is tautologically equivalent to $\phi \wedge \neg\psi$. Instead of assuming the former, it is sometimes more convenient to assume the latter, in a special format. One assumes ϕ, as in the start of a conditional proof, and then assumes $\neg\psi$, the negation of that which one would try to derive in the conditional proof. Suppose that a contradiction is now derived. By ContraPrm, infer ψ. Conditionalize (use Rule C) to obtain $\neg\psi \rightarrow \psi$. By Clav, obtain ψ. Now conditionalize again to get $\phi \rightarrow \psi$, the desired conditional goal.

As previously mentioned in the discussion of disjunctive goals, RAA is sometimes helpful in deriving a disjunction, $\phi \vee \psi$. Assume its negation,

$$\neg(\phi \vee \psi),$$

then infer $\neg\phi \wedge \neg\psi$, by DeM. Now exploit this conjunction and try to derive a contradiction.

RAA can also be helpful with a goal that is a negation, such as $\neg\phi$. Assume its negation,

$$\neg\neg\phi,$$

or equivalently, ϕ. Try to derive a contradiction. If one is obtained, use ContraPrm to get $\neg\phi$. By C, we then obtain $\phi \rightarrow \neg\phi$. From this, one can use CDis to obtain $\neg\phi \vee \neg\phi$, whence $\neg\phi$ by Taut; or one can go directly to $\neg\phi$ by invoking Clav.

This completes the strategies for $\mathcal{SC}$ proofs. They are not guaranteed to produce a proof when one exists. Yet typical $\mathcal{SC}$ derivations can usually be analyzed into the combined use of one or more of these strategies. It is often helpful to work backwards from the goal. Ask yourself: From what other sentences could I derive this goal sentence? If you find some, repeat this process. Eventually, you may get lucky and end up with the original, given premise(s). If not, then look at some consequences of these premises. Do they match any of the sentences from which the goal can be derived? If so, you might be able to work from both ends (the original premises and the goal) toward the middle and connect them together into a good proof.

A word of caution. A few beginners try to construct all of their proofs just using the tautological equivalence rules. This is not a good strategy— try to derive the conclusion Q from the premise $P \wedge Q$ by this method. The equivalence rules are very useful, but they are intended to be used together with the other rules in our system.

Another word of caution. Rule P gives total freedom to introduce new premises into a derivation. Yet one should not introduce new premises without a clear strategy for their use. If a new premise is not used in the derivation of the desired goal, then it was not needed in the first place. If it is used, so it becomes a member of the set of premises of the desired conclusion, then it must be removed. Our system of rules has only one way to remove a premise name from a set of premise names, and that is by Rule C. Thus, if a new premise is added to a derivation, one should expect to apply Rule C to this premise (conditionalize it) at some point in the proof. The only strategies that readily permit effective conditionalization of this kind are CP and RAA. So, in *SC* derivations, one is well-advised *not* to add a new premise except for CP or RAA. (Of course, the RAA strategy makes use of CP.)

If a conclusion ϕ is a tautological consequence of premises Γ, a proof of ϕ from Γ does exist (by the Completeness Theorem). In my experience, a proof can almost always be constructed using the aforementioned strategies together with our rules of derivation. It is astonishing that so much can be accomplished with so few heuristic rules. However, most people need a fair amount of practice to develop the following abilities to (1) recognize which strategies are appropriate in a given situation, (2) decompose a proof into smaller subproofs, (3) avoid some common misuses of the rules of derivation, and (4) recognize when a particular approach to a proof is not making progress toward its goal. Initially, many people find it rather difficult to construct sentential calculus derivations. It is necessary to keep practicing to become familiar with the rules of derivation and the proof strategies and to develop the required abilities. Eventually, proofs that at first appeared formidable will be viewed as simple. More importantly, the knowledge and skills that are acquired will be useful in constructing informal arguments, such as the arguments used in the next chapter.

1.6.3 Sentential Calculus Derivation Examples

Example.

Problem Statement. Derive $\neg A \lor \neg B$ from the set $\{ A \rightarrow (B \rightarrow C), \neg C \}$.

Discussion. The goal is a disjunction, so I transform the disjunction into an equivalent conditional, $\neg\neg A \rightarrow \neg B$, which is equivalent to $A \rightarrow \neg B$. Use CP to prove the latter conditional. To do this, A is assumed as a premise. The rest is easy.

Written Solution.

$\{\ Pr_1\ \}$	(1)	$A \rightarrow (B \rightarrow C)$	P
$\{\ Pr_2\ \}$	(2)	$\neg C$	P
$\{\ Pr_3\ \}$	(3)	A	P (for CP)
$\{\ Pr_1, Pr_3\ \}$	(4)	$B \rightarrow C$	MP (1), (3)
$\{\ Pr_1, Pr_2, Pr_3\ \}$	(5)	$\neg B$	MT (2), (4)
$\{\ Pr_1, Pr_2\ \}$	(6)	$A \rightarrow \neg B$	C (3), (5)
$\{\ Pr_1, Pr_2\ \}$	(7)	$\neg A \vee \neg B$	CDis (6) ∎

Example.

Problem Statement. Derive R from $Q \vee P$, $(T \rightarrow P) \rightarrow R$, and $\neg R \rightarrow (S \wedge \neg Q)$.

Discussion. Although R is embedded in the second and third premises, trying to extract it appears to require some effort, so I use RAA. $\neg R$ is assumed as a new, fourth, premise. If we could somehow get $T \rightarrow P$, then from the second premise we could get R, and therefore a contradiction. Moreover, if we could obtain P, then by the first part of Rule C, we could get $T \rightarrow P$. We might be able to obtain P from the first premise, so let us try to use the new premise $\neg R$. From it and the given third premise, we can obtain $(S \wedge \neg Q)$ by MP. We can then use Simp to get $\neg Q$. Using DS, $\neg Q$ and the first premise yield P. We can therefore obtain $T \rightarrow P$, and the rest is straightforward.

Written Solution.

$\{\ Pr_1\ \}$	(1)	$Q \vee P$	P
$\{\ Pr_2\ \}$	(2)	$(T \rightarrow P) \rightarrow R$	P
$\{\ Pr_3\ \}$	(3)	$\neg R \rightarrow (S \wedge \neg Q)$	P
$\{\ Pr_4\ \}$	(4)	$\neg R$	P (for RAA)
$\{\ Pr_3, Pr_4\ \}$	(5)	$S \wedge \neg Q$	MP (3), (4)
$\{\ Pr_3, Pr_4\ \}$	(6)	$\neg Q$	Simp (5)
$\{\ Pr_1, Pr_3, Pr_4\ \}$	(7)	P	DS (1), (6)
$\{\ Pr_1, Pr_3, Pr_4\ \}$	(8)	$T \rightarrow P$	C (7)
$\{\ Pr_1, Pr_2, Pr_3, Pr_4\ \}$	(9)	R	MP (2), (8)
$\{\ Pr_1, Pr_2, Pr_3\ \}$	(10)	$\neg R \rightarrow R$	C (4), (9)
$\{\ Pr_1, Pr_2, Pr_3\ \}$	(11)	R	Clav (10)

Remark. As mentioned before, the antecedent of (2) was obtained by conditionalizing the P on line (7). In this type of situation, I am often asked, "Why not just assume $T \to P$ by Rule P?" As already pointed out in the discussion of the strategy for Conditional Goals, it is usually *not* wise to assume the antecedent of a conditional premise. If we had simply entered (8) by Rule P, this line would have the premise names, $\{ Pr_8 \}$. We could have then derived R, as desired, but the premise names of R would contain Pr_8, and we would be stuck with this unwanted premise name. One usually gets stuck with unwanted premise names if one assumes the antecedent of a conditional premise. ■

Example.

Problem Statement. Write an $\mathcal{SC}$ derivation of ϕ, the sentence $C \vee A$, from the set of premises $\Delta = \{ \neg C \to (\neg A \to B),\ B \to C \}$.

Discussion. The goal sentence ϕ is a disjunction, but it is not obvious how to obtain it. Here is one way: By repeated use of CDis and DN, the first premise is seen to be equivalent to $C \vee (A \vee B)$, and the second premise to $\neg B \vee C$. By Assoc, $C \vee (A \vee B)$ is equivalent to $(C \vee A) \vee B$. Applying Cut to $(C \vee A) \vee B$ and $\neg B \vee C$, we obtain $(C \vee A) \vee C$. After some additional work, one can obtain ϕ. The reader should write this proof. It works, but is rather tedious. Perhaps we can do better.

Fortunately, there are several strategies for disjunctive goals. One approach is to derive a conditional that is equivalent to the disjunction. For instance, ϕ is equivalent to $\neg C \to A$, so one might try to use CP to derive the latter sentence. The reader should also write this proof. But also notice this: Suppose one assumes the additional premise $\neg C$ for CP. One then tries to derive A. But perhaps directly deriving A is not easy. Then one might try RAA, by assuming $\neg A$ and trying to derive a contradiction. This method uses two extra premises and will eventually require two separate applications of Rule C to remove these extra premises. Why not just assume both of these premises together? This would amount to adding $\neg C \wedge \neg A$ as the sole extra premise. But $\neg C \wedge \neg A$ is equivalent to the negation of ϕ, so this method is equivalent to using RAA on the original goal, which is one of the methods previously mentioned for disjunctive goals.. The next written solution uses RAA in this manner. Although it works well for this problem, RAA is not always the best approach to use. Nevertheless, it is useful when other approaches are not succeeding or appear to be very complicated.

Written Solution.

{ Pr_1 }	(1) $\neg C \rightarrow (\neg A \rightarrow B)$	P
{ Pr_2 }	(2) $B \rightarrow C$	P
{ Pr_3 }	(3) $\neg(C \vee A)$	P (for RAA)
{ Pr_3 }	(4) $\neg C \wedge \neg A$	DeM (3)
{ Pr_3 }	(5) $\neg C$	Simp (4)
{ Pr_1, Pr_3 }	(6) $\neg A \rightarrow B$	MP (1), (5)
{ Pr_3 }	(7) $\neg A$	Simp (4)
{ Pr_1, Pr_3 }	(8) B	MP (6), (7)
{ Pr_1, Pr_2, Pr_3 }	(9) C	MP (2), (8)
{ Pr_1, Pr_2, Pr_3 }	(10) $C \vee A$	ContraPrm (5), (9)
{ Pr_1, Pr_2 }	(11) $\neg(C \vee A) \rightarrow (C \vee A)$	C (3), (10)
{ Pr_1, Pr_2 }	(12) $C \vee A$	Clav (11) ∎

Example.

Problem Statement. Let $\Gamma = \{\ \neg P,\ \ R \rightarrow P,\ \ (P \wedge Q) \rightarrow (P \vee R)\ \}$ and let ϕ be the sentence $Q \rightarrow P$. Does $\Gamma \vdash_T \phi$? Either prove this by a derivation, or give a counterexample.

Discussion. It is not obvious whether ϕ is $\mathcal{SC}$ provable from Γ. From $\neg P$ and $R \rightarrow P$, we obtain $\neg R$ by MT. Thus, we have $\neg P \wedge \neg R$, which is equivalent to $\neg(P \vee R)$. This, together with the third premise and MT, yields $\neg(P \wedge Q)$, which is equivalent to $\neg P \vee \neg Q$. Now, $\neg P \vee \neg Q$ is tautologically consistent with $Q \rightarrow P$ (Definition 1-10), but the fact that $\neg P \vee \neg Q$ is consistent with $Q \rightarrow P$ does *not* imply that $\neg P \vee \neg Q$ tautologically implies $Q \rightarrow P$. Indeed, it is hard to see why $Q \rightarrow P$ must be true, given Γ. Therefore, I now try to construct a counterexample interpretation **I**. We need an **I** that satisfies Γ and makes ϕ false. Clearly, we need Q: **T**, P: **F**. But since we need $R \rightarrow P$ to be true, and P is false, we need R: **F**. The written solution shows that this works.

Written Solution. Let $\Gamma = \{\ \neg P,\ \ R \rightarrow P,\ \ (P \wedge Q) \rightarrow (P \vee R)\ \}$ and let ϕ be the sentence $Q \rightarrow P$. Then not $\Gamma \vdash_T \phi$.

Proof. The following counterexample interpretation establishes that it is not the case that $\Gamma \vDash_T \phi$. Let Q: **T**, P: **F**, and R: **F**. Then $\neg P$ is **T**, and, since R is **F**, the truth value of $R \rightarrow P$ is **T**. Finally, since

$$\mathbf{V_I}((P \wedge Q) \rightarrow (P \vee R)) = ((\mathbf{F} \wedge \mathbf{T}) \rightarrow (\mathbf{F} \vee \mathbf{F})) = (\mathbf{F} \rightarrow \mathbf{F}) = \mathbf{T},$$

I satisfies Γ. But also $Q \rightarrow P$ has the value $(\mathbf{T} \rightarrow \mathbf{F}) = \mathbf{F}$ under **I**. By the

Soundness Metatheorem, if $\Gamma \vdash_T \phi$ is true, then $\Gamma \vDash_T \phi$. Since the latter is false, we also have not $\Gamma \vdash_T \phi$. ***Q.E.D.*** ■

Example.

Problem Statement. $\mathcal{SC}$ Theorem. $((X \to Y) \to X) \leftrightarrow X$

Discussion. We will follow the usual convention that if a problem states a theorem, the task is to prove it. One way would be to show that it is tautologous, and then use completeness, i.e., invoke Metatheorem 1-27. However, normally, one is expected to write a formal derivation to establish an $\mathcal{SC}$ theorem. Following the definition of $\mathcal{SC}$ theorem, we must write a derivation of the sentence from the empty set (of sentences). Because it is a biconditional, I shall divide the derivation into two subproofs, one proving the conditional in one direction, the other proving it in the opposite direction. Of course, any proof in our formal system must begin with one or more premises. In a proof of an $\mathcal{SC}$ theorem, these premises may not occur in the set of premises of the last line of the derivation because we are deriving this line from the empty set of premises. In so far as the goal sentence of each subproof is a conditional, I shall try the CP strategy. First I shall assume the antecedent $(X \to Y) \to X$ as a premise. Then by CDis we obtain $\neg(X \to Y) \vee X$. Applying the NC rule yields $(X \wedge \neg Y) \vee X$. It is now easy to obtain X. The second subproof is even easier.

Written Solution.

$\{ Pr_1 \}$	(1)	$(X \to Y) \to X$	P
$\{ Pr_1 \}$	(2)	$\neg(X \to Y) \vee X$	CDis (1)
$\{ Pr_1 \}$	(3)	$(X \wedge \neg Y) \vee X$	NC (2)
$\{ Pr_1 \}$	(4)	$X \vee (X \wedge \neg Y)$	Comm (3)
$\{ Pr_1 \}$	(5)	$(X \vee X) \wedge (X \vee \neg Y)$	Dist (4)
$\{ Pr_1 \}$	(6)	$X \vee X$	Simp (5)
$\{ Pr_1 \}$	(7)	X	Taut (6)
$\varnothing$	(8)	$((X \to Y) \to X) \to X$	C (1), (7)
$\{ Pr_9 \}$	(9)	X	P
$\{ Pr_9 \}$	(10)	$(X \to Y) \to X$	C (1st part) (9)
$\varnothing$	(11)	$X \to ((X \to Y) \to X)$	C (9), (10)
$\varnothing$	(12)	$(((X \to Y) \to X) \to X) \wedge$ $(X \to ((X \to Y) \to X))$	Conj (8), (11)
$\varnothing$	(13)	$((X \to Y) \to X) \leftrightarrow X$	Equiv (12) ■

There are many interesting and useful $\mathcal{SC}$ theorems. Line (18) of the next example states one such theorem. The derivation illustrates the same standard strategy for biconditionals that is used in the previous example. The proof also illustrates the use of nested applications of CP. For this theorem, a shorter proof can be obtained by converting the conditionals to disjunctions, and using Tautological Equivalence Rules. This is left for practice.

Example.

Problem Statement. Derive $(P \to (Q \to R)) \leftrightarrow ((P \wedge Q) \to R)$ from the empty set of premises.

Discussion. A derivation of this sentence from $\varnothing$ will prove that it is an $\mathcal{SC}$ theorem. Because it is a biconditional, I try dividing the proof into two subproofs: First derive one side of the biconditional from the other side, then derive the converse conditional, then combine these two results. I shall also use the Conditional Proof (CP) strategy for each of the subproofs.

Written Solution.

$\{ Pr_1 \}$	(1)	$P \to (Q \to R)$	P
$\{ Pr_2 \}$	(2)	$P \wedge Q$	P
$\{ Pr_2 \}$	(3)	P	Simp (2)
$\{ Pr_1, Pr_2 \}$	(4)	$Q \to R$	MP (1), (3)
$\{ Pr_2 \}$	(5)	Q	Simp (2)
$\{ Pr_1, Pr_2 \}$	(6)	R	MP (4), (5)
$\{ Pr_1 \}$	(7)	$(P \wedge Q) \to R$	C (2), (6)
$\varnothing$	(8)	$(P \to (Q \to R)) \to ((P \wedge Q) \to R)$	C (1), (7)
$\{ Pr_9 \}$	(9)	$(P \wedge Q) \to R$	P
$\{ Pr_{10} \}$	(10)	P	P
$\{ Pr_{11} \}$	(11)	Q	P
$\{ Pr_{10}, Pr_{11} \}$	(12)	$P \wedge Q$	Conj (10), (11)
$\{ Pr_9, Pr_{10}, Pr_{11} \}$	(13)	R	MP (9), (12)
$\{ Pr_9, Pr_{10} \}$	(14)	$Q \to R$	C (11), (13)
$\{ Pr_9 \}$	(15)	$P \to (Q \to R)$	C (10), (14)
$\varnothing$	(16)	$((P \wedge Q) \to R) \to (P \to (Q \to R))$	C (9), (15)
$\varnothing$	(17)	$((P \to (Q \to R)) \to ((P \wedge Q) \to R))$ $\wedge (((P \wedge Q) \to R) \to (P \to (Q \to R)))$	Conj (8), (16)
$\varnothing$	(18)	$(P \to (Q \to R)) \leftrightarrow ((P \wedge Q) \to R)$	Equiv (17)

Remark. The theorem on line (18) is sometimes called the *export–import law* or the *principle of exportation*. Knowing that the right and left sides of (18) are equivalent can be useful in situations where one of the two forms is more convenient to use than the other. This principle can also be iterated, for example, to show that

$$((A \land B) \land C) \to R$$

is equivalent to

$$A \to (B \to (C \to R))$$

and so on. Caution: Do not confuse $P \to (Q \to R)$ with $(P \to Q) \to R$; they are not tautologically equivalent. ■

The most often used $\mathcal{SC}$ theorems are simple consequences of our derivation rules. Here is a typical example:

$\{ Pr_1 \}$	(1)	$(P \land \neg Q) \lor R$	P
$\{ Pr_1 \}$	(2)	$(\neg Q \land P) \lor R$	Comm (1)
$\varnothing$	(3)	$((P \land \neg Q) \lor R) \to ((\neg Q \land P) \lor R)$	C (1), (2)

Instead of trying to remember the theorem on line (3), and scores of other specialized theorems, it is easier to remember the rules of inference. However, there are some $\mathcal{SC}$ theorems that do not so closely match the rules. I have listed a few interesting and useful ones next; they are well-known and can be found in a number of logic texts. The proofs are left as exercises; some require fairly long derivations.

Selected Sentential Calculus Theorems

1. $(A \land B) \to (A \leftrightarrow B)$

2. $(A \leftrightarrow B) \leftrightarrow ((A \leftrightarrow C) \leftrightarrow (B \leftrightarrow C))$

3. $((A \lor B) \to C) \leftrightarrow ((A \to C) \land (B \to C))$

4. $((A \leftrightarrow B) \leftrightarrow C) \leftrightarrow (A \leftrightarrow (B \leftrightarrow C))$

5. $\neg(A \leftrightarrow B) \leftrightarrow (\neg A \leftrightarrow B)$

6. $(A \lor (B \to C)) \leftrightarrow ((A \lor B) \to (A \lor C))$

7. $(A \lor (B \leftrightarrow C)) \leftrightarrow ((A \lor B) \leftrightarrow (A \lor C))$

8. $(A \wedge (B \to C)) \to ((A \wedge B) \to (A \wedge C))$

9. $(A \wedge (B \leftrightarrow C)) \to ((A \wedge B) \leftrightarrow (A \wedge C))$

10. $(A \wedge \neg(B \leftrightarrow C)) \leftrightarrow \neg((A \wedge B) \leftrightarrow (A \wedge C))$

Several of these theorems state important facts about conditionals and biconditionals. In particular, Theorems 6–9 state some distribution principles that are useful in some applications. Theorem 4 states that the biconditional is associative. Associativity is a very useful property; it will appear again, in several different contexts, in later chapters. Among other things, Dijkstra and Scholten (1990) put heavy emphasis on the use of biconditionals and show the value of the associativity of the biconditional.

Exercises

General Instructions

Exercises for which answers are provided at the back of the book are marked with "Answer Provided."

Please note that exercises that introduce important principles or useful applications are especially difficult, or are special in some way, are indicated by **bold** print. In some cases, parenthetical comments describe the special features of the problem. If such comments are missing, the exercise is nonetheless of special value in introducing a new concept, principle, or technique.

E 1-53 (Answers Provided for 1 and 4) For each of the following arguments, give an $\mathcal{SC}$ derivation of the conclusion from the premise(s). The conclusion is indicated by / $\therefore$. Before starting, please also review and follow the more detailed instructions for E 1-47.

1. $A \vee (B \wedge C)$ / $\therefore$ $(A \to C) \to C$

2. $A \to B$, $A \vee C$, $\neg(C \wedge D)$ / $\therefore$ $(A \to B) \to (B \vee \neg D)$

3. $\neg(A \wedge B) \to (\neg A \wedge \neg B)$ / $\therefore$ $A \leftrightarrow B$

4. $A \leftrightarrow \neg B$, $B \vee \neg C$, $C \to A$ / $\therefore$ $\neg C$

5. $J \to (K \to L)$, $\neg N \to (J \wedge M)$, $(K \to N) \to \neg P$, $P \wedge \neg Q /$ $\therefore$ $L \vee Q$

6. $(A_1 \to B_1) \vee (A_2 \to B_2)$ / $\therefore$ $(A_1 \to B_2) \vee (A_2 \to B_1)$

7. $R \vee (S \vee T)$, $T \to P$, $Q \wedge (\neg P \wedge N)$, $R \to S$ $/ \therefore S$

8. $(F \to G) \to H$, $G \wedge (I \to J)$, $\neg (I \wedge \neg J) \to \neg K$ $/ \therefore \neg (H \to K)$

9. $A \vee (B \vee C)$, $P \to \neg A$ $/ \therefore (P \wedge \neg B) \to C$

10. $A \leftrightarrow B$, $\neg B$ $/ \therefore (P \vee A) \to \neg A$

11. $\neg A \to C$, $B \to P$, $(P \vee C) \to H$ $/ \therefore (A \to B) \to (Q \to H)$

12. $(A \vee B) \to \neg(\neg D \wedge \neg C)$ $/ \therefore B \to (\neg C \to D)$

13. $\neg A \to C$, $B \to Q$, $S \vee A$, $C \to B$ $/ \therefore \neg A \to (Q \wedge (B \wedge C))$

14. $\neg C$, $(B \to \neg C) \to A$ $/ \therefore (A \to C) \to F$

15. $(P \to Q) \wedge (Q \to R)$, $(R \to P) \wedge (T \wedge U)$, $(P \leftrightarrow R) \to S$, $(T \wedge S) \to (V \vee W)$ $/ \therefore (U \wedge V) \vee (U \wedge W)$ (Hint. Try to derive U and $(V \vee W)$.)

16. (Lengthy) D, $A \vee \neg B$, $A \to P$, $(\neg B \wedge D) \to C$, $C \to B$, $\neg C \to (\neg P \vee \neg D)$ $/ \therefore A \wedge (B \wedge (C \wedge D))$

E 1-54 (Answer Provided for 3) For each of the following arguments, either give an SC derivation of the conclusion from the premises, or give a counterexample interpretation. If you construct an interpretation, show the truth values of the sentences under this interpretation.

1. $A \vee B$, $C \to \neg A$ $/ \therefore C \to B$

2. $(P \vee Q)$, $(P \to R)$, $(Q \to S)$ $/ \therefore T \to (R \vee S)$

3. $A \vee E$, $A \to ((P \vee Q) \to C)$, $\neg E \vee C$, $(A \wedge \neg E) \to F$ $/ \therefore C \vee P$

4. $E \to (F \vee G)$, $G \to (H_1 \wedge I)$, $\neg H_1$ $/ \therefore (E \to I)$

5. $P \to (Q \vee R)$, $(R \to P)$, $\neg(P \wedge Q)$ $/ \therefore (P \to S)$

6. $D \leftrightarrow B$, $\neg B \to C$ $/ \therefore \neg D \to B$

7. $T \vee Q$, $\neg R \leftrightarrow S$, $S \vee \neg Q$ $/ \therefore R \to (T \to Q)$

8. $A \vee D$, $B \leftrightarrow C$, $A \vee \neg C$ $/ \therefore B \to D$

9. $(A \leftrightarrow B) \wedge (C \leftrightarrow D)$ $/ \therefore (A \wedge C) \leftrightarrow (B \wedge D)$

10. $P \vee Q, \quad \neg R \leftrightarrow Q, \quad S \vee Q \quad / \therefore \ P \to (R \to (Q \vee \neg R))$

11. $C \to Q, \quad (A \vee B) \to C, \quad Q \to \neg C, \quad A \leftrightarrow \neg B \quad / \therefore \ Q \vee B$

E 1-55 Give two $\mathcal{SC}$ derivations of $(C \to B) \to B$ from the premises: $A \vee B$, $B \vee (A \to C)$. First use CP by assuming $C \to B$; second, use RAA.

E 1-56 Prove each of the following $\mathcal{SC}$ theorems by deriving it from the empty set of premises.

1. $\neg Q \to (P \to (Q \to R))$

2. $((R \wedge \neg Q) \wedge (Q \vee \neg R)) \to (R \leftrightarrow Q)$

3. $A \leftrightarrow (A \vee (A \wedge X))$

4. $(A \leftrightarrow B) \to ((A \wedge X) \leftrightarrow (B \wedge X))$

E 1-57 Derive each of the following from the empty set of premises:

$$((A \vee \neg A) \wedge B) \leftrightarrow B$$

$$((A \wedge \neg A) \vee B) \leftrightarrow B$$

(Hint. $(A \vee \neg A)$ is derivable from $\varnothing$.)

E 1-58 Either prove that $\neg((A \vee B) \to C)$ is tautologically equivalent to

$$(C \to A) \vee (C \to B)$$

by deriving each from the other, or else show by a counterexample interpretation that they are not tautologically equivalent.

E 1-59 For each of the following sentences, either prove that it is an $\mathcal{SC}$ theorem by deriving it from the empty set of premises, or else show by a suitable counterexample interpretation that it is not an $\mathcal{SC}$ theorem.

1. $((A \wedge X) \leftrightarrow (B \wedge X)) \to (A \leftrightarrow B)$

2. $((P \to Q) \wedge (\neg P \to Q)) \leftrightarrow Q$

3. $(P \to Q) \to (\neg(Q \vee R) \to (\neg P \wedge \neg R))$

4. $(A \leftrightarrow B) \leftrightarrow ((A \wedge X) \leftrightarrow (B \wedge X))$

E 1-60 Let $\Delta = \{\ P_1 \rightarrow Q_1,\ \ P_2 \rightarrow Q_2,\ \ P_1 \vee P_2,\ \ \neg(Q_1 \wedge Q_2)\ \}$.

1. Show that Δ is satisfiable.

2. Give $\mathcal{SC}$ derivations of each of the following from Δ:

$$Q_1 \rightarrow P_1$$

$$Q_2 \rightarrow P_2$$

$$\neg(P_1 \wedge P_2)$$

E 1-61 **(Lengthy**; Answer Provided for the first theorem**)** A list of selected $\mathcal{SC}$ theorems is given just before the beginning of this exercise set. Derive each of these theorems from the empty set of premises. (Remark. Since there are several theorems, this exercise requires several derivations. Some of these derivations are rather long.)

E 1-62 For each of the following statements, either write an informal metatheoretical proof, or give a suitable counterexample. As usual, you may use theorems and metatheorems that are stated in the text before this problem, including the Soundness and Completeness metatheorems.

1. Let ϕ, ψ, λ be any $\mathcal{SC}$ sentences. Then $(\phi \rightarrow \psi) \vee (\lambda \rightarrow \phi)$ is an $\mathcal{SC}$ theorem.

2. Let Γ be a set of $\mathcal{SC}$ theorems and ϕ and ψ be any $\mathcal{SC}$ sentences. If $\Gamma \vdash_T \phi \rightarrow \psi$, then ψ must be an $\mathcal{SC}$ theorem.

3. Let ϕ, ψ be any $\mathcal{SC}$ sentences. Then $\{\neg\phi \vee \psi, \phi \vee \neg\psi\}$ is satisfied by every $\mathcal{SC}$ interpretation.

4. Let τ be an $\mathcal{SC}$ theorem and ϕ be any $\mathcal{SC}$ sentence. Then

$$(\tau \rightarrow \phi) \leftrightarrow (\phi \wedge \tau)$$

is an $\mathcal{SC}$ theorem.

5. Let ϕ, ψ be any $\mathcal{SC}$ sentences. Then $\neg(\phi \rightarrow \psi) \rightarrow (\psi \rightarrow \phi)$ is an $\mathcal{SC}$ theorem.

E 1-63 (Illustrates some application techniques; lengthy) Suppose that we wish to achieve some goal A. Suppose that, if we achieve B and C and D, then we will also achieve A. One can also think of A as some problem to solve where A has the property that it will be solved if each of B, C, and D is solved. When this relation holds, it is said that the goal or problem A can be *reduced* to the subgoals or subproblems B, C, and D. In sentential calculus, this situation can be represented by

$$(B \land (C \land D)) \rightarrow A$$

We can also represent this situation by the tree in Figure 1.2. In this figure, B, C, D are the *child nodes* of A. The curved arc indicates that A is reducible to the *conjunction* of B, C, D.

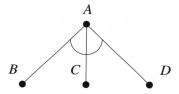

Figure 1.2.

Now suppose that we achieve B if we achieve E, so

$$E \rightarrow B.$$

Suppose that we will also achieve B if we achieve F, so

$$F \rightarrow B.$$

We need B *and* C *and* D to get A, but either child node E *or* F will get us B. This situation is represented in Figure 1.3, where there is no arc between the lines from B to E and from B to F.

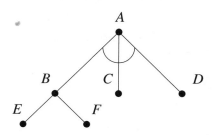

Figure 1.3.

Suppose further that

$$(G \land H) \rightarrow E.$$

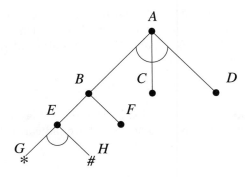

Figure 1.4.

Also suppose that G is a "good" node, i.e., something that we can simply do (or a problem that we can directly solve), without depending on any further action (or further solutions). In sentential calculus, we represent this by

$$G.$$

Suppose that H is a "no-good" node, i.e., something that we cannot achieve or realize or solve; it is a dead end. In sentential calculus, we represent this by

$$\neg H.$$

In the tree diagram, terminal nodes that are good are represented by * and terminal nodes that are no-good are represented by #. The situation is now represented by Figure 1.4.

Suppose we have the following additional information:

$$(I \wedge S) \to F$$
$$I$$
$$(J \wedge K) \to S$$
$$J$$
$$K$$
$$L \to C$$

$$M \to C$$
$$T \to C$$
$$\neg L$$
$$M$$
$$T$$
$$(N \wedge (O \wedge P)) \to D$$
$$N$$
$$O$$
$$Q \to P$$
$$R \to P$$
$$U \to P$$
$$\neg Q$$
$$R$$
$$U$$

Now please do the following:

1. Draw the complete tree representing all of the preceding data. This type of diagram is called an "AND/OR" tree, because each branch point corresponds to an 'AND' or to an 'OR'.

A tree of this kind is helpful in planning problems. To obtain a plan for achieving A, find a subtree τ such that every node in τ is achievable and the root of τ is A. In particular, if ν is an AND node in τ, then each child node of ν must also be achievable. If ν is an OR node, then one achievable child node of ν must be in τ. All terminal nodes of τ must be good nodes. Such a subtree gives us a plan. Some AND/OR trees have no solution plans; some have multiple solutions.

2. By trial-and-error search, find one solution subtree in your complete tree. Draw this subtree.

3. From all of the preceding $\mathcal{SC}$ sentences, select a subset that corresponds to the nodes in your solution subtree from 2. Terminal nodes should correspond to atomic sentences; other nodes should correspond to antecedents of conditionals. Order these sentences in a systematic way starting with terminal nodes and working upwards toward A. Give a sentential calculus derivation of A to show that your subtree plan really works. Your subtree accomplishes two things: It selects an appropriate set of premises, and it systematically determines the derivation.

1.7   Conjunctive Normal Form and Resolution Proofs

Given an $\mathcal{SC}$ sentence, it is sometimes convenient to find a tautologically equivalent sentence that is in a special form. Various special forms may be used and are often called *normal forms*. We will consider one kind of normal form.

Definition 1-29 A *literal* is a sentential letter or the negation of a sentential letter.

Thus, A, B_1, $\neg A_1$, $\neg C$ are literals.

Definition 1-30 A *clause* is an $\mathcal{SC}$ sentence satisfying the following:

A literal is a clause.

If ϕ and ψ are clauses, then $(\phi \vee \psi)$ is a clause.

Any clause must be obtainable by use of one or both of the previous rules.

This is another recursive definition; it states how clauses are constructed. Here are some clauses:

$$A, \quad B, \quad C, \quad \neg A, \quad \neg B, \quad \neg C,$$

$$(A \vee B), \quad (A \vee \neg A), \quad (B \vee C),$$

$$((A \vee \neg A) \vee B), \quad (B \vee (B \vee C)),$$

$$((A \vee \neg A) \vee B) \vee (B \vee (B \vee \neg C)).$$

In general, a clause is either a single literal or else a disjunction of literals, where *disjunction* is understood to mean the recursively constructed multiple disjunctions just now defined. Chapter 3 will develop techniques for proving that the parentheses in such multiple disjunctions can be omitted; for now we will continue to work with the parentheses.

Definition 1-31 An $\mathcal{SC}$ sentence satisfying the following is said to be in *conjunctive normal form* (CNF):

A clause is in CNF.

If δ and η are in CNF, then $(\delta \wedge \eta)$ is in CNF.

Any sentence in CNF must be obtainable by use of one or both of the previous rules.

The following are in CNF:

$$P, \quad \neg P, \quad (P \vee \neg Q),$$

$$((C \vee \neg A) \vee B) \wedge ((P \vee \neg Q) \wedge (\neg Q \vee \neg P)).$$

In this example, $(P \vee \neg Q)$ and $(\neg Q \vee \neg P)$ are in CNF, since they are clauses. Therefore, their conjunction,

$$((P \vee \neg Q) \wedge (\neg Q \vee \neg P)),$$

is in CNF. Also,

$$((C \vee \neg A) \vee B)$$

is a clause, so it is in CNF. Thus, the conjunction of the two preceding sentences is also in CNF. Speaking loosely, we can say that a sentence in conjunctive normal form is a "conjunction of clauses." In this description, "conjunction of clauses" means that the sentence can be built up from clauses by recursively conjoining parts according to the previous definition.

It can be proved that, for any $\mathcal{SC}$ sentence ϕ, there is a tautologically equivalent sentence that is in CNF. There is also a procedure for constructing an equivalent CNF. I will not give the proof; instead, the procedure will be demonstrated by an example. After the steps of the procedure are understood, it should be clear how a corresponding metatheorem could be stated and proved. The procedure is useful because some theorem-proving computer programs require that the sentences the programs manipulate be in CNF. The logical operations used in a program of this kind will be described later in this section. The present task is to see how the procedure for conversion to CNF is performed.

I will transform

$$\neg(A \leftrightarrow B) \tag{1.4}$$

into a tautologically equivalent CNF. *I will first do this with a fairly lengthy procedure, and then describe a shorter method that works well for this particular example.* Both methods will be exhibited in an informal manner, but all of the steps can be expanded into a formal $\mathcal{SC}$ derivation. Also, each of the steps uses a Tautological Equivalence Rule, so each intermediate sentence obtained, as well as the final result, is equivalent to the original sentence.

CNF Procedure

1. We first eliminate occurrences of $\leftrightarrow$. By repeatedly using the Equivalence Biconditional, all occurrences of $\leftrightarrow$ can be replaced by occurrences of $\wedge$ and $\rightarrow$. If we do this to sentence (1.4), we obtain

$$\neg((A \rightarrow B) \wedge (B \rightarrow A)).$$

2. By repeatedly using the Conditional–Disjunction Rule, we convert occurrences of $\rightarrow$ into occurrences of disjunctions. The previous sentence becomes

$$\neg((\neg A \vee B) \wedge (\neg B \vee A)).$$

3. We now have a sentence in which the only connectives that occur are $\neg$, $\wedge$, $\vee$. De Morgan's theorem enables us to move negations from the outside of parentheses to the inside. Applying DeM to the preceding, we obtain

$$\neg(\neg A \vee B) \vee \neg(\neg B \vee A).$$

Applying DeM two more times yields

$$(\neg\neg A \wedge \neg B) \vee (\neg\neg B \wedge \neg A).$$

It can be seen that repeated use of DeM eventually produces a sentence in which no negations are outside parentheses.

4. By repeated use of DN, multiple negations can be eliminated. This produces literals. In our example, we obtain

$$(A \wedge \neg B) \vee (B \wedge \neg A). \tag{1.5}$$

Sentence (1.5) is a disjunction of conjunctions; it is what is called a *Disjunctive Normal Form*. It can be converted into a CNF.

5. First we use the Distribution Rule to distribute $(A \wedge \neg B)$ over the $\wedge$ in the right-hand disjunct of (1.5):

$$((A \wedge \neg B) \vee B) \wedge ((A \wedge \neg B) \vee \neg A).$$

For easier reading and handling, the two disjunctions are commuted:

$$(B \vee (A \wedge \neg B)) \wedge (\neg A \vee (A \wedge \neg B)).$$

We can now distribute the left-hand disjuncts over the $\wedge$s to produce a conjunction of disjunctions,

$$((B \vee A) \wedge (B \vee \neg B)) \wedge ((\neg A \vee A) \wedge (\neg A \vee \neg B)). \tag{1.6}$$

Sentence (1.6) is a conjunction of clauses, so it is a CNF sentence that is tautologically equivalent to (1.4). This completes the procedure.

In the case of this particular example, it is possible to obtain a simpler CNF. Although (1.6) is a perfectly good CNF, we might as well go for the simpler one. In order to justify the additional simplification procedure, I shall prove the following metatheorem.

Metatheorem 1-32   Let τ and ϕ be $\mathcal{SC}$ sentences. If τ is a tautology, then $(\tau \wedge \phi)$ is tautologically equivalent to ϕ.

Proof. Let **I** be any interpretation that satisfies $(\tau \wedge \phi)$. Then **I** satisfies each conjunct, so ϕ is true under **I**. Hence, $(\tau \wedge \phi) \vDash_T \phi$.

Conversely, let **I** be any interpretation that satisfies ϕ. Since τ is tautologous, it is true under **I**. Therefore, **I** satisfies $(\tau \wedge \phi)$, so $\phi \vDash_T (\tau \wedge \phi)$. **Q.E.D.**

Sentence (1.6) contains two conjuncts that are tautologies, namely,

$$(B \vee \neg B) \text{ and } (\neg A \vee A).$$

By Metatheorem 1-32,

$$((B \vee A) \wedge (B \vee \neg B))$$

is tautologically equivalent to

$$(B \vee A),$$

and similarly for the right-hand conjunct of (1.6). Consequently, any interpretation that satisfies (1.6) must satisfy (1.7), and conversely

$$(B \vee A) \wedge (\neg A \vee \neg B). \tag{1.7}$$

Roughly speaking, we can remove ("cancel out") the two tautologies in (1.6), and obtain (1.7), which is also tautologically equivalent to the starting sentence, (1.4). Notice that all of the steps, except the last, were justified by application of Tautological Equivalence Rules, so those steps can be directly used in an $\mathcal{SC}$ derivation. The justification of the last step depended on the preceding metatheorem, which is a semantical metatheorem. It does not show how to give an $\mathcal{SC}$ *proof* of

$$(\tau \wedge \phi) \leftrightarrow \phi$$

when τ is a tautology. By the Completeness theorem, such a proof must exist. It is left as an exercise to show how to construct the general schematic form of such a derivation.

Now let us briefly consider a shorter method that could have been applied to this problem. Our starting sentence, $\neg(A \leftrightarrow B)$, is the negation of a biconditional. I used the first of our two Equivalence rules. This was done because it led to an illustration of a general method that applies to biconditionals, conditionals, negated compound subsentences, etc. In other words, the previous, long method illustrates a general procedure that applies to any form of SC sentence. However, for this particular problem we could have applied the second Equivalence rule and directly moved to

$$\neg((A \wedge B) \vee (\neg A \wedge \neg B)).$$

Applying DeM, we obtain

$$\neg(A \wedge B) \wedge \neg(\neg A \wedge \neg B),$$

and from this further applications of DeM and DN soon yield

$$(\neg A \vee \neg B) \wedge (A \vee B),$$

which is essentially the previous result (1.7). Therefore, for this example, the second part of Equiv is more convenient. When treating a negated biconditional, this second part of Equiv is useful; yet the first part of Equiv is often useful for a biconditional that is not negated.

As previously mentioned, CNF is used in some theorem-proving computer programs. Suppose that Γ is a set of SC sentences and ϕ is an SC sentence. We would like to write a computer program that might be able to construct a proof of ϕ from Γ if ϕ is in fact derivable from Γ. We could try to write a program that uses all of the rules and heuristic proof strategies introduced in this chapter. As a long-run project, this might be worth trying, but it would require a major effort. Another approach is to develop a simpler and more "mechanical" set of derivation rules that could be implemented in the program. The vague term "mechanical" means that the rules can be used in a derivation according to strict algorithms or procedures for their application, rather than being guided by a set of subtle heuristic strategies.

One of the standard methods of "mechanical" computer deduction is the *Resolution Method*. For sentential calculus, it amounts to this: Suppose that we have a set of premises Γ and a desired conclusion ϕ. Put each premise in Γ into CNF. Also put the negation of the conclusion, i.e., $\neg\phi$, into CNF. Each of the resulting CNFs is, of course, a conjunction of clauses. Let Δ be the set of all of these clauses. Δ may be called a set of *clausal premises*. Now use Δ as a new set of premises of a derivation, i.e., each clause in Δ becomes one premise. Try to derive a contradiction from Δ. If a contradiction is obtained, stop.

It is obvious that the sentences in Δ are implied by $\Gamma \cup \{\neg\phi\}$, because of the way Δ was constructed using CNFs. Thus, if a contradiction can be derived from Δ, then a contradiction can be derived from $\Gamma \cup \{\neg\phi\}$. By applying the RAA strategy, we could then obtain a derivation of ϕ from Γ. In practice, one does not finish off this entire RAA proof; as soon as a contradiction is obtained from Δ, the resolution proof is done, for we now know that ϕ is derivable from Γ.

The sentences of $\Gamma \cup \{\neg\phi\}$ are converted into Δ so that the rules of derivation can be greatly restricted and applied "mechanically." For sentential calculus, after the original sentences are converted into a set of clausal premises, a resolution proof uses *only* the following rules: DS, Cut, Comm, Assoc, and Taut. Of course, Rule P is used in stating the premises Δ, but it is used *only* for this purpose. It can be proved that, for a set of clausal premises, this set of rules is *complete with respect to contradiction*, i.e., if a set of SC clauses Δ is unsatisfiable, then a contradiction can be derived from Δ using these rules. Consequently, if $\Gamma \vDash_T \phi$, and Δ is obtained as described here, then there exists a resolution derivation of a contradiction from Δ.

Here is a simple example. Suppose we want to derive $\neg B$ from

$$\neg A, \quad (B \to A) \vee C, \quad C \to A.$$

There is a short direct proof using MT, DS, and MT. If we assume $\neg\neg B$, or equivalently, B, there is also a short derivation that produces a contradiction for RAA. To construct a Resolution Proof, we convert the premises to CNF, obtaining the following clauses:

$$\neg A,$$

$$(\neg B \vee A) \vee C,$$

$$\neg C \vee A.$$

For the negation of the conclusion, we may use B, which is also in CNF, with the one clause B. We use B, together with the previous clauses, to derive a contradiction, as follows:

Example.

SC Resolution Derivation.

$\{\, Pr_1 \,\}$	(1)	$\neg A$	P
$\{\, Pr_2 \,\}$	(2)	$(\neg B \vee A) \vee C$	P
$\{\, Pr_3 \,\}$	(3)	$\neg C \vee A$	P
$\{\, Pr_4 \,\}$	(4)	B	P
$\{\, Pr_1, Pr_3 \,\}$	(5)	$\neg C$	DS (1), (3)
$\{\, Pr_1, Pr_2, Pr_3 \,\}$	(6)	$\neg B \vee A$	DS (2), (5)
$\{\, Pr_1, Pr_2, Pr_3, Pr_4 \,\}$	(7)	A	DS (4), (6)

As previously mentioned, a resolution proof is considered done as soon as a contradiction is obtained; we have a contradiction between lines (1) and (7). ■

This derivation was especially simple. With more complicated CNF premises, one often needs to use Cut, Comm, etc. Humans can look at the sentences and "see" when DS or Cut will yield simpler sentences or new combinations of sentences. For instance, I saw immediately that DS could be applied to lines (1) and (3). A "dumb" computer program typically tries all combinations of available sentences and laboriously searches for matches between literals of the forms ϕ and $\neg\phi$. This is computationally expensive, so various procedures have been invented to produce more efficient Resolution Proofs. Resolution for predicate calculus is much more complex; a good introduction to the subject is in Genesereth and Nilsson (1987).

Exercises

E 1-64 (Answer Provided for 1) Transform the following sentences into Conjunctive Normal Form. Use the informal procedure described in Section 1.7. Cancel out tautologies (as described in the section), and simplify expressions of the forms $\phi \vee \phi$ and $\phi \wedge \phi$ by using the Tautology Laws.

1. $(A \to B) \to (C \wedge A)$

2. $\neg(P \to \neg(Q \vee (\neg R \wedge S)))$

E 1-65 Let τ be a tautology and ϕ be any $\mathcal{SC}$ sentence. Show that it is always possible to construct an $\mathcal{SC}$ derivation of $(\tau \wedge \phi) \leftrightarrow \phi$ from the empty set of premises. (Hint. Suppose that $(\tau \wedge \phi)$ is placed on the first line of a derivation by Rule P. Then one can infer ϕ on the second line by Simp. Now use C to obtain one-half of the biconditional. Now suppose that ϕ is entered on a line by P. You will need to get $(\tau \wedge \phi)$. You may assume Completeness.)

E 1-66 Let $\Gamma = \{ A \to (C \vee D), B \to A, D \to C, B, \neg(B \wedge C) \}$. Show that Γ is unsatisfiable by deriving a contradiction from it, i.e., construct a standard $\mathcal{SC}$ derivation with some sentence ϕ on one line and with $\neg\phi$ on another line, or else with their conjunction on a line.

E 1-67 (Answer Provided) Let the premises of an argument be

$$\Gamma = \{A \to (C \vee D),\ B \to A,\ D \to C,\ B\},$$

and let the conclusion be $(B \wedge C)$. Put each premise and the negation of the conclusion into CNF, form the corresponding set of clausal premises, and construct a Resolution Proof for this argument.

E 1-68 Use Resolution (Proof) to derive a contradiction from the following set of clauses:

$$(A \vee \neg B) \vee C, \quad B \vee \neg D, \quad \neg C \vee D,$$

$$B \vee (C \vee D), \quad \neg A \vee \neg B, \quad \neg D \vee \neg B$$

E 1-69 For each of the following arguments, convert each premise and the negation of the conclusion into equivalent CNFs, form the corresponding set of clausal premises, and construct a Resolution Proof for the argument.

1. $C \rightarrow D, \quad A \vee C, \quad \neg(B \wedge D) \quad / \therefore \ B \rightarrow A$

2. $D \leftrightarrow B, \quad \neg B \rightarrow C, \quad \neg C \quad / \therefore \ \neg D \rightarrow B$

3. $M \vee (B \wedge X), \quad (M \leftrightarrow \neg B), \quad \neg X \quad / \therefore \ B \rightarrow T$

4. $(A \leftrightarrow B) \wedge (C \leftrightarrow D) \quad / \therefore \ (A \wedge C) \leftrightarrow (B \wedge D)$

Chapter 2

Basic Set Theory

2.1 Sets

2.1.1 Extensionality, Predicates, and Abstraction

Intuitively, a *set* S is a collection of distinct objects considered just as a collection, without regard to the order or manner in which these objects may be named or described. The objects that S comprises are called the *element*s or *members* of S. Our idea of a set includes the case in which there may be no elements in the collection. Here are descriptions of some sets:

- The set of people on the Island of Trinidad at noon, January 1, 1992.

- The set of positive integers.

- The set of all integers.

- The set of real numbers greater than $\pi = 3.14159 \cdots$.

In this chapter, and later chapters, I will denote sets and other objects with some of the same symbols used in the sentential calculus. For instance, 'A' may denote a set; if it does, then it is *not* a sentential letter. The uses of the symbols will always be stated explicitly or be clear from the context of the discussion.

The basic concept of set theory is the *membership relation*, $\in$. If A is a set and b is any object, then $b \in A$ means that b is an element (or member) of A. For instance, Thomas Edison $\in$ the set of famous inventors. If *Int* is the set of integers, $-13 \in Int$. In the foundations of mathematics, $\in$ is characterized by several axioms stated in a formalized language.

115

We will not do this. In this book, set theory is treated informally, using mathematical English, which is just ordinary English plus some mathematical terminology and stylistic features. We will assume the existence of some simple and familiar kinds of sets. We will develop principles for defining new sets in terms of given ones, and will prove some theorems regarding basic properties of sets. We begin with a fundamental axiom, formulated in informal language; it implies that the identity of a set is determined *only* by what is in it (and thus depends on nothing else).

Axiom of Extensionality. Let A and B denote any sets. If A and B denote the same set, we write $A = B$, and

$$A = B \text{ iff for every } x, (x \in A \text{ iff } x \in B).$$

This axiom states the necessary and sufficient condition for $A = B$, namely, that for any object x, x is an element of A iff it is an element of B. If $A = B$, then anything in A is in B and conversely; so A and B contain exactly the same elements. In order to have $A \neq B$, there must be some element of A that is not in B, or some element of B that is not in A, or both.

Assume that Raquel is a member of the set of people on the Island of Trinidad at noon, January 1, 1992. Ignoring English verb tenses, under our assumption, the expression 'Raquel is a person and Raquel is on the Island of Trinidad at noon, January 1, 1992.' is a true English sentence. The expression 'x is a person, and x is on the Island of Trinidad at noon, January 1, 1992' is called an *English-language predicate*, and x is called a *variable*. We could have used a different variable, say, y, in place of each occurrence of x in this predicate. In this chapter, the lowercase letters s through z, possibly with subscripts, will be used as the variables in English-language predicates. We will also sometimes use k, m, n as variables, especially when discussing numerical examples. We can say that Raquel *satisfies* the predicate because a true sentence is obtained if the name 'Raquel' is substituted for x throughout this predicate. This use of the word 'satisfies' is different from the previous use of this word in the phrase 'sentential calculus interpretation **I** satisfies sentence ϕ'. The context of discussion should always make clear which use of 'satisfies' is intended.

Suppose that we use $trin(x)$ as an abbreviation for the predicate 'x is a person and x is on the Island of Trinidad at noon, January 1, 1992.' Now we may use $trin(\text{Raquel})$ as an abbreviation for 'Raquel is a person, and Raquel is on the Island of Trinidad at noon, January 1, 1992.' If this sentence is true, then Raquel satisfies the abbreviated predicate $trin(x)$. There may be other people who also satisfy this predicate, for instance, $trin(\text{Rafael})$ may be true. Try to think of *all* the people who satisfy this predicate. For a given person, say, Jane Doe, we may not *know* whether $trin(\text{Jane Doe})$ is true or false, but it must be one or the other. If *trin*(Jane Doe) is true, then she is a member of the set of all people on the Island of Trinidad at noon, January 1, 1992. If *trin*(Jane Doe) is not true, then she is not an element of this set. Thus, even if we have no *practical way* of naming all of the people who satisfy the predicate, *trin*(x), we can still say that the set of these people is the same as the set of people on the Island of Trinidad at noon, January 1, 1992. This set is called the *extension* of the predicate *trin*(x). This suggests that, for any predicate with one variable, there exists a corresponding set that is the extension of this predicate. This vague idea requires important clarifications and qualifications, but first, here are some more examples:

Examples. Some predicates, with descriptions of their extensions:

- x is an even positive integer. Extension is the set of even positive integers.

- y is a lion in Africa on January 1, 1992. Extension is the set of all lions in Africa on January 1, 1992.

- u is a real number such that $u^2 < 0$. Extension has no elements.

- s is an $\mathcal{SC}$ sentence that is true under every interpretation. Extension is the set of all sentential calculus tautologies.

- x is an integer such that $x^2 = 9$. Extension is the set whose members are 3 and –3.

Each of these predicates uses only one variable, but this variable may occur more than once in the predicate, just as x does in *trin*(x). Using the appropriate substitutions, the last example leads to the two true sentences: '3 is an integer such that $3^2 = 9$' and '–3 is an integer such that $(-3)^2 = 9$'. If we substitute for x in this predicate any number other than 3 or –3, we do not obtain a true sentence. Hence, the extension of the predicate contains only 3 and –3. ■

We often use the expressions 'for every x' and 'for any x'. As used here, these expressions mean the same thing, and each is called *the informal universal quantifier*. There are *also informal existential quantifiers* such as 'there exists x' in 'there exists x such that x is a real number greater than 100,000.' Notice that this expression is a complete sentence, i.e., a declarative sentence that has a truth value. If we have a predicate such as 'y is rich' (where 'rich' is short for 'rich person'), we can make a complete sentence from this predicate by substituting a name for the occurrences of the variable in it. For example, we can form the sentence, 'John Doe is rich.', which may or may not be true depending on who John Doe is. We can also form a sentence from a predicate by putting an appropriate quantifier in front of the predicate to form a *quantified sentence*. Thus, we can obtain 'There exists x such that x is rich', which means 'There is a rich person.' We can also obtain, 'For any x, x is rich.', which means, 'Everyone is rich.' In Chapter 4, we will examine some technical differences between the occurrences of variables in quantified sentences and their occurrences in predicates. For now let us say that the occurrence of x in 'x is rich' is a *free occurrence* because there is no quantifier (such as 'for any x') affecting it. This can be seen by noticing that 'x is rich' is not a complete sentence because 'x' does not denote any particular thing, so the expression has no truth value. It can be transformed into a sentence in two ways. We can substitute a name for the free occurrence of the variable 'x', as in 'Jane Roe is rich.' We can also quantify the variable, as in 'Everyone is rich.' ('For every x, x is rich.') and in 'There is someone who is rich.' ('There exists an x such that x is rich.') Occurrences of variables inside of quantifiers, and occurrences of variables affected by quantifiers, are called *bound occurrences*. These concepts will be made precise in Chapter 4. For now let us introduce the following definition. It is not highly precise, and it will be subjected to further clarification as needed in certain contexts. In the statement of this definition, 'names' refers to any proper name of a person, thing, place, etc., and 'constants' refers to symbols that denote specific mathematical objects such as numbers. These symbols may be numerals (like '23' and '23.04') or may be other special symbols that denote particular objects (such as π, which denotes $3.14159\cdots$). The kinds of symbols that count as names or constants may be expanded later as needed.

Definition 2-1 Let $\sigma(n_1, n_2, \cdots, n_k)$ be a sentence of mathematical English containing occurrences of names or constants $n_1, \cdots, n_k$. Suppose that the variables $x_1, \cdots, x_k$ do not occur in this sentence. Then an expression $\sigma(x_1, x_2, \cdots, x_k)$ is an *English predicate in the variables* $x_1, \cdots, x_k$ iff it is obtainable from $\sigma(n_1, n_2, \cdots, n_k)$ by substituting occurrences of $x_1, \cdots, x_k$ for corresponding occurrences of $n_1, \cdots, n_k$ in this sentence. If desired, other new variables that do not occur in the sentence, such as s, u_3, y, etc., may be used in place of the x_i. The number of different new variables that are substituted into the sentence is called the *arity* of the predicate. A predicate of arity k is called a *k-ary* predicate. A 1-ary predicate is also called a *unary* predicate and a 2-ary predicate is called a *binary* predicate.

Thus, 'x is rich' is obtainable from 'Bill is rich' by substituting the variable x for the name Bill. This predicate has an arity of 1. The (English) predicate, '$x < y$' is obtainable by substituting the variables x and y for the constants 2 and 3 in the sentence '$2 < 3$'. The arity of '$x < y$' is 2. The predicate 'There exists x such that $y < x$' is obtainable from 'There exists x such that $5 < x$' by substituting the new variable y for the constant 5. Since only one new variable, y, is used, this predicate has arity of 1. In Section 2.2 we will begin to use predicates, such as '$x < y$', with occurrences of more than one variable. Until then the predicates we consider will have only one variable, although this variable may occur more than once in the predicate.

The Axiom of Extensionality provides a criterion of identity for sets, but it does not say which sets exist. It is tempting to assume that there is a *universal set* containing *everything*, so that anything that exists (including any set) is an element of this big set. Unfortunately, assuming the existence of an absolutely universal set can lead to a contradiction, so we will not make this assumption. There is a need to augment the Axiom of Extensionality with other axioms such that the entire system of axioms produces a satisfactory characterization of sets. Since the late nineteenth century, much effort has been devoted to this task, and several good systems of axioms have been developed. As already mentioned, we will not take this approach in this book, but we still need a suitable way to define sets.

One approach is this: Suppose that we are given a predicate such as $trin(x)$ in the variable x. We just assume that its extension exists. We obtain the set of all people on Trinidad at noon, January 1, 1992. This result seems reasonable. Suppose that we were to introduce the following as an axiom: Let $\rho(x)$ be any predicate. Then there is a set S such that, for any u, $u \in S$ iff $\rho(u)$ (i.e., u satisfies ρ). This principle is called the *Axiom of Comprehension* or the *Axiom of Abstraction*. In order to make it precise, we would need to specify a precise language to be used for formulating the allowable predicates. The good news is that this can be done; normally by using first-order predicate calculus, which is the topic of Chapter 4. The bad news is that, even if one does use such a formal language, the axiom of comprehension can lead to a contradiction. Luckily, it turns out that a limited form of comprehension can be used safely. It can be stated informally as follows, where it is understood that we are using English predicates in our mathematical English language.

Axiom of Separation. Let D be a set and let $\rho(x)$ be a predicate in the one variable x. Assume that this predicate can be meaningfully applied to the elements of D. Then there exists a set S such that, for any x, $x \in S$ iff ($x \in D$ and $\rho(x)$).

This axiom, and especially the term 'meaningfully applied', requires some explanation. Notice that the Axiom of Separation is really an infinite set of axioms corresponding to the infinite set of predicates that can be formulated. It is what is called an *axiom schema*. It has the name 'separation' because it states that we can separate out of D exactly those elements that satisfy the predicate $\rho(x)$. But we really do not apply ρ directly to the *elements* of D; we substitute *names*, or other expressions denoting these elements, into the predicate $\rho(x)$ to form sentences. It is possible that an element of $\rho(x)$ may have more than one name or expression that denotes (refers to) it. For example, if we are talking about numbers, the expressions '12', '5 + 7', and '3 × 4' all denote the same number. Consider the predicate, '$8 < x$', and notice that '$8 < 12$', '$8 < (5 + 7)$', and '$8 < (3 \text{ x } 4)$' are all true. If we use the same predicate and substitute for x the expressions, '2' and '$5 - 3$', in both cases the resulting sentence is false. In general, we must allow the possibility that elements of D may be denoted by more than one name or other referring expression. In order for a predicate $\rho(x)$ to separate out a definite set of elements of D, $\rho(\nu_1)$ and $\rho(\nu_2)$ must have the same truth value whenever ν_1 and ν_2 are two expressions denoting the same element of D. If one of these sentences were true and the other false, it would not be determined whether the corresponding element of D is in S or not in S.

Saying that $\rho(x)$ can be *meaningfully applied* to elements of D means this: Let β be a name (or referring expression) for any element of D. Then if we substitute β for the occurrences of x in $\rho(x)$ to obtain $\rho(\beta)$, it will be the case that $\rho(\beta)$ is a complete declarative sentence and has the same truth value as $\rho(\delta)$, where δ is any other name (or referring expression) for that element of D denoted by β. The predicate, '$8 < x$', can be meaningfully applied to real numbers in ordinary mathematical English. The predicate, 'x is in India at noon on July 2, 2004', can be meaningfully applied to the set of human beings.

It might be thought that any predicate can be meaningfully applied to any set, but this is not true. Consider the expression $\eta(x)$: 'Betty believes that x is one of the two tallest women on the basketball team.' Suppose there is someone on the team named 'Miranda', Betty knows this, and the sentence, 'Betty believes that Miranda is one of the two tallest women on the basketball team' is true. Yet Betty is not aware that Miranda is the same person who is also named 'Miraculous Mira' (as a nickname). Then the sentence 'Betty believes that Miraculous Mira is one of the two tallest women on the basketball team' may be false. Hence, the expression $\eta(x)$ can yield different truth values when different names for the same object (in this case, same person) are substituted for x. According to the characterization of 'meaningfully applied', $\eta(x)$ cannot be meaningfully applied to the set of persons on the basketball team. Recall that Chapter 0, the Introduction, mentioned that Chapter 2 would show an exception to the Principle of Substitutivity of Identicals. This is it.

We use *meaningfully applied* in a special, technical way, and it is important that we do so. According to our intuitive idea of a set, it is just a collection of objects without regard to the order or manner in which these objects may be named or described. If an object has more than one name, such as 'Miranda' and 'Miraculous Mira', it should not matter which name we use in determining whether or not this object is in a given set. Since we will usually use predicates to help us define sets, we need the following to hold: If any name for an object is substituted for the variable in a predicate $\rho(x)$, the truth value of the resulting sentence should be the same as the truth value of any other sentence obtained by substituting any other name for this object into $\rho(x)$. The expression 'Betty believes that x is one of the two tallest women on the basketball team' fails to satisfy this condition, so according to our terminology, it cannot be meaningfully applied to the set of women on the basketball team. It can be seen that the peculiar behavior of $\eta(x)$ results from the occurrence of 'believes' in it. A person may believe something about Miranda but not necessarily believe the same thing about Miraculous Mira. Expressions that can yield different truth values when different names for the same object are substituted into them are called *referentially opaque*. Referential opacity has been extensively discussed in the philosophical literature. A good beginning source on the subject is Quine (1960).

Without resorting to a formal language (such as predicate calculus), it is difficult to state all of these ideas, or the Axiom of Separation, much more precisely. On the other hand, terms such as 'believes', 'knows', 'desires', etc., that are troublesome, are presumably not in mathematical English and (one hopes) are not required in many application problems. For our purposes, the previous statement of the axiom should be adequate, provided that one proceeds with caution. In order to use the Separation Axiom, the set D must already be given. In formalized set theory, one uses additional axioms that permit the construction of a huge variety of sets. We will not do this, although one additional axiom will be introduced later. For now, we adopt the following:

Modus Operandi. In any use of set theory, it will be assumed that we are given some specified nonempty set of objects called the *domain of discourse* or *universe of discourse*. Any additional sets that are used in an application problem must be defined by applying the Axiom of Separation to this domain or else be definable in terms of such sets by using methods developed in the remainder of Chapter 2.

There is an old recipe for rabbit stew that begins, "First catch your rabbit." The *Modus Operandi* says, in effect, if you want to define sets, first catch a domain of discourse. This is a metaphorical way of saying that the domain is assumed to exist (i.e., it does not lead to contradictions), and that our mathematical English has suitable predicates for defining sets of elements of this domain. But the *Modus Operandi* goes beyond this: After we have defined some special sets of elements of the domain of discourse, we may then use these sets in definitions of additional sets. The members of these additional sets need not be restricted to being elements of the *original domain*. It will turn out that, although we start with a given domain, D, we may eventually define a new, extended domain D_1. The details of how this is done will be given later when the Power Set Axiom is introduced. Fortunately, most of the sets used in applications will be separated out of the original domain. The *Modus Operandi* is vague, but if we tried to make it very precise, we would be starting down the road toward axiomatic set theory, which is exactly what our informal approach is intended to avoid. Our aim is not to develop a systematic theory about the existence of sets. Instead, we will assume the existence of some sets and develop ways of working usefully with these assumptions.

To illustrate a simple use of the *Modus Operandi*, suppose the domain of discourse is the set of all people who work for the business organization, the Omega Corporation. Let D denote this set of people. Then, by using separation, there exists a set M such that, for any x, $x \in M$ iff ($x \in D$ and x is male). The predicate used here is 'x is male' and the set M is the set of male employees of the Omega Corporation. (It is possible there are none.) If it is clearly understood what the domain of discourse is, we will apply the Axiom of Separation without directly referring to this domain. Thus, we may simply write this: Let M be the set of x such that x is male. This amounts to using the Axiom of Comprehension (Abstraction) *restricted to the domain D*.

Definition 2-2 Let D be a domain of discourse and $\rho(x)$ be a predicate that can be meaningfully applied to the elements of D. Then $\{x \mid \rho(x)\}$ denotes the set of elements x, such that $x \in D$ and $\rho(x)$. Defining a set in this manner is called *definition by abstraction*.

The justification for definition by abstraction is the Separation Axiom because this axiom asserts the existence of the set that is defined. It is important to remember that we will *only* use the notation $\{x \mid \rho(x)\}$ when some domain is specified and that this notation is just an abbreviated form of application of the Separation Axiom. For example, suppose that the domain is the set of integers. Then we may define the following:

$$A = \{ x \mid x > 21 \}$$

and

$$B = \{ x \mid \text{ there exists an integer } y > 5 \text{ such that } x = y^2 \} .$$

Let D be any domain of discourse. Applying the Axiom of Extensionality, we obtain $D = \{ x \mid x = x \}$, because $x \in D$ iff ($x \in D$ and $x = x$). The biconditional is true because anything is identical to itself.

Some related notations will also be used. Given a domain D, if we can list the elements (within D) of some set, we can describe it in terms of these elements, e.g., $\{ 0, 1, 2 \}$. It is also customary to denote certain infinite sets by indicating an infinite list of elements, e.g., $\{ 0, 1, 2, \cdots \}$. The three-dot notation (ellipsis) should be used only when it is perfectly clear how the list is to be extended; the next chapter will provide a clear meaning for the three dots, at least in many practical contexts. This text will not attempt to define *finite* or *infinite* in a precise, general way. But we do assume familiarity with the basic number systems described in the Introduction. Any particular natural number, such as 0, 13, 2398474, is considered to be a *finite number*. If a set S has a finite number of elements, we call it a *finite set*. Moreover, if the finite number of elements in S is n, then we say that the *cardinality of S* is n, which we will write as $|S| = n$. For example, $|\varnothing| = 0$, and, if

$$B = \{ \, a, \ b, \ 1, \ 2, \ \{ \, s, \ t, \ 15 \, \}, \ \sqrt{2} \, \},$$

then the elements of B are a, b, 1, 2, $\{ \, s, \ t, \ 15 \, \}$, and $\sqrt{2}$, so $|B| = 6$. It is important to see that $\{ \, s, \ t, \ 15 \, \}$ is only one element of B, even though $\{ \, s, \ t, \ 15 \, \}$ has three elements. More generally, we denote the cardinality of any set S by $|S|$. If $|S| = n$, for some $n \in Nat$, then S is a *finite set* (has finite cardinality); if, for all $n \in Nat$, $|S| > n$, the S is an *infinite set*. The concept of cardinality in general set theory (including infinite sets) is rather complex and difficult to define, and beyond the scope of this book. However, for a finite set S we can just use the intuitive idea of the number of elements in S.

We now proceed using our *Modus Operandi*. In any particular application problem, the domain of discourse will be specified. For general discussions and statements of theorems, it is assumed that a domain of discourse exists, even if this assumption is not explicitly mentioned. Suppose that we have a domain D and a predicate $\rho(x)$ that can be meaningfully applied to the elements of D. Also suppose that we define a set $S = \{ \, x \mid \rho(x) \, \}$. If S is defined by applying the Axiom of Separation to D, we have

$$x \in S \text{ iff } (x \in D \text{ and } \rho(x)).$$

Definitions of this form will arise repeatedly, and it becomes tiresome always to include '$x \in D$'. Therefore, in place of the preceding biconditional, we may simply write

$$x \in S \text{ iff } \rho(x),$$

where it is understood that the possible values of x are the elements of D. Correspondingly, we can write the explicit definition of the set S in the simple form,

$$S = \{ \, x \mid \rho(x) \, \},$$

without including '$x \in D$'. From now on we will make extensive use of definitions by abstraction and the corresponding biconditionals in the forms specified here.

Example.

Problem Statement. The domain of discourse is *Nat* (as described in the Introduction), and you are given the following predicates that apply to natural numbers:

   even(x): x is an even number,

   odd(x): x is an odd number.

Also, assume that you can form the square of a number, so that if $x \in Nat$, then x^2 denotes the square of x. Finally, let H be the set of natural numbers starting with 0 and ranging through 100. In set theory, it is normally taken for granted that we may also use the identity predicate, $=$, and we shall do this here, along with using informal sentential connectives. Using (definition by) abstraction, write definitions of the following sets:

1. $E =$ the set of even (natural) numbers.

2. $S =$ the set of all odd numbers that are squares.

3. $H_S =$ the set of all numbers that are squares and equal to or less than 100. Also, list these particular numbers in braces.

4. $O =$ the set of all numbers that are either odd or are not in H.

Briefly justify your definitions.

Discussion. Our task here is to translate the somewhat informal descriptions of sets that are given into more precise definitions using abstraction. In doing this we are limited to standard set theory notation and the specified terms that are given in the problem.

   Consider the set, E, of even numbers. Because E is the set that contains only even numbers and all even numbers, we have $x \in E$ iff x is an even number. Using the given predicate, we then have $x \in E$ iff *even(x)*. Thus, the definition by abstraction can be written as $E = \{ x \mid even(x) \}$. The other definitions can be handled in a similar manner.

Written Solutions.

1. An element is in E iff it is an even number. Therefore,

$$E = \{x \mid even(x)\}.$$

2. An element of *Nat* is in S iff it is a square, i.e., has the form of a square (number), and is also odd. Let x be in *Nat*. To say that x is odd is equivalent

to having $odd(x)$ true, which can simply be stated as $odd(x)$. To say that x has the form of a square means that there is some number y such that $x = y^2$. The elements of S are all of the natural numbers that are both odd and square. Therefore,

$$S = \{x \mid \text{there exists } y \text{ such that } x = y^2 \ \& \ odd(x)\}.$$

3. If we had a predicate for 'less than', we could use it here, but we are not given this predicate, and we must work with what we are given. Fortunately, we can use the given predicate H, because a square equal to or less than 100 is also a square in H. We just need to conjoin the *square*-condition with the H-condition. The result is similar to the previous one:

$$H_S = \{x \mid \text{there exists } y \text{ such that } x = y^2 \ \& \ x \in H\} \ .$$

It is easy to compute the elements of H_S, obtaining

$$H_S = \{0, \ 1, \ 4, \ 9, \ 16, \ 25, \ 36, \ 49, \ 64, \ 81, \ 100\}.$$

4. This predicate is more complex than the previous ones, but we can easily express O as follows:

$$O = \{x \mid odd(x) \text{ or not } x \in H\}.$$

By applying the $\mathcal{SC}$ Comm and CDis rules, this is equivalent to

$$O = \{x \mid x \in H \text{ only if } odd(x)\}. \ \blacksquare$$

2.1.2   Some Special Sets and Set Operations

I will now introduce some definitions of familiar set theoretical concepts and state some elementary theorems. Keep in mind that we make the general assumption that all sets under consideration satisfy the *Modus Operandi*. In the case of the following definition, we assume that A and B are obtained from a domain of discourse by use of separation. Also, the possible values of variables, such as x, are the elements of the domain.

Definition 2-3 Let A, B be sets. A is a *subset* of B, written $A \subseteq B$, iff for every x, if $x \in A$, then $x \in B$. A and B may be identical. If they are not identical, then A is a *proper subset* of B, written $A \subset B$.

If D is the domain of discourse, then $A \subseteq D$ and $B \subseteq D$, because both A and B are obtained from D by separation. The concept of $\subseteq$ must not be confused with the concept of $\in$. For example, let Int be the set of all integers and Int^+ be the set of all positive integers. Then $-13 \in Int$, but it is not the case that $-13 \in Int^+$. This negation is usually expressed by $-13 \notin Int^+$. Also, $Int^+ \subseteq Int$ in as much as, for any x, if x is an element of Int^+ ($x \in Int^+$), then x is an element of Int ($x \in Int$). When $A \subseteq B$ it is sometimes said that A is *included* in B.

Example.

Problem Statement. Let the domain be *Re*, the set of real numbers (as described in the Introduction). We are given the predicates

$$x < y : \; x \text{ is less than y,}$$

$$y = \frac{1}{x} : \; y \text{ is the reciprocal of } x,$$

rational(x): x is a rational number (i.e., a member of *Rat*, as described in the Introduction).

Using these predicates and definition by abstraction, write definitions of these sets:

1. $A =$ the set of real numbers that are both greater than 0 and less than 0.

2. $B =$ the set of real numbers less than 100.

3. $C =$ the set of reciprocals of all *positive* rational numbers.

4. $E =$ the set whose elements are each sets, with cardinality equal to 7 or to 11, of rational numbers.

Discussion. These definitions are fairly straightforward except that we are not given a predicate for *positive*. But this is not a serious concern, because we can use the standard definition:

$$positive(x) \text{ iff } 0 < x.$$

Written Solutions.

1. $A = \{x \mid 0 < x \; \& \; x < 0\}$. We know from elementary mathematics that A is empty.

2. $B = \{x \mid x < 100\}$.

3. We use the standard definition,

$$positive(x) \text{ iff } 0 < x.$$

Then a *positive* rational number is specified by the predicate,

$$0 < x \ \& \ rational(x).$$

The number y is a reciprocal of a *positive* rational number iff

$$\text{there exists } x \text{ such that, } 0 < x \ \& \ rational(x) \ \& \ y = \frac{1}{x}.$$

Thus, we write

$$C = \left\{ y \mid \text{there exists } x \text{ such that, } 0 < x \ \& \ rational(x) \ \& \ y = \frac{1}{x} \right\}.$$

4. Define $R = \{r \mid rational(r)\}$. Then R is the set of all rational numbers. Then S is a set of rational numbers iff $S \subseteq R$. Furthermore, S is a set with cardinality equal to 7 or 11, of rational numbers iff

$$S \subseteq R \ \& \ (|S| = 7 \text{ or } |S| = 11),$$

or

$$S \subseteq \{r \mid rational(r)\} \ \& \ (|S| = 7 \text{ or } |S| = 11).$$

Therefore, we can define

$$E = \{S \mid S \subseteq \{r \mid rational(r)\} \ \& \ (|S| = 7 \text{ or } |S| = 11)\}.$$

(*Remark.* Notice that E is not a subset of $D = Re$, but instead is a set of subsets of D. Yet the definition of E is allowable under our *Modus Operandi*.)

∎

The concept of *subset* is related to set identity by the following simple theorem. This theorem is proved in an informal but rigorous mathematical style of proof. Again we suppose that the possible values of the variable x are the elements of some domain of discourse. I will also follow the common convention that an expression of the form 'ϕ iff ψ and χ' is an abbreviation for 'ϕ iff (ψ and χ)'. 'ϕ iff ψ or χ' may also be used to abbreviate 'ϕ iff (ψ or χ)'. Similar abbreviations may also be used for sentences containing 'if–then' instead of 'iff', for instance, sentences of the form 'If ϕ, then (ψ and χ)'.

Theorem 2-4 Let A, B be sets. Then $A = B$ iff ($A \subseteq B \ \& \ B \subseteq A$).

Proof. (*Remark.* Since the theorem states a biconditional, I will first prove that the left side implies the right side, and then prove that the right side implies the left.)

Suppose that $A = B$. By (the Axiom of) Extensionality, $x \in A$ iff $x \in B$. This 'iff' statement is tautologically equivalent to

$$(\text{ if } x \in A, \text{ then } x \in B) \ \& \ (\text{ if } x \in B, \text{ then } x \in A).$$

By the definition of subset, we obtain $A \subseteq B \ \& \ B \subseteq A$. By informal use of the Conditionalization Rule, we obtain: If $A = B$, then $A \subseteq B \ \& \ B \subseteq A$.

Conversely, suppose that $A \subseteq B \ \& \ B \subseteq A$. By the definition of subset, we have for any x, if $x \in A$, then $x \in B$, and, for any x, if $x \in B$, then $x \in A$. So, if x is any element of the domain of discourse,

$$(\text{if } x \in A, \text{ then } x \in B) \ \& \ (\text{if } x \in B, \text{ then } x \in A).$$

So, for any x, we have $x \in A$ iff $x \in B$. Whence $A = B$ by extensionality. Conditionalizing, we obtain: If $A \subseteq B \ \& \ B \subseteq A$, then $A = B$. **Q.E.D.**

This proof was written in considerable detail to show how definitions and the informal use of sentential calculus principles are used. Later proofs will be presented in a more concise style. In particular, conditionalization steps like those used in this proof typically are not even stated. In an informal proof, as soon as the consequent of a conditional is deduced from the antecedent, the reader is expected to see that the conditional has been proved. The remark that I will prove that each side implies the other is also unnecessary, in as much as it is understood that a typical way of proving a biconditional is to give separate proofs that each side implies the other.

Suppose that A and B are sets. The Axiom of Extensionality provides one way to prove that they are identical: We try to show directly that, for any x, $x \in A$ iff $x \in B$. Theorem 2-4 provides another way: First try to prove that $A \subseteq B$, and then try to prove that $B \subseteq A$. The two methods are essentially equivalent, but in practice it is sometimes easier to break the problem into two parts according to the second method. Here is another simple theorem about subsets; the proof is brief.

Theorem 2-5 Let A, B, C be sets. Then

$$\text{if } (A \subseteq B \ \& \ B \subseteq C), \text{ then } A \subseteq C.$$

Proof. Suppose that $A \subseteq B \ \& \ B \subseteq C$. Then, for any x in the domain of discourse, by the definition of subset, we have

$$\text{if } x \in A, \text{ then } x \in B,$$

and

$$\text{if } x \in B, \text{ then } x \in C.$$

Applying the $\mathcal{SC}$ Hypothetical Syllogism Rule, we obtain the following:

$$\text{if } x \in A, \text{ then } x \in C,$$

$$\text{so } A \subseteq C. \ \boldsymbol{Q.E.D.}$$

In Chapter 1 we used the empty set in an intuitive way. We also assumed the meaning of 'nonempty' in the statement of the *Modus Operandi*. In that context, 'nonempty' means that there is at least one element in the domain of discourse. We can now define 'empty set' using abstraction.

Definition 2-6 $\varnothing = \{\, x \mid x \neq x \,\}$. $\varnothing$ is called *the empty set.*

The predicate '$x \neq x$' is not satisfied by any object, so $\varnothing$ contains no elements. By (the Axiom of) Extensionality, any two empty sets are identical, so there is exactly one empty set, *the* empty set, that we denote by $\varnothing$. Also, recall that $D = \{\, x \mid x = x \,\}$, so we now have useful characterizations of both the empty set and the domain of discourse. These characterizations will be used in some later proofs. We turn now to some basic operations that can be performed on sets.

Definition 2-7 Let A, B be sets. The *union* of A and B is

$$A \cup B = \{\, x \mid x \in A \text{ or } x \in B \,\}.$$

The *intersection* of A and B is

$$A \cap B = \{\, x \mid x \in A \ \& \ x \in B \,\}.$$

If $A \cap B = \varnothing$, we say that A and B are *disjoint.*

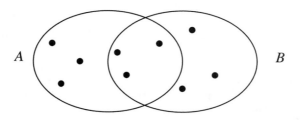

Figure 2.1. $A \cup B$

Set union and intersection are familiar concepts; in fact, we have already used them in a limited way in the first chapter. When working with sets, it is often helpful to represent situations with pictures called *Venn diagrams*, named

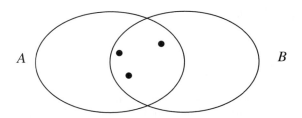

Figure 2.2. $A \cap B$

for the nineteenth-century philosopher John Venn. They were also used by the eighteenth-century mathematician Leonhard Euler and earlier writers (see Gardner (1982), Chapters 1–2). Figure 2.1 is a Venn diagram of the union of two sets, and Figure 2.2 is a Venn diagram of their intersection. The dots represent the elements of the union and intersection in each diagram. Instead of using dots, one often just shades in the areas in question. Other representational techniques are used for special purposes. Here are some simple facts that use the concepts defined above.

Theorem 2-8 Let D be the domain of discourse, and let A be any subset of this domain. Then the following statements are true:

1. $\varnothing \subseteq A$.	2. $A \subseteq A$.
3. $A \cup \varnothing = A$.	4. $A \cap \varnothing = \varnothing$.
5. $A \cup D = D$.	6. $A \cap D = A$.

Proof.

1. By the definition of $\subseteq$ (subset), $\varnothing \subseteq A$ means if $x \in \varnothing$, then $x \in A$. But $x \in \varnothing$ is false for all x, so the conditional is true. Therefore, $\varnothing \subseteq A$.

2. 'If $x \in A$, then $x \in A$' is true for any x and A, since it is a tautology. Hence, $A \subseteq A$.

3. Let x be any element of the domain of discourse such that $x \in A \cup \varnothing$. By the definition of $\cup$, $x \in A$ or $x \in \varnothing$. But $\varnothing$ has no elements, so $x \notin \varnothing$, so $x \in A$, by the $\mathcal{SC}$ DS rule. Therefore, if $x \in A \cup \varnothing$, then $x \in A$, so (by the definition of subset) $A \cup \varnothing \subseteq A$.

Conversely, suppose that $x \in A$. Then $x \in A$ or $x \in \varnothing$, by the $\mathcal{SC}$ Add rule. Thus, $A \subseteq A \cup \varnothing$ (again using the definition of subset).

Now, by Theorem 2-4, $A \cup \varnothing = A$.

(*Remark.* Notice the proof strategy that we use here. By (the Axiom of) Extensionality, $A \cup \varnothing = A$ is equivalent to the biconditional

$$x \in (A \cup \varnothing) \text{ iff } x \in A.$$

We use one of the most common $\mathcal{SC}$ strategies for biconditional goals: Reduce the problem to proving two (single) conditionals:

$$\text{if } x \in (A \cup \varnothing), \text{ then } x \in A,$$

and

$$\text{if } x \in A, \text{ then } x \in (A \cup \varnothing).$$

Once these are proved, we conclude that each set is a subset of the other and then use Theorem 2-4 to conclude that they are identical. If one forgets about Theorem 2-4, one can simply use the Equiv rule to infer the biconditional from the two single conditionals, then invoke extensionality.)

4. We use the RAA (*reductio ad absurdum*) proof strategy. Suppose that

$$A \cap \varnothing \neq \varnothing.$$

By extensionality, either $\varnothing$ has a member not in $A \cap \varnothing$, or $A \cap \varnothing$ has a member not in $\varnothing$. But $\varnothing$ has no members, so suppose there is some x such that $x \in A \cap \varnothing$. Then $x \in A$ & $x \in \varnothing$, so $x \in \varnothing$. This is a contradiction, since $\varnothing$ has no members.

5. If $x \in D$, then by the $\mathcal{SC}$ Add rule, we have ($x \in A$ or $x \in D$). Hence, $x \in A \cup D$ by the definition of $\cup$. Therefore, $D \subseteq A \cup D$, by definition of $\subseteq$.

Conversely, suppose that $x \in A \cup D$. By the definition of $\cup$, $x \in A$ or $x \in D$. Since $A \subseteq D$, by the definition of $\subseteq$, if $x \in A$, then $x \in D$. Also, 'if $x \in D$, then $x \in D$' is tautologous. By the $\mathcal{SC}$ CD and Taut rules, it follows that $x \in D$. Hence, if $x \in A \cup D$, then $x \in D$, so $A \cup D \subseteq D$ (again using the definition of $\subseteq$).

Therefore, $A \cup D = D$, by Theorem 2-4.

6. Let $x \in A \cap D$. By the definition of intersection, we have $x \in A$ and $x \in D$. Then, by the $\mathcal{SC}$ Simp rule, $x \in A$. Thus, $A \cap D \subseteq A$ (using the definition of $\subseteq$).

Conversely, let $x \in A$. Since $A \subseteq D$, $x \in D$. By the $\mathcal{SC}$ Conj rule, $x \in A$ and $x \in D$, so, by the definition of intersection, $x \in A \cap D$. Therefore, $A \subseteq A \cap D$.

Hence, by Theorem 2-4, $A \cap D = A$. *Q.E.D.*

Here are some important comments about informal proofs and, in particular, about set theory proofs. The proofs just given illustrate four very common features:

1. In general, we assume that the variable x is an *arbitrary* element of the domain of discourse. In particular, if we assume that $x \in A$, that is *all* we assume about x, namely, that it is an element of A. Similarly, if we assume that $x \in \varnothing$, then that is *all* we assume about x and so on.

2. We may use the axioms and the definitions of the relevant terms, for instance, *subset*.

3. We informally use tautological consequences or tautologies.

4. We may use previously proved theorems.

Most theorems of basic set theory can be proved with these techniques. Furthermore, the various SC proof strategies are usually very helpful in constructing the proofs. Mathematicians usually do not cite every obvious axiom, definition, or tautological consequence they use in a proof. However, it is generally expected that a proof written by a beginning student will justify every step, or nearly every step, used in the proof.

When writing set theory proofs, it is important not to confuse set theoretical notation with logical notation. The symbols $\cup$ and $\cap$ are *different from* $\vee$ and $\wedge$. The former two are operations on sets; the latter two are sentential connectives of the sentential calculus. In this book I do not use $\vee$ and $\wedge$ in informal proofs. Instead, I use their correlates in mathematical English, namely, 'or' and 'and', as well as the abbreviation '&'. When convenient, I will also use $\Rightarrow$ and $\Leftrightarrow$ for the informal conditional and biconditional. Thus, where ϕ and ψ are informal predicates, we may write

$$\phi \Rightarrow \psi \text{ as an abbreviation for if } \phi, \text{ then } \psi,$$

and

$$\phi \Leftrightarrow \psi \text{ as an abbreviation for } \phi \text{ iff } \psi.$$

For example, instead of writing

$$\text{if } x \in A \cup \varnothing, \text{ then } x \in A,$$

or

$$x \in A \cup \varnothing \text{ only if } x \in A,$$

we may write

$$x \in A \cup \varnothing \Rightarrow x \in A.$$

In set theory, it is vitally important not to confuse a *set* with a *statement* or *sentence*. For sets, one can write expressions like: "Let $A = B$" or "Let $A \subseteq B$." These quoted expressions are *sentences* in mathematical English. But it makes *no sense* to write: "Let A" or "Suppose A," when A is a set. Because sets are not sentences, it makes no sense to give them truth values or to assert them. To illustrate this further, let a be an element of the domain of discourse and let S and T denote sets. Then, 'S or T' is meaningless, in so far as S and T are not sentences. On the other hand, $S \cup T$ is meaningful, in that this expression denotes the set that is the union of the set S with the set T. In addition, '$a \in S$' and '$a \in T$' are sentences meaning 'a is an element of S' and 'a is an element of T'. Because they are sentences, it is meaningful to write, for example, '$a \in S$ or $a \in T$', but it is meaningless to write '$a \in S \cup a \in T$'. *To summarize: Names or variables for sets may be combined with set operators (such as $\cup$) but may not be combined with sentential operators (such as 'or'). Names or variables for sentences may be combined with sentential operators (such as 'or') but may not be combined with set operators (such as $\cup$) .*

Example.

Problem Statement. Theorem. If A, B are any sets, then

$$A \subseteq B \text{ iff } A \cup B = B.$$

Discussion. This theorem is not a simple statement of identity of sets; it is a biconditional asserting that a certain identity holds iff A is a subset of B. There are different approaches one might apply to this kind of problem, but the most straightforward strategy is this: First prove that one side of the biconditional implies the other side, and then prove the converse implication. This approach has the same form as one of the $\mathcal{SC}$ strategies described in Chapter 1. Of course, we also use the general Conditional Proof (CP) strategy to prove each one-way conditional. The proof to be developed here is further complicated by the fact that the right side of the biconditional (the identity statement) is equivalent to another biconditional statement by virtue of the Axiom of Extensionality. Thus, we will need to be very cautious in constructing the proof, and be careful to write it clearly.

Written Solution. Theorem. If A, B are any sets, then

$$A \subseteq B \text{ iff } A \cup B = B.$$

Proof.

Part 1. First suppose that $A \subseteq B$. Then, by the definition of $\subseteq$, for any element x of the domain,

$$x \in A \Rightarrow x \in B,$$

so

$$x \notin A \text{ or } x \in B,$$

by the $\mathcal{SC}$ CDis rule.

We now need to prove that $A \cup B = B$. I shall prove that each side of this identity is a subset of the other side.

> Suppose that $x \in A \cup B$. By the definition of union, $x \in A$ or $x \in B$. This, together with $x \notin A$ or $x \in B$, tautologically implies $x \in B$ (using Cut and Taut). We have thus proved that $x \in A \cup B \Rightarrow x \in B$. Hence, using the definition of subset, $A \cup B \subseteq B$ (under the main assumption that $A \subseteq B$).

> Conversely, suppose that $x \in B$. Then, $x \in A$ or $x \in B$ by the $\mathcal{SC}$ Add rule. Thus, using the definitions of union and subset, we have that $B \subseteq A \cup B$.

We now have $A \cup B \subseteq B$ and $B \subseteq A \cup B$. By Theorem 2-4, we infer that $A \cup B = B$. Therefore, $A \subseteq B \Rightarrow A \cup B = B$, which proves Part 1.

Part 2. Conversely, assume that $A \cup B = B$. We need to prove that $A \subseteq B$, so let $x \in A$. Then $x \in A$ or $x \in B$ by the $\mathcal{SC}$ Add rule, so $x \in A \cup B$, by the definition of $\cup$. Since $A \cup B = B$, $x \in B$. Hence, $x \in A$ only if $x \in B$, so by the definition of subset, $A \subseteq B$. Therefore, $A \cup B = B \Rightarrow A \subseteq B$, which proves Part 2. ***Q.E.D.*** ▪

Let us now consider some additional set operations.

Definition 2-9 Let A, B be sets. The *set theoretic difference between* A and B, also called *the complement of B relative to A*, is

$$A - B = \{x \mid x \in A \ \& \ x \notin B\}.$$

Example. If

$$A = \{\text{Pythagoras, Aristotle, Omar, Descartes, Leibniz, Frege}\},$$

and

$$B = \{\text{Pythagoras, Descartes, Frege, Russell, Turing}\},$$

then

$$A - B = \{\text{Aristotle, Omar, Leibniz}\},$$

and

$$B - A = \{\text{Russell, Turing}\}. \blacksquare$$

A diagram of set theoretic difference is given in Figure 2.3. Although the complement of B relative to A is useful for an arbitrary set A, one is often especially interested in the case where A is the domain of discourse. This motivates the next definition.

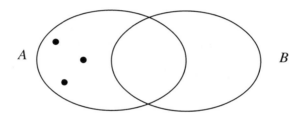

Figure 2.3. $A - B$

Definition 2-10 Let D be a domain of discourse, and let B be a subset of D. Then the complement of B relative to D is called the *complement of B* and is denoted by B'. Thus, $B' = (D - B)$.

If D is a given domain, then any element under consideration is in D, hence $x \in D$ is true for any x. But $x \in D - B$ is true iff $(x \in D \ \& \ x \notin B)$, and the latter is true iff $x \notin B$ (since $x \in D$ is true for any x). Thus, we can simplify the definition of B' to

$$B' = \{\, x \mid x \notin B \,\}. \tag{2.1}$$

From (2.1), $x \in B' \Leftrightarrow x \notin B$, and similarly, $x \in A' \Leftrightarrow x \notin A$. For any sets A, B, by (the Axiom of) Extensionality,

$$A = B \ \Leftrightarrow \ (x \in A \Leftrightarrow x \in B).$$

But, tautologically, $(x \in A \Leftrightarrow x \in B)$ holds iff $(x \notin A \Leftrightarrow x \notin B)$, so

$$A = B \ \Leftrightarrow \ (x \notin A \Leftrightarrow x \notin B).$$

But since $x \in B' \Leftrightarrow x \notin B$, $x \in B'$ and $x \notin B$ always have the same truth value so they can be substituted for each other. The corresponding fact holds for $x \in A'$ and $x \notin A$. Thus, by substitution we can obtain

$$A = B \Leftrightarrow A' = B'. \tag{2.2}$$

The argument used to obtain (2.2) used several biconditionals. We will frequently work with sequences of biconditionals. Consider any sequence of two or more biconditionals, for instance, this sequence of three:

$$\phi_1 \quad \Leftrightarrow \quad \phi_2,$$
$$\phi_2 \quad \Leftrightarrow \quad \phi_3,$$
$$\phi_3 \quad \Leftrightarrow \quad \phi_4.$$

By repeated use of the $\mathcal{SC}$ rules Equiv, Simp, HS, and Conj, we can obtain

$$\phi_1 \quad \Leftrightarrow \quad \phi_4.$$

But there is a simpler, more intuitive way to obtain this result. Consider the second biconditional. It states that ϕ_2 is equivalent to ϕ_3, so they always have the same truth value, so can be substituted for one another. If we substitute ϕ_3 for ϕ_2 in the first biconditional, we obtain

$$\phi_1 \quad \Leftrightarrow \quad \phi_3.$$

We can replace the second line by the above to obtain

$$\phi_1 \quad \Leftrightarrow \quad \phi_2,$$
$$\phi_1 \quad \Leftrightarrow \quad \phi_3.$$

We can now use the third original line to substitute ϕ_4 for ϕ_3 in $\phi_1 \Leftrightarrow \phi_3$ to obtain

$$\phi_1 \quad \Leftrightarrow \quad \phi_4,$$

and this result can replace the third original line. Thus, the original three lines can be transformed into

$$\phi_1 \quad \Leftrightarrow \quad \phi_2,$$
$$\phi_1 \quad \Leftrightarrow \quad \phi_3,$$
$$\phi_1 \quad \Leftrightarrow \quad \phi_4.$$

It is customary to write this new sequence of lines in the form

$$\phi_1 \quad \Leftrightarrow \quad \phi_2$$
$$\Leftrightarrow \quad \phi_3$$
$$\Leftrightarrow \quad \phi_4,$$

where it is understood that ϕ_2 is equivalent to ϕ_3, ϕ_3 is equivalent to ϕ_4, and ϕ_1 is then successively equivalent to the right-hand side of $\Leftrightarrow$. This type of array thus shows that $\phi_1 \Leftrightarrow \phi_4$. Of course, in an actual proof, each of the equivalences must be justified. An example of such a proof is in the next theorem.

Complements correspond, in a special way, to negation in sentential calculus. Since the tautologies called 'De Morgan's theorem' are very important in sentential calculus, it is natural to seek an analog in set theory. Here it is.

Theorem 2-11 (De Morgan). Let A and B be sets. Then

$$(A \cup B)' = A' \cap B',$$

$$(A \cap B)' = A' \cup B'.$$

Proof. Let x be any element of the domain of discourse. Then the following equivalences result from definitions and a tautological equivalence, namely, De Morgan's theorem for sentential calculus. The justification for each line is stated in an abbreviated form at the start of the line.

Def. of $'$	$x \in (A \cup B)'$	$\Leftrightarrow$ not $(x \in A \cup B)$,
Def. of $\cup$		$\Leftrightarrow$ not $(x \in A$ or $x \in B)$,
$\mathcal{SC}$ DeM		$\Leftrightarrow$ $x \notin A$ & $x \notin B$,
Def. of $'$		$\Leftrightarrow$ $x \in A'$ & $x \in B'$,
Def. of $\cap$		$\Leftrightarrow$ $x \in (A' \cap B')$.

We now have a sequence of biconditional statements that yield

$$x \in (A \cup B)' \Leftrightarrow x \in (A' \cap B').$$

Therefore, by (the Axiom of) Extensionality, $(A \cup B)' = (A' \cap B')$.

The proof of the second equality is similar and is left for practice. **Q.E.D.**

It should be apparent that all of the previous definitions and theorems of this section apply to subsets of a domain of discourse D. The elements of these subsets are just elements of D. But, as the members, or elements, of a set can be any type of object, they can also be sets themselves. For some purposes, it is necessary to consider sets of sets. One special case is the set of all subsets of a given set. Suppose that $A \subseteq D$. Then, by Theorem 2-5, any subset of A is also a subset of D. Intuitively, it seems that we should be able to work with the set of all subsets of A. But these subsets are not elements of D; they are subsets of D. Thus, simple application of separation does not provide a definition of the set of subsets of A. Instead, we will use a third axiom. In the statement of this axiom, it is assumed that $A \subseteq D$.

> ***Power Set Axiom.*** Let A be a set. Then there exists a set, the *power set* of A, denoted by $\mathcal{P}(A)$, satisfying the following:
>
> $$\text{For any } S, \ S \in \mathcal{P}(A) \ \Leftrightarrow \ S \subseteq A.$$

It follows from the Axiom of Extensionality that the power set of A is unique, and we can define it as follows:

Definition 2-12 For any set A, the *power set* of A is

$$\mathcal{P}(A) = \{S \mid S \subseteq A\}.$$

In this definition, it is assumed that A is defined by separation from D. We also assume that the subsets of A can be so defined. Notice, however, that $\mathcal{P}(A)$ is not defined in this manner. Yet it does satisfy the *Modus Operandi*, because it is defined in terms of sets, A and its subsets, that can be separated from D. This is the case even if A happens to be the same as the domain of discourse D. Thus, $\mathcal{P}(D)$ also exists, and its elements are all of the subsets of D. One can think of $\mathcal{P}(D)$ as an extended domain D_1 to which the Axiom of Separation may be applied if desired. So, we may now work with sets of subsets of D, as well as with subsets of D. If we wish, we can also extend $\mathcal{P}(D)$ to $\mathcal{P}(\mathcal{P}(D))$, and so on, although we will have little need for such extensions.

From the definition of power set, we have, for any set S,

$$S \in \mathcal{P}(A) \ \Leftrightarrow \ S \subseteq A.$$

By Theorem 2-8, for any set A, $A \subseteq A$ and $\varnothing \subseteq A$. Thus, for any set A, A and $\varnothing$ are elements of $\mathcal{P}(A)$. If $A = \{\,0, 1\,\}$, then

$$\mathcal{P}(A) = \{\varnothing, \{0\}, \{1\}, A\}.$$

Given an original domain D, we can now extend it to $\mathcal{P}(D)$, so we can work with subsets of $\mathcal{P}(D)$, which are just sets of sets of elements of D. If we have such a set A of sets, we sometimes want to form the intersection or the union of all of these sets. We want this operation to be defined even if A has an infinite number of sets in it. It is easy to define such operations.

Definition 2-13 Let A be a set of sets, then

$$\bigcup A = \{x \mid \text{there exists } S \in A \ \& \ x \in S\}.$$

Also, if A is a nonempty set of sets, then

$$\bigcap A = \{x \mid \text{for any } S, \text{ if } S \in A, \text{ then } x \in S\}.$$

$\bigcap A$ is called the *intersection* of A and $\bigcup A$ the *union* of A. Some writers call $\bigcap A$ the *generalized intersection* of A and $\bigcup A$ the *generalized union*.

It can be seen that x is a member of $\bigcap A$ iff it is a member of *every* set in A. Also, x is a member of $\bigcup A$ iff it is a member of *at least one* set in A.

Example.

Let $A = \{\{ 0, 1 \}, \{ 2, 1 \}, \{ 17, 1 \}, \{ 1 \}, \{ 1, 0, 13 \}\}$. Then

$$\bigcap A = \{1\}$$

and

$$\bigcup A = \{0, 1, 2, 13, 17\}. \ \blacksquare$$

The previous definition of generalized intersection is restricted to nonempty A. The reason for this restriction is examined in a problem in the next exercise set. It takes considerable practice to make effective use of the terminology developed in this section of the book. The next example provides some additional illustrations.

Example.

Problem Statement. This is related to an earlier example. The domain of discourse is *Nat*, and you are given the following predicates that apply to natural numbers:

$$natural(x): x \text{ is a natural number},$$
$$even(x): x \text{ is an even number},$$
$$odd(x): x \text{ is an odd number},$$
$$x < y.$$

Using (definition by) abstraction, and set operators defined previously in this chapter, write definitions of the following sets:

1. S = the set of natural numbers comprising 0, all odd natural numbers less than 100, and all even natural numbers greater than 1000.

2. S' = the complement of S. Don't just write 'S''; define it in terms of the given predicates.

3. H_5 = the set of all subsets of *Nat* with cardinality = 5.

4. S_E = the set of all nonempty subsets of *Nat* whose elements are even numbers.

5. A = the set of even numbers equal to or greater than 2 and equal to or less than 6. Also, list the elements of A in braces.

6. B = List the elements of $\mathcal{P}(A)$ in braces.

Written Solutions.

1. S has a curious mixture of numbers, but it can be defined by

$$S = \{x \mid x = 0 \text{ or } (odd(x) \ \& \ x < 100) \text{ or } (even(x) \ \& \ 1000 < x)\}.$$

2. We have $x \in S' \Leftrightarrow$ not $(x \in S)$. But also from the previous answer,

$$x \in S \Leftrightarrow x = 0 \text{ or } (odd(x) \ \& \ x < 100) \text{ or } (even(x) \ \& \ 1000 < x).$$

By repeated use of $\mathcal{SC}$ DeM, we obtain

$$x \in S' \Leftrightarrow \text{not } (x \in S) \Leftrightarrow$$
$$x \neq 0 \ \& \ \text{not } (odd(x) \ \& \ x < 100) \ \& \ \text{not } (even(x) \ \& \ 1000 < x).$$

We could apply DeM again to the negations of the conjunctions in this formula, but it is a suitable solution as is, so we write

$$S' = \{x \mid x \neq 0 \ \&$$
$$\text{not } (odd(x) \ \& \ x < 100) \ \& \ \text{not } (even(x) \ \& \ 1000 < x)\}.$$

3. Obviously we have $Nat = \{x \mid natural(x)\}$, so S is a subset of *Nat* iff

$$S \subseteq Nat \text{ or } S \subseteq \{x \mid natural(x)\}.$$

We just need to add to this the condition that the cardinality of S is 5. Thus,

$$H_5 = \{S \mid S \subseteq \{x \mid natural(x)\} \ \& \ |S| = 5\}.$$

There are other approaches to the same end. First define $Nat = \{x \mid natural(x)\}$ to obtain a notational simplification. We can now write

$$H_5 = \{S \mid S \subseteq Nat \ \& \ |S| = 5\},$$

which, according to a standard notational convention, can be rewritten as

$$H_5 = \{S \subseteq Nat \mid |S| = 5\}.$$

But the power set of Nat is $\mathcal{P}(Nat)$, which is the set of all subsets of Nat, so the previous two equations can also be expressed by

$$H_5 = \{S \mid S \in \mathcal{P}(Nat) \;\&\; |S| = 5\}$$

and

$$H_5 = \{S \in \mathcal{P}(Nat) \mid |S| = 5\}.$$

4. Using the characterization of Nat in the previous answer, one way of stating that X is a nonempty subset of Nat whose elements are even numbers is this:

$$X \subseteq Nat \;\&\; X \neq \varnothing \;\&\; \text{all elements of } X \text{ are even.}$$

But to say that all elements of X are even means that "for any y, if y is an element of X, then y is even." A shorter way of saying the latter is this: "For all $y \in X$, $even(y)$." Therefore, we may write

$$S_E = \{X \mid X \subseteq Nat \;\&\; X \neq \varnothing \;\&\; \text{for all } y \in X, \; even(y)\},$$

or

$$S_E = \{X \subseteq Nat \mid X \neq \varnothing \;\&\; \text{for all } y \in X, \; even(y)\}.$$

Using the power set of Nat, we obtain

$$S_E = \{X_{\textbf{.}} \mid X \in \mathcal{P}(Nat) \;\&\; X \neq \varnothing \;\&\; \text{for all } y \in X, \; even(y)\},$$

and

$$S_E = \{X \in \mathcal{P}(Nat) \mid X \neq \varnothing \;\&\; \text{for all } y \in X, \; even(y)\}.$$

5. We can define $x \leqslant y$ by: $x < y$ or $x = y$. Then

$$A = \{x \mid even(x) \;\&\; 2 \leqslant x \leqslant 6\} = \{2, 4, 6\}.$$

6. $\mathcal{P}(A) = \{\varnothing, \{2, 4, 6\}, \{2, 4\}, \{2, 6\}, \{4, 6\}, \{2\}, \{4\}, \{6\}\}.$ ∎

Using the preceding definitions, one can prove many elementary facts about sets. For reference and practice, several of these facts are summarized in the following theorem. Two of the proofs are given as examples. The other proofs are similar and left as exercises. They provide good practice in using informal tautological inferences. Note that A'' means the same as $(A')'$.

Theorem 2-14 Let D be a nonempty domain of discourse and let A, B, C be subsets of D. Then

1. $\emptyset' = D$.

2. $A'' = A$.

3. $D' = \emptyset$.

4. $(A - B) = A \cap B'$.

5. $A - \emptyset = A$.

6. $A \cup A' = D$ and $A \cap A' = \emptyset$.

7. $A \cup B = B \cup A$ [Commutation].

8. $A \cap B = B \cap A$ [Commutation].

9. $A \cup (B \cup C) = (A \cup B) \cup C$ [Association].

10. $A \cap (B \cap C) = (A \cap B) \cap C$ [Association].

11. $A \cup (B \cap C) = (A \cup B) \cap (A \cup C)$ [Distribution].

12. $A \cap (B \cup C) = (A \cap B) \cup (A \cap C)$ [Distribution].

Proof.

1. By the definition of the empty set, $x \in \emptyset \Leftrightarrow x \neq x$. Since each side of this biconditional has the same truth value as the opposite side, the same is true of the negations of each side. One can also verify that it follows tautologically by Equiv, Simp, ContraPos, and Conj, that $x \notin \emptyset \Leftrightarrow x = x$. Hence, by the definition of complement, $x \in \emptyset' \Leftrightarrow x = x$. But the domain $D = \{x \mid x = x\}$, and thus $x \in D \Leftrightarrow x = x$. The last two biconditionals tautologically imply that $x \in \emptyset' \Leftrightarrow x \in D$, so by extensionality, $\emptyset' = D$.

11. Let $x \in D$. Then

Defs. of $\cup$, $\cap$	$x \in A \cup (B \cap C) \Leftrightarrow x \in A$ or $(x \in B$ & $x \in C)$,
$\mathcal{SC}$ Dist	$\Leftrightarrow (x \in A$ or $x \in B)$ & $(x \in A$ or $x \in C)$,
Def. $\cup$	$\Leftrightarrow (x \in A \cup B)$ & $(x \in A \cup C)$,
Def. $\cap$	$\Leftrightarrow x \in (A \cup B) \cap (A \cup C)$.

Hence, by extensionality, $A \cup (B \cap C) = (A \cup B) \cap (A \cup C)$.

The other proofs are left as exercises. ***Q.E.D.***

These basic identities imply a multitude of others that can often be proved largely or entirely by "set algebra," i.e., by substitution into some of the set identities in Theorems 2-8, 2-11 (De Morgan), and 2-14. The next example illustrates this kind of proof.

Example.

Problem Statement. Theorem. For any sets, A, B,

$$A - (A \cap B) = A - B.$$

Discussion. The problem is to prove this simple set identity theorem. A simple Venn diagram shows that it is obviously true, but a Venn diagram is not a proof. We need to prove it by using axioms, definitions, previous theorems, and logic. Clearly, one needs to be familiar with these facts and rules.

There are three common methods of proving set identities:

- The first is to apply the Axiom of Extensionality directly to a sequence of biconditional statements. An example of this method is the proof of De Morgan's Theorem 2-11.

- Sometimes it is not easy to construct a usable sequence of biconditionals. Suppose that one is trying to prove that *Expression*1 = *Expression*2, where *Expression*1 and *Expression*2 are set theoretical expressions (like those in the current problem statement). One can try to prove: *Expression*1 $\subseteq$ *Expression*2 and *Expression*2 $\subseteq$ *Expression*1. If both of these are proved, then the desired identity follows by Theorem 2-4. This method has been used in several proofs earlier in this section.

- A third method of proving some set identities is by applying previous theorems and working "algebraically." The next solution mainly uses this method.

Written Solution. Theorem. For any sets, A, B,

$$A - (A \cap B) = A - B.$$

Proof. By Theorem 2-14, for any sets X, Y, $X - Y = X \cap Y'$. Starting with this, we proceed "algebraically":

Thm. 2-14	$A - (A \cap B)$	$=$	$A \cap (A \cap B)'$
Set DeM		$=$	$A \cap (A' \cup B')$
Set Dist.		$=$	$(A \cap A') \cup (A \cap B')$
Thm. 2-14		$=$	$(A \cap A') \cup (A - B)$.

But notice that $A \cap A' = \{\, x \mid x \in A \; \& \; x \notin A \,\} = \varnothing$. Substituting into the previous equation yields

By subst.	$A - (A \cap B)$	$=$	$\varnothing \cup (A - B)$
Comm		$=$	$(A - B) \cup \varnothing$
Thm. 2-8		$=$	$(A - B)$ ***Q.E.D.*** ■

EXERCISES

General Instructions

Throughout this book, if an exercise is simply the statement of a theorem or a metatheorem, then the task is to prove this theorem or metatheorem. In your proof, you may use the definitions given in the text. Unless stated to the contrary, you may also use theorems proved in the text.

Exercises for which answers are provided at the back of the book are marked with "Answer Provided."

In this chapter, unless instructed otherwise, you may assume elementary facts of arithmetic and algebra in these exercises.

Finally, please note that exercises that introduce important principles, useful applications, are especially difficult, or are special in some way, are indicated by bold print. In some cases, parenthetical comments describe the special features of the problem. If such comments are missing, the exercise is nonetheless of special value in introducing a new concept, principle, or technique.

E 2-1 Let $A = \{\, a, \; b \,\}$, $B = \{\, b, \; c \,\}$, where a, b, c are mutually distinct. Let the domain of discourse be $D = A \cup B$. List the elements of each of the following sets in braces.

$$
\begin{array}{ll}
A \cap B & A - B \\
B - A & A' \\
B' & (B - A)' \\
(A \cap B)' & (A \cup B)' \\
(A - B) \cup (B - A) & (A \cup B) - (A \cap B)
\end{array}
$$

E 2-2 *Theorem.* For any sets A, B, $(A \cap B) \cup B = B$.

E 2-3 (Answer Provided) *Theorem.* For any sets, A, B, C, D,

$$\text{if } A \subseteq B \ \& \ C \subseteq D, \text{ then } A \cup C \subseteq B \cup D.$$

E 2-4 *Theorem.* Let A, B be any sets; then

1. $A \cap B \subseteq A$ and $B \subseteq A \cup B$. If $x \in A \wedge B$, then $x \in A$ (2-3)
   $$\text{If } x \in B, \text{ then } x \in A \cup B.$$
2. $A \cap B = A \Leftrightarrow A \cup B = B$.

E 2-5 *Theorem.* For any sets A, B, $A \subseteq B$ iff $B' \subseteq A'$.

E 2-6 (Answer Provided) *Theorem.* For any sets A, B,

$$A = B \Leftrightarrow (A \cup B) \subseteq (A \cap B).$$

E 2-7 Let W be the set of all words in a large English dictionary. If x is a word in W, let

$len(x)$ denote the length of x (e.g., $len(\text{apple}) = 5$),

$verb(x)$ mean that x is a verb (e.g., $verb(\text{run})$),

$has\text{-}z(x)$ mean that x contains a 'z' (e.g., $has\text{-}z \ (\text{crazy})$).

Assuming that W is the domain of discourse, and using definition by abstraction, write definitions of the following sets:

1. $V = $ the set of all verbs.

2. $V_z = $ the set of all verbs containing a 'z'.

3. $L_7 = $ the set of all words with length $= 7$.

4. $V_5 = $ the set of all verbs of length 5.

5. $V_{7z} = $ the set of all verbs that have length 7 and contain a 'z'.

E 2-8 *Theorem.* For any sets, A, B, C,

$$A - (B \cap C) = (A - B) \cup (A - C).$$

E 2-9 Complete the proof of De Morgan's theorem, Theorem 2-11.

E 2-10 Complete the proof of Theorem 2-14.

E 2-11 *Theorem.* For any sets, A, B, $A' - B = B' - A = (A \cup B)'$.

E 2-12 *Theorem.* Let X, Y, Z be any sets. Then

$$(X \subseteq Z \text{ or } Y \subseteq Z) \Rightarrow X \cap Y \subseteq Z.$$

E 2-13 *Theorem.* Let A, B, C be any sets, and let $C \subseteq B$. Then

$$(A \cap B) - (A \cap C) \subseteq B.$$

E 2-14 *Theorem.* For any sets X, Y, $X \subseteq Y \Leftrightarrow X - Y = \emptyset$.

E 2-15 *Theorem.* Let B, C be any sets. Then, $B \cap C = \emptyset \Leftrightarrow C - B = C$.

E 2-16 *Theorem.* For any sets A, B, $(A - B)' = A' \cup B$.

E 2-17 *Theorem.* For any sets A, B, $A - (A - B) = A \cap B$.

E 2-18 *Theorem.* Let A, B, C be sets; then

$$(A - B) \cup C = (C \cup A) \cap (C \cup B').$$

E 2-19 *Theorem.* For any sets A, B,

$$(A - B) \cup (B - A) = (A \cup B) - (A \cap B).$$

E 2-20 *Theorem.* Let C, A, B be any sets. If $B \subseteq A$, then $B \cap C \subseteq A \cap C$.

E 2-21 *Theorem.* Let D be the domain of discourse, and A, B_1, B_2, B_3 be any subsets of D. Then the following hold:

1. $A = A \cap D$.
2. $D = B_1 \cup (B_2 \cup B_3) \Rightarrow A = (A \cap B_1) \cup ((A \cap B_2) \cup (A \cap B_3))$.
3. $B_1 \cap B_2 = \emptyset \Rightarrow (A \cap B_1) \cap (A \cap B_2) = \emptyset$.

E 2-22 *Theorem.* Let D be a domain of discourse, and A, B be any subsets of D. Then

$$A \cap B = \emptyset \text{ iff } A \subseteq B'$$

and

$$A \cup B = D \text{ iff } A' \subseteq B.$$

E 2-23 Let X, Y, Z be any sets on some domain (of discourse). For each of the following statements, give either a proof or a counterexample.

1. $(X \cup Y) - Z = (X - Z) \cup (Y - Z)$.

2. $Z - (X \cup Y) = (Z - X) \cup (Z - Y)$.

3. $Z - (X \cap Y) = (Z - X) \cap (Z - Y)$.

E 2-24 *Theorem.* If X, Y, Z are any sets, then

$$(Z \subseteq X \ \& \ Z \subseteq Y) \Rightarrow Z \subseteq X \cap Y,$$

and

$$(X \subseteq Z \ \& \ Y \subseteq Z) \Rightarrow X \cup Y \subseteq Z.$$

E 2-25 Let A, B, C be any sets. For each of the following statements, give either a proof or a counterexample.

1. $(A \cup B) - B = A$.

2. $C \subseteq B \Rightarrow (A - B \subseteq A - C)$.

3. $(A - B) \cap (B - C) \cap (C - A) = \varnothing$.

4. $(A \cap B) \cup (A' \cap B') = (A' \cap B) \cup (A \cap B')$.

E 2-26 *Theorem.* For any sets A, B, C,

$$(A \cap B) \subseteq C$$

iff

$$(A \cap B) \subseteq (B \cap C) \text{ and } (A \cap B) \subseteq (A \cap C) .$$

E 2-27 Using the predicates and domain (W) given in E 2-7, abstraction, and the operators defined earlier in this chapter, write definitions of the following:

1. P_v = the set of all sets of verbs.

2. P_{7v} = the set of all cardinality = 7 sets of verbs.

3. $V_{(7 \text{ or } z)}$ = the set of all verbs that are either of length 7 or contain a 'z'.

E 2-28 Let $M = \{ 0, 1, 2, 11, 3 \}$, $N = \{7, 1, 3, 2 \}$, $S = \{ 8, 3, 0, 2, 11 \}$, and $A = \{ M, N, S \}$.

1. What is $\bigcup A$?

2. What is $\bigcap A$?

E 2-29 *Theorem.* Let S, T be any sets. Then $S = \bigcap\{S\}$ and

$$S \cap T = \bigcap\{S, T\}.$$

E 2-30 Let S, T be any sets. What is $\bigcup\{S\}$ and what is $\bigcup\{S, T\}$? Carefully state and prove your answers. (Hint. Compare with the previous exercise.)

E 2-31 1. What is $\bigcup\varnothing$? 2. Definition 2-13 defines

$$\bigcap A = \{\, x \mid \text{for any } S, \text{ if } S \in A, \text{ then } x \in S \,\}$$

for nonempty sets A. What would be in $\bigcap A$ if we let A be empty?

E 2-32 *Theorem.* For any set A, $A = \bigcup \mathcal{P}(A)$. (Hint. If $x \in A$, then $\{x\} \subseteq A$.)

E 2-33 (Answer Provided) *Theorem.* For any sets, A, B,

$$\mathcal{P}(A \cap B) = \mathcal{P}(A) \cap \mathcal{P}(B).$$

E 2-34 *Theorem.* For any sets, A, B, $A \subseteq B$ iff $\mathcal{P}(A) \subseteq \mathcal{P}(B)$.

2.2   Relations

2.2.1   General Features

The predicates used to define sets in the previous section were all formulated in terms of a single variable, so each had an arity of 1. For a unary predicate, such as $trin(x)$ from Section 2.1.1, it makes sense to ask which objects, e.g., Raquel, Rafael, etc., satisfy it. Let us extend this idea of satisfaction to predicates with arity greater than 1. Suppose the sentence, 'Miranda is the wife of George', is true. From it we can obtain the 2-ary predicate, 'y is the wife of x.' Of course, if we substitute 'Miranda' for y and 'George' for x in this predicate, we obtain the original sentence, which is true. One might suggest that we say that the set, { Miranda, George }, satisfies this predicate. But a set imposes no order on its elements, so

{ Miranda, George } = { George, Miranda } .

If we now substitute 'George' for y and 'Miranda' for x, we obtain 'George is the wife of Miranda', which is false. Therefore, the order in which we consider the objects is important, and we encounter problems if we merely talk about the set of objects satisfying the predicate.

I will abbreviate the predicate, 'y is the wife of x', by the shorter form, $WIFExy$. The expression 'WIFE' abbreviates the longer English phrase, and the variables have been placed after $WIFE$, but they have the same significance. Thus, if g abbreviates 'George' and m abbreviates 'Miranda', then $WIFEgm$ abbreviates 'Miranda is the wife of George.' The crucial point is that the *order* of x and y in $WIFExy$ is significant. The position held by x can be replaced by names of husbands and the position held by y can be replaced by names of wives. Now let $< x, y >$ denote a *pair* of objects, x and y, considered as ordered with x coming before y. So, $< g, m >$ denotes the pair George and Miranda, considered in the order given. Unlike sets, in which order is irrelevant, the pair $< g, m >$ is not identical to the pair $< m, g >$. We could have defined $WIFExy$ so that it would mean 'x is the wife of y'; the order specified in the definition is purely conventional. But once the order has been set by a definition, then further uses of the predicate must be in accordance with this order. I will continue to use the order in which the second-variable position of $WIFE$ corresponds to the wife.

We can now discuss whether or not such an *ordered pair* satisfies a 2-ary (binary) predicate. If $< a, b >$ is an ordered pair, where a and b are names, then it satisfies the binary predicate $\rho(x, y)$ iff a true sentence results when a is substituted for x and b is substituted for y. The sentence one obtains by substituting $< a, b >$ into a binary predicate is, in general, different from the sentence obtained by substituting $< b, a >$. Consider the example of the predicate $WIFExy$. The pair $< g, m >$ satisfies this predicate, in so far as $WIFEgm$ is true. The pair $< m, g >$ does not satisfy this predicate because $WIFEmg$ is not true. Although the order of the names is important, we assume that, if an object has more than one name, the particular name that is used in a pair is irrelevant to the truth value of the resulting sentence. In the previous section, we assumed that we would use only unary predicates that can be meaningfully applied to a domain of discourse; thus, the particular name (of an object) that is substituted into a predicate is irrelevant to the truth value of the resulting sentence. We are now also restricting our treatment to binary predicates that can be meaningfully applied. This assumption will be made for all the predicates considered in this book, unless explicitly stated otherwise.

To work with ordered pairs, we need a criterion of identity for such pairs. In the case of sets, the criterion of identity was stated in the Axiom of Extensionality. This is too weak for pairs, in that extensionality yields no ordering of the elements. A little thought leads to a natural criterion for pairs. Suppose that George, in addition to being named g, also has another name, g_1. Also suppose that Miranda, in addition to being named m, has another name, m_1. If George is married to Miranda, then the predicate $WIFExy$ should yield true sentences regardless of what names for the persons George and Miranda are substituted for the variables x and y. In that case, both pairs, $< g, m >$ and $< g_1, m_1 >$, should satisfy $WIFExy$. Think of these two pairs of names as each denoting the pair of persons George followed by Miranda. These pairs of names are identical pairs because the names g and g_1 in the first position each denote the person George, and the names m and m_1 in the second position each denote the person Miranda. So we are assuming that the identity of an ordered pair does not depend on the particular names that occur in the pair, but only on what these names denote and the order in which the names occur in the pair.

These considerations motivate the following general criterion of identity for ordered pairs: Let s, t, x, y denote elements of a domain of discourse, then $< s, t >=< x, y>$ iff $s = x$ & $t = y$. Stated simply, ordered pairs are identical iff their first elements are identical and their second elements are identical. This criterion does not actually say what an ordered pair is. The idea of an ordered pair is still a primitive concept, i.e., it has not been defined in terms of concepts that are already familiar to us. We could leave the concept primitive, and go ahead and work with pairs, using the criterion of identity. Because this chapter is developing basic mathematical concepts in terms of sets, it would be better to define ordered pairs as special kinds of sets. Fortunately, N. Wiener and K. Kuratowski showed how to do this in the early part of the twentieth century. Using their idea, we replace the previous informal introduction of ordered pairs with the following definition. From now on '$x, y \in S$' abbreviates '$x \in S$ and $y \in S$' and similarly for additional variables.

Definition 2-15 Let D be a domain of discourse and let $x, y \in D$. The *ordered pair* of x and y is denoted by $< x, y >$, and

$$< x, y >= \{\{x\}, \{x, y\}\}.$$

The clever idea behind this definition is to give one of the elements, x, some property not possessed by the other element, y. In the definition, this feat is accomplished by letting x occur in a set by itself, $\{x\}$. But the same feat could have been accomplished by other definitions as well. It is sometimes asked: How do we know that this is a *good* definition? An interesting philosophical discussion of this question is Hochberg (1981). For now, it is enough to ask whether the definition is *good enough* for our purposes of developing the basic concepts of mathematics. We already have an intuitively motivated criterion of identity for ordered pairs. If the definition is to satisfy our needs, then the concept of an ordered pair that it yields must at least satisfy this criterion. Fortunately, this can be proved from the definition using principles of set theory.

Theorem 2-16 For any s, t, x, y,

$$< s, t >=< x, y > \Leftrightarrow s = x \& t = y.$$

Proof. Although the proof is important, it is not very instructive because it does not introduce any new ideas or techniques. I will indicate the main steps, leaving the details to be filled in for practice.

1. By the Principle of Substitutivity of Identicals (in the Introduction), the right side implies the left side.

2. Conversely, suppose that $< s, t > = < x, y >$, so that

$$\{\{s\}, \{s, t\}\} = \{\{x\}, \{x, y\}\}.$$

Since $s = t$ or $s \neq t$, we argue by cases (see in Chapter 1 the "Using a Disjunction" $\mathcal{SC}$ proof strategy):

Case 1. If $s = t$, then

$$< s, t > = \{\{s\}, \{s, t\}\} = \{\{s\}\} = \{\{x\}, \{x, y\}\}.$$

Whence, $s = x$ and $t = y$.

Case 2. Suppose that $s \neq t$. If we had $x = y$, we could apply an argument of the type used in Case 1 to show that $s = t$, contrary to the supposition. So $x \neq y$. Since

$$\{\{s\}, \{s, t\}\} = \{\{x\}, \{x, y\}\},$$

we must have $s = x$. Whence, $t = y$. **Q.E.D.**

The Wiener–Kuratowski definition of ordered pair is mainly of theoretical interest because it enables us to *define* ordered pairs as special kinds of sets. If D is a domain of discourse, and $x, y \in D$, then $\{x\}$ and $\{x, y\}$ are both subsets of D. Therefore, $< x, y >$ is defined in terms of subsets of D. Hence, ordered pairs and sets of ordered pairs satisfy our *Modus Operandi*. Having established this, from now on we will simply work with pairs of the form, $< x, y >$, keeping in mind that they are just special kinds of sets of sets and that they satisfy the criterion of identity for pairs.

Although the present development is abstract and general, ordered pairs of numbers have been used for some time. The great seventeenth-century philosopher, mathematician, and scientist, René Descartes, made systematic use of pairs of numbers in analytic geometry, which he invented. Although it was also a great invention, the geometry of the ancient Greeks was largely limited to variations on certain types of figures, e.g., lines, polygons, and conic sections (obtained by slicing a cone with a plane). Descartes' geometry enabled one to represent *arbitrary* figures by algebraic relationships. This was an important new method of *representing* geometric information. In current artificial intelligence research, there is much interest in "knowledge representation." Historically, Descartes' invention of analytic geometry was one of the most important advances in this field. The invention of modern symbolic logic and set theory is another important advance. Descartes' idea is now generalized and abstracted in the following definition, which plays a fundamental role in the representation of information.

Definition 2-17 Let A, B be sets. The *Cartesian product* of A, B is

$$A \times B =$$

$\{\, u \mid \text{there exist } x, y, \text{ such that } x \in A \,\&\, y \in B \,\&\, u = < x, y > \,\}.$

Since $A \times B$ is a set of ordered pairs, it is customary to write the preceding equation in the simpler form,

$$A \times B = \{\, < x, y > \mid x \in A \,\&\, y \in B \,\}.$$

$A \times B$ is just the set of all pairs whose first element is from A and whose second element is from B. For instance, if Re is the set of all real numbers, then $Re \times Re$ is the set of all pairs corresponding to the points in the two-dimensional Cartesian coordinate system. If $A = \{\, 0, 1 \,\}$ and $B = \{\, 3, 5 \,\}$, then

$$A \times B = \{\, < 0, 3 >, < 0, 5 >, < 1, 3 >, < 1, 5 > \,\}.$$

In general, we are not especially interested in $A \times B$ itself, but we are interested in its subsets.

Definition 2-18 Let A, B be sets. Then a *binary (2-ary) relation R between A and B* is a subset of $A \times B$. For any x and y, we may write

$$Rxy \iff < x, y > \in R.$$

Furthermore,

$$\{\, x \mid x \in A \,\&\, \text{there exists } y \in B \text{ such that } Rxy \,\}$$

is called the *domain of R* and

$$\{\, y \mid y \in B \,\&\, \text{there exists } x \in A \text{ such that } Rxy \,\}$$

is called the *range of R*.

If A is nonempty and $A = B$, then we say that R is a *relation on A*.

This definition does not require that R be nonempty, so even $\varnothing$ is a binary relation, although it is not a very interesting one. The domain of R must not be confused with the domain of discourse. If D is the domain of discourse, then, in general, $A \subseteq D$ and the domain of R is a subset of A. In the special case where $A = B = D$ and $R = D \times D$, the domain of R is the same as the domain of discourse.

In everyday life we often mention relations between various kinds of things, such as: 'Antony *loves* Cleopatra', 'Sam *is an employee of* the Omega Corporation', and 'Robot A3XW is *in* the nuclear power plant.' We can also consider relations between abstract objects, e.g., '$(P \wedge Q)$ *tautologically implies* Q', which is a relation between $\mathcal{SC}$ sentences. Definition 2-18 may not appear to have any connection with these examples of relations, but it works fine in a certain way. Consider the predicate 'x loves y' applied to the domain of human beings, both past and present. Let H be the set of all humans. Then $H \times H$ is the set of all ordered pairs of humans including, for example,

<David, Bathsheba>, <Bathsheba, David>, <Antony, Cleopatra>,

<Cleopatra, Antony>, <Napoleon, Wellington>, <Wellington, Napoleon>,

<David, David>, <Cleopatra, Cleopatra>, etc.

Consider such a pair, say <Antony, Cleopatra>. This pair satisfies the predicate 'x loves y' iff the person Antony loves the person Cleopatra. Similarly, the pair <Bathsheba, David> satisfies this predicate iff Bathsheba loves David. Consider the set of *all* pairs that satisfy this predicate. This is the set of all pairs of people such that the first one loves the second one. This is a special subset of $H \times H$. To use the previous definition of relation, we can identify the relation expressed by 'x loves y' with the set of all pairs of people such that the first loves the second. This definition of relation is not concerned with the exact *meaning* of 'x loves y', nor with the physical, psychological, or social aspects of this relation. Although these aspects may be important for some purposes, our definition abstracts away from them to the set of ordered pairs of people that satisfy the predicate 'x loves y'. One can say that Definition 2-18 of a *binary relation between A and B* identifies such a binary relation with the set of all ordered pairs that satisfy a predicate in two variables. In fact, the definition goes farther than this, because it does not require that a specific predicate be formulated; it requires only that there exist a suitable subset of $A \times B$. In practical applications, such a subset will usually be specifiable by some predicate in two variables. This idea will now be illustrated with several examples.

Examples.

Let Re be the set of real numbers. Then $Re \times Re$ contains the points (ordered pairs) in the two-dimensional Cartesian coordinate system.

$$E = \{< x, y > \mid x, y \in Re \ \& \ x = y\}$$

is the relation that holds between x and y when $x = y$. We may also define E in "relational notation" by: For any $x, y \in Re$, $Exy \Leftrightarrow x = y$. Geometrically, this relation consists of all points on the line obtained by plotting $y = x$ in the Cartesian coordinate system. The domain and range of E are both equal to Re.

$$L = \{< x, y > \mid x, y \in Re \ \& \ x < y\}$$

is the relation that holds between x and y when $x < y$. In relational notation, we may also define L by: For any $x, y \in Re$, $Lxy \Leftrightarrow x < y$. Geometrically, this relation comprises all points "above" the line given by $y = x$.

$$C = \{< x, y > \mid x, y \in Re \ \& \ x^2 + y^2 = 25\}$$

is also a relation on Re. Geometrically, it comprises all points on the circle of radius 5 obtained by plotting $x^2 + y^2 = 25$. We may also define C by: For any $x, y \in Re$,

$$Cxy \Leftrightarrow x^2 + y^2 = 25.$$

The domain of C is

$$\{x \mid x \in Re \ \& \ -5 \leqslant x \leqslant 5\}.$$

The range of C is the same set of reals.

Whereas relations are sets, we can also operate on them with union, intersection, difference, and complementation. For example, $E \cup C$ contains all points that are either on the line or on the circle. L' (relative to $Re \times Re$) contains all points that are on the line corresponding to E or "below" it. We may use expressions like $E \cup C$ and $E \cap C$ to represent relations, as in '$(E \cap C)xy$', and in 'For any $x, y \in Re$, $(C \cap L)xy$ iff $< x, y > \in (C \cap L)$'. From the latter, because

$$< x, y > \in (C \cap L) \Leftrightarrow (< x, y > \in C \ \& \ < x, y > \in L),$$

and

$$< x, y > \in C \Leftrightarrow Cxy,$$

and

$$< x, y > \in L \Leftrightarrow Lxy,$$

we may also infer

$$(C \cap L)xy \iff (Cxy \ \& \ Lxy),$$

and hence,

$$(C \cap L)xy \iff (x^2 + y^2 = 25 \ \& \ x < y).$$

Thus, the relation $C \cap L$ corresponds to the arc of the circle that is above the line corresponding to E.

When an '&' or an 'or' connects two expressions in a conjunction or disjunction that is on one side of a $\Rightarrow$ or a $\iff$, it is customary to abbreviate the expression by dropping the parentheses around the conjunction or disjunction. For example, we may abbreviate the previous equivalence in this form:

$$(C \cap L)xy \iff x^2 + y^2 = 25 \ \& \ x < y,$$

where it is understood that the conjuncts are to be grouped together by parentheses. Note, however, that the left side of this expression is not a conjunction and the parentheses in it are necessary. They indicate that the relation between x and y is the intersection, $(C \cap L)$. ∎

The preceding examples are relations on Re. Here is a non-numerical example.

Example.

The domain of discourse is the set of all human beings. Let

$$W = \{\text{Alice, Jane, Miranda}\} = \{a, j, m\},$$

where the letters a, j, m are abbreviations for the longer names. Let

$$M = \{ \text{Bill, George, Hal, Rafael}\} = \{b, g, h, r\}.$$

Suppose that the wife of Bill is Alice, the wife of George is Miranda, and the wife of Rafael is Jane. Hal is a bachelor. The Cartesian product $M \times W$ is displayed in Table 2.1. The pairs that are members of the *WIFE* relation are printed in a special bold font, where *WIFExy* iff y is the wife of x.
In set theoretical notation we write

$$M \times W = \{ \ <x, y> \ | \ x \in M \ \& \ y \in W \ \},$$

and

$$WIFE = \{ \ <b, a>, <g, m>, <r, j> \ \}. \ ∎$$

	Alice	Jane	Miranda
Bill	$<$ **b**, **a** $>$	$< b, j >$	$< b, m >$
George	$< g, a >$	$< g, j >$	$<$ **g**, **m** $>$
Hal	$< h, a >$	$< h, j >$	$< h, m >$
Rafael	$< r, a >$	$<$ **r**, **j** $>$	$< r, m >$

Table 2.1. $M \times W$

A diagram consisting of a set of points, *(nodes)*, some of which are connected by lines (*arcs, edges,* or *links*), is called a *graph*. An arc may have an arrowhead on it indicating a direction from one node to another. A graph that has arcs with such arrowheads is called a *directed graph*; a graph, all of whose arcs have no arrowheads, is an *undirected* or *symmetric graph*. This use of the word 'graph' is different from its meaning in analytic geometry, where it refers to a plot of points in the Cartesian coordinate system. In this book, 'graph' refers to a diagram of nodes connected by arcs; I use the word 'plot' to refer to a sketch of points in the Cartesian coordinate system. Graphs are very convenient for illustrating binary relations. Here is a typical example.

Example.

Suppose there is a company, the Omega Corporation, also called 'Omega-Corp', and we want to represent its organization. Let O be the domain of discourse, where O is the set of all people in OmegaCorp. Let D be a binary relation such that Dxy holds iff y is the direct superior of x, i.e., y is the immediate boss of x.

Suppose that

$$O = \{ a, b, c, d, e, f, g, h, i, j, k, l, m \},$$

where a, b, etc., denote the people in OmegaCorp. The graph of OmegaCorp is shown in Figure 2.4. The dots are the nodes, which represent the people in O. An arc directed from node x to node y indicates that Dxy, i.e., the arc points toward y, who is the direct superior of x. Strictly speaking, the graph consists only of the nodes and arcs. However, this graph has also been enhanced to represent important subsets of O.

In OmegaCorp there are four departments:

M = middle management = $\{ b, c, d \}$.

R = research and development = $\{ e, f \}$.

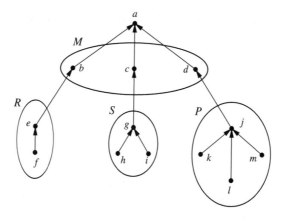

Figure 2.4. *The Omega Corporation*

$S = $ shipping $= \{\, g, h, i \,\}$.

$P = $ production $= \{\, j, k, l, m \,\}$.

Notice that a is not in any department. She is the owner and company president. Graphs of this type can help one to visualize binary relations. For
example, a quick glance at Figure 2.4 shows that we have, among other things,
Dfe, *Deb*, and *Dba*. One can therefore see that there is a 'chain of command'
from f up to a. In Chapter 4 I will show how the information about Omega-
Corp can be represented so that a computer may deduce, i.e., prove, that this
chain of command exists. The computer can do this by using symbolic logic.
However, most humans find that sketching graphs can be helpful in solving
problems about relations. ■

   The next section discusses some special kinds of relations. Before proceeding to it, we will generalize from binary relations to relations of higher arity.

Definition 2-19 For any x, y, z in the domain of discourse, the *ordered triple*
$< x, y, z >$ is

$$< x, y, z >=<< x, y >, z > .$$

For sets A, B, C, the *Cartesian product*, $A \times B \times C$, is

$$A \times B \times C = \{< x, y, z > \mid (x \in A \ \& \ y \in B) \ \& \ z \in C \}.$$

Ordered triples satisfy the same kind of criterion of identity as ordered pairs. Suppose that

$$< x,\, y,\, z >=< u,\, v,\, w > .$$

Then

$$<< x,\, y >,\, z >=<< u,\, v >,\, w >,$$

so $z = w$ and $< x,\, y >=< u,\, v >$, so also $x = u$ and $y = v$. Using ordered triples corresponds to three dimensions in geometry. Corresponding to the circle relation C given in a previous example, we can define a sphere relation by

$$Sxyz \Leftrightarrow x^2 + y^2 + z^2 = 25,$$

where x, y, z are real numbers. If desired, the previous definition for triples can be extended to ordered 4-tuples by defining

$$< x,\, y,\, z,\, u >=<< x,\, y,\, z >,\, u > .$$

This same idea can be extended to any positive integer k to define ordered k-tuples. An ordered $k + 1$-tuple is an ordered pair whose first element is an ordered k-tuple, for $k \geqslant 1$. We will consider sets to be 1-tuples. A subset of $A \times B \times C$ is called a 3-ary relation, a subset of $A \times B \times C \times D$ is a 4-ary relation, etc.

Ordered k-tuples are very useful for representing information. Suppose, for example, that we are storing personnel information about university professors. We could use a 4-ary predicate, FACULTY, which is used in the form, FACULTY(name, employee-ID-number, department, rank), where the parentheses and commas serve to separate the items and avoid confusion. The data would then look like this:

FACULTY(John Doe, 3857, chemical engin., instructor)

FACULTY(Jane Doe, 9813, philosophy, assistant professor)

FACULTY(Bill Roe, 2003, classics, professor)

FACULTY(Sally Roe, 7733, geology, associate professor)

FACULTY(Roza Roe, 1005, physics, professor), etc.

There might be additional data about the physical plant, e.g., department names and locations of department offices:

DEPARTMENT(geology, Rockhound Hall 100)

DEPARTMENT(physics, Newton Bldg. 3.24)

DEPARTMENT(philosophy, Wisdom Hall 8.905), etc.

For printed display, it is customary to present such data in the form of *relation tables*. The FACULTY data are shown in Table 2.2.

NAME	EMPLOYEE-ID-NO.	DEPT.	RANK
John Doe	3857	chemical engin.	instructor
Jane Doe	9813	philosophy	assistant prof.
Bill Roe	2003	classics	professor
Sally Roe	7733	geology	associate prof.
Roza Roe	1005	physics	professor

Table 2.2. FACULTY

Relation tables can be applied in many contexts. They can be used for airline flight schedules, inventories of goods, mailing lists, library catalogs, poisons and antidotes, etc. Storing data in relational form makes it easy to retrieve information and to make logical deductions from the data. This will be discussed further in Chapter 4.

2.2.2 Special Kinds of Relations

In this section, 'relation' means 'binary relation', unless otherwise stated. I will define many new terms. It is difficult to remember all of the definitions right away, so one should expect to refer back to them from time to time. As always, we assume there is a domain of discourse lurking in the background.

Definition 2-20 Let R be a binary relation. The *converse* of R is

$$R^{-1} = \{<x,y> \mid <y,x> \in R\}.$$

R is a set of ordered pairs; R^{-1} is obtained from R by reversing the order of each pair in R. The superscript -1 is not an exponent; it is merely one of several conventional notations used for the converse of a relation. From the above definition by abstraction, the following holds for any x and y:

$$R^{-1}xy \Leftrightarrow Ryx.$$

Ordinary language contains special names for the converses of some familiar relations. For example, the converse of *above* is *below*, and the converse of *before* is *after*. For any objects x and y, x is below y iff y is above x, and similarly, for events u and v, u is after v iff v is before u. Notice that for any relation R, $(R^{-1})^{-1} = R$. Therefore, R and R^{-1} are converses of each other. Thus, *above* and *below* (also, *before* and *after*) are converses of each other.

Example.

Problem Statement. Let L be the less-than relation, i.e.,

$$L = \{ <x, y> \mid x < y \},$$

on *Re*. Assuming simple facts about $<$, do the following: Write definitions by abstraction for L', L^{-1} (converse), $R = L' - L^{-1}$, $S = L \cup L^{-1}$. We will consider L' to be the complement of L relative to $Re \times Re$, i.e., $L' = (Re \times Re) - L$. Wherever possible, replace occurrences of L with $<$ or similar symbols. Also, show how R and S are related in simple ways to the identity relation on *Re*.

Discussion. L is a binary relation on *Re*, so it is a subset of $Re \times Re$. The complement of L is just the subset of $Re \times Re$ containing those ordered pairs that are not in L. This consideration will make it easy to do the first task. The converse is just as easy to handle. We can define R and S by applying set theory operations to the sets L' and L^{-1}. The rest of the problem requires the written justifications given in the solutions.

Written Solutions. Let L be the less-than relation, i.e.,

$$L = \{ <x, y> \mid x < y \},$$

on *Re*. Then

1. Using the specified definition of complement, we have

$$L' = \{ <x, y> \mid <x, y> \notin L \} = \{ <x, y> \mid \text{not } Lxy \}.$$

The two formulations are equivalent. By replacing L with $<$, and using facts about $<$, we can simplify to

$$L' = \{ <x, y> \mid <x, y> \notin < \} = \{ <x, y> \mid \text{not } x < y \}$$
$$= \{ <x, y> \mid y \leqslant x \}.$$

2. From the definition of converse,

$$L^{-1} = \{ <x,y> \mid <y,x> \in L \} = \{ <x,y> \mid Lyx \}$$
$$= \{ <x,y> \mid y < x \}.$$

3. $R = L' - L^{-1} = \{ <x,y> \mid y \leqslant x \} - \{ <x,y> \mid y < x \}$
$$= (\{ <x,y> \mid y = x \} \cup \{ <x,y> \mid y < x \}) - \{ <x,y> \mid y < x \}$$
$$= \{ <x,y> \mid y = x \} = \{ <x,x> \mid x \in Re \},$$

so R is just the identity relation on Re.

4. $S = L \cup L^{-1} = \{ <x,y> \mid x < y \} \cup \{ <x,y> \mid y < x \}$. But we know that either $x < y$ or $x = y$ or $y < x$, and that these are mutually exclusive possibilities. Therefore, $S = \{ <x,y> \mid x \neq y \}$, so S is just the nonidentity relation on Re. ∎

Every relation has a converse; it is just another relation and does not necessarily have any special properties. The following definition introduces several special features of relations. Some relations have none of these features; others may have only one of them or perhaps a few.

Definition 2-21 Let R be a binary relation on the set A, i.e., $R \subseteq A \times A$.

R is *reflexive* on A iff
    for every x in A, Rxx.

R is *irreflexive* on A iff
    for every x in A, not Rxx.

R is *symmetric* on A iff
    for every x, y in A, if Rxy, then Ryx.

R is *asymmetric* on A iff
    for every x, y in A, if Rxy, then not Ryx.

R is *transitive on* A iff
    for every x, y, z in A, if $(Rxy$ & $Ryz)$, then Rxz.

R is *antisymmetric* on A iff
    for every x, y in A, if $(Rxy$ & $Ryx)$, then $x = y$.

R is *strongly connected* on A iff
    for every x, y in A, $(Rxy$ or $Ryx)$.

R is *connected* on A iff
    for every x, y in A, if $x \neq y$, then $(Rxy$ or $Ryx)$.

Some of these properties are familiar. For instance, the relation specified by 'x is a sibling of y' is symmetric. The relation specified by 'x is a sister of y' on the set of all humans is not symmetric (in general), but it is also not asymmetric. The relation 'x is taller than y' is asymmetric. It is easy (and left for practice) to prove that if R is symmetric, then $R = R^{-1}$. Any arc in the graph of a symmetric relation would have arrowheads pointing in both directions. For this reason, the arrowheads are usually omitted in such a graph. It is called a *symmetric* or *undirected* graph, as was mentioned in the previous section. Each node in the graph of a reflexive relation has a little arc going out from that node and turning back into it. The relation $x \leqslant y$ on the real numbers is reflexive. But $x < y$ is irreflexive, as there is no number x such that $x < x$. This relation $x < y$ is also asymmetric, transitive, and connected. We will return to it a little later.

It is important to remember that a relation on a set A is a subset of $A \times A$. In ordinary language, we may use a predicate like 'x likes y' to denote a relation, without specifying its domain. This is risky because it can lead to confusion and errors. Suppose there is a group of people, $G_1 = \{a, b, c\}$, such that, for any two distinct people in this group, one of them likes the other. Then *LIKESxy* is connected on G_1. Further suppose that $G_1 \subseteq G_2$, where

$$G_2 = \{a, b, c, d\}.$$

Also suppose that b does not like d and that d does not like b. Then *LIKESxy* is not connected on G_2. When working with relations, it is important to specify the domain, or domains, of these relations.

Example.

Problem Statement. Let R and S be nonempty binary relations on a set A. Either prove or disprove each of these assertions:

1. If R, S are symmetric on A, then $R \cup S$ is symmetric on A.

2. If R, S are transitive on A, then $R \cup S$ is transitive on A.

Discussion.

1. Since $R \cup S$ is a union of two sets of ordered pairs, we have, by the definition of union,

$$< x, y > \in R \cup S \Leftrightarrow (< x, y > \in R \quad \text{or} \quad < x, y > \in S).$$

I shall also write $< x, y > \in R \cup S$, using relational notation, in the form $(R \cup S)xy$. For the first question, we need to consider whether $(R \cup S)xy$ implies $(R \cup S)yx$. If we assume $(R \cup S)xy$, then we have Rxy or Sxy. I shall now consider the cases (again using the familiar "case argument" proof strategy).

Case 1. Suppose that we have Rxy. Since R is symmetric, we also have $Rxy \Leftrightarrow Ryx$. Thus, by the $\mathcal{SC}$ MP rule, Ryx. Hence, by the Add rule, we obtain $(Ryx$ or $Syx)$.

Case 2. Similarly, if we have Sxy, then we can obtain $(Ryx$ or $Syx)$.

In either case, we end up with Ryx or Syx, so $R \cup S$ is symmetric. Case arguments like this one are related to the Constructive Dilemma rule. I will use that rule in the written solution to shorten the proof a little.

2. Now consider the second question, in which it is given that both relations are transitive. Does this imply that

$$((R \cup S)xy \ \& \ (R \cup S)yz) \Rightarrow (R \cup S)xz?$$

The antecedent of this conditional is equivalent to

$$(Rxy \text{ or } Sxy) \ \& \ (Ryz \text{ or } Syz).$$

Does this imply the consequent, $(Rxz$ or $Sxz)$? We have some cases to consider, so I shall write the conjunction in this format:

$$(Rxy \text{ or } Sxy)$$

$$\&$$

$$(Ryz \text{ or } Syz).$$

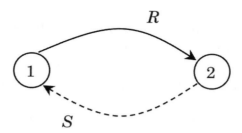

Figure 2.5.

Now quickly consider the cases: From the first conjunct we have Rxy or Sxy. Suppose we happen to have Rxy and also, from the second conjunct, we have Ryz. Then, because R is transitive, we obtain Rxz. By Add, we can infer Rxz or Sxz, which suggests that $R \cup S$ *possibly* is transitive. But other cases could occur instead of this one. Suppose that we have Rxy from the first conjunct and Syz from the second? We have no justification for inferring transitivity from this pair of facts. This *suggests* that we might be able to construct a counterexample and that it might be useful to look for one at this point.

When working with binary relations, it is often very helpful to sketch graphs representing, or at least partially representing, the problem. It is usually advisable to begin with simple graphs in domains with only a few elements. Sometimes it is necessary to use more complicated graphs, perhaps with infinite domains. Nevertheless, if it is not immediately obvious that a large domain is required, then it is a good strategy to begin with simple possibilities.

In the current problem, a domain with only one element, say a, leads to Raa and Saa, which is not a counterexample. Let us try a domain of two elements, say $A = \{1, 2\}$. A graph that easily comes to mind is Figure 2.5.

From the figure, we have $R12$ and no other elements of A related by R. Let z be any element of A and consider $(R12 \ \& \ R2z)$. This conjunction is false for both $z = 1$ and $z = 2$. Hence, R is "vacuously" transitive because the antecedent in the conditional in the definition of transitivity is always false. Similarly, S is vacuously transitive. Now, since we have $R12$ and $S21$, using the Add rule, we also have $(R12$ or $S12)$ and $(R21$ or $S21)$. In that event, $(R \cup S)12$ and $(R \cup S)21$. But $(R \cup S)11$ is false.

Written Solution.

1. *Theorem.* If R, S are nonempty symmetric relations on A, then $R \cup S$ is symmetric on A.

Proof. Let x, y be any elements of A and assume that $(R \cup S)xy$. Then Rxy or Sxy. Since the given relations are symmetric,

$$Rxy \Rightarrow Ryx$$

and

$$Sxy \Rightarrow Syx.$$

Applying the $\mathcal{SC}$ CD rule to the disjunction and the two conditionals, we obtain

$$Ryx \text{ or } Syx,$$

and hence $(R \cup S)yx$. Therefore, $R \cup S$ is symmetric on A. **Q.E.D.**

2. It is not the case that if R, S are transitive on A, then $R \cup S$ is transitive on A.

Counterexample. Let

$$A = \{1, 2\}, \quad R = \{< 1, 2 >\}, \quad S = \{< 2, 1 >\}.$$

From the definition of R, we have $R12$ and no other elements of A related by R. Let z be any element of A, and consider $(R12 \ \& \ R2z)$. This conjunction is false for both $z = 1$ and $z = 2$, so it is false for every element of A. As a result, R is "vacuously" transitive because the antecedent of the conditional in the definition of transitivity is always false. Similarly, S is vacuously transitive. Now, since we have $R12$ and $S21$, using the SC Add rule, we obtain $(R12$ or $S12)$ and $(R21$ or $S21)$. Hence, $(R \cup S)12$ and $(R \cup S)21$. But $(R \cup S)11$ is false, so $(R \cup S)$ is not transitive. ∎

Identity is a very special relation. It is reflexive, for anything is identical to itself. It is also symmetric, because, if $x = y$, then $y = x$. Finally, it is transitive, as one can easily check. There are many other relations that are also reflexive, symmetric, and transitive. Suppose A is a set of uniformly colored marbles; some are solid green, some solid red, some yellow, some blue, etc. Consider the relation Sxy on A which holds iff x has the same color as y. Each marble has the same color as itself, so Sxx for every x in A. Thus, S is reflexive. Also, if Sxy, then Syx, so S is symmetric. Finally, if $Sxy \ \& \ Syz$, then Sxz, so S is transitive. We can use the relation S as a basis for classifying the marbles according to color: We put all green marbles into one subset, all reds into another subset, all yellows into a third subset, and so on. When done, the set A is divided (or partitioned) into disjoint subsets such that each marble is in one of these subsets, and such that all of the marbles in any given subset have the same color (so are related by S). This manner of classification can often be performed and is widely used in mathematics and science. The conjunction of the three properties, reflexivity, symmetry, and transitivity, is what enables this kind of classification. Identity has these three properties but also satisfies a more general principle of substitutivity (which should now be reviewed in the Introduction). The next few pages are devoted to the very important kind of relation defined next.

Definition 2-22 Let R be a binary relation on the set A. R is an *equivalence relation* on A iff R is reflexive, symmetric, and transitive on A.

In order to relate equivalence relations to classification, we also use

Definition 2-23 A *partition* Π of a nonempty set A is a set of subsets of A, satisfying all of the following:

1. Each set in Π is nonempty.

2. For any x, if $x \in A$, there is $C \in \Pi$ such that $x \in C$.

3. If $C, D \in \Pi$ and $C \neq D$, then $C \cap D = \varnothing$.

The sets in Π are called *cells*.

In order to partition A, we divide it into nonempty disjoint subsets (cells) such that every element of A is a member of one and only one cell. A partition Π of A is *not* the same as the power set, $\mathcal{P}(A)$. This can be seen by recalling that $\varnothing \in \mathcal{P}(A)$, but no element (cell) of Π is empty. Moreover, a set A can often be partitioned in many different ways, but A has only one power set.

Definition 2-24 Let R be an equivalence relation on A and let $u \in A$. The *R-equivalence class* of u is

$$[u]_R = \{x \mid x \in A \ \& \ Rux\}.$$

Notice that $[u] \subseteq A$. The subscript R is normally used when more than one equivalence relation is under consideration. If R is the only one, then we may write $[u]$ in place of $[u]_R$. The set of all equivalence classes of R is called the *quotient set of A modulo R* and is denoted by A/R. Using abstraction, this set of sets is defined by

$$A/R = \{ [u] \mid u \in A\}.$$

On first reading, this definition may be difficult to understand. The main ideas can be paraphrased as follows: A is a given set, R is an equivalence relation on A, and u is an element in A. Then $[u]$ is the *subset* of all elements of A that are related to the element u by the relation R. Remember that $[u]$ is a subset of A. 'Class' is another word that is sometimes used for some kinds of sets, and it is traditional to call $[u]$ an *equivalence class*. Now every element in A will have an equivalence class. If $a, b, c, \cdots \in A$, then there exist the equivalence classes, $[a], [b], [c], \cdots$, and each of these is a subset of A. Finally, A/R is just the set of all of these equivalence classes, so

$$[a], [b], [c], \cdots \in A/R.$$

The notation

$$\{ [u] \mid u \in A\}$$

just denotes all of these classes, when u takes the values $a, b, c, \cdots$ for all of the elements in A.

Consider a set A of objects that have different shapes. Define E by Exy iff (x and y have the same shape). E is very much like the previous example of the same color relation. E is also an equivalence relation. Figure 2.6 depicts E on a domain of several objects with the shapes heart, circle, and diamond. There are two sizes, small and large. Ignoring the sizes, all of the hearts are contained in the topmost two rectangles stretching horizontally across the figure. All of these hearts together form one equivalence class. Pick any heart h in that bunch; then any other heart has the same shape as h. The circles are contained in the middle two rectangles. They form another equivalence class, just as all of the diamonds do. The three subsets of hearts, circles, and diamonds make up A/E.

Now define another relation, F by Fxy iff x and y have the same size. There are two sizes, small and large. All of the small objects are contained in the three rectangles in the leftmost vertical column. They are one equivalence class of F. All of the large elements are in the three rectangles in the right vertical column. They are the other equivalence class of F. These two sets make up A/F.

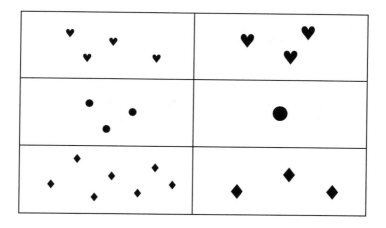

Figure 2.6. *Two Equivalence Relations*

E partitions the domain into three equivalence classes according to the *same shape* relation. F partitions it according to the *same size* relation. These two partitions cut across each other. It can be seen that the intersection, $E \cap F$, partitions the domain into smaller equivalence classes according to *same shape and same size*. Thus, $A/(E \cap F)$ is the set of six subsets represented by the smaller, single rectangles. Of course, I have not proved any of this. We need a fundamental theorem that relates equivalence relations to partitions. In order to prove this theorem, I first prove a couple of lemmas; a *lemma* is a little theorem that is used to help prove a bigger one. The first lemma says that, if x and y are in the same R-equivalence class, then they are related by R. For instance, in Figure 2.6, if we pick out any two of the hearts, they are in the same E-equivalence class. According to the next lemma, they must be related by E. Of course, they are; they have the same shape!

Lemma 2-25   Let R be an equivalence relation on set A.   Then for any $x, y, u \in A$,

$$\text{if } x \in [u] \ \& \ y \in [u], \text{ then } Rxy.$$

Proof.   Let $x, y \in [u]$.   Then, by Definition 2-24, Rux & Ruy.   Because R is symmetric, if Rux, then Rxu.   Because we also have Rux, we infer (by MP) Rxu. Now we have Rxu & Ruy.   Because R is transitive, we infer that Rxy. **Q.E.D.**

This is a typical proof about relations. It makes use of the definitions of special properties of the relations involved. Such a proof may also use some facts about sets, and it will use informal logic, especially tautological inferences. It is often helpful to sketch a graph of the situation described by the antecedent conditions of such a theorem. For practice, one should sketch the graph corresponding to this lemma. Whereas equivalence relations are symmetric, the graph will be a symmetric, i.e., undirected graph; the arcs can be drawn without arrowheads. Because $x, y \in [u]$, we have Rux & Ruy, so there is an arc from u to x and an arc from u to y. These are undirected. In so far as R is transitive, it follows that there is an (undirected) arc from x to y. By sketching this graph first, it becomes obvious how to prove the lemma. Notice that this lemma used only the transitivity and symmetry properties of R. This is also true of the next lemma.

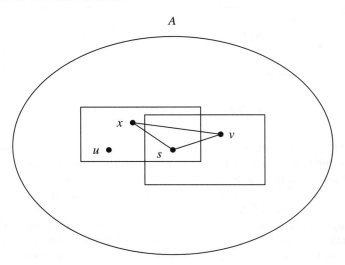

Figure 2.7. *Lemma 2-26*

Lemma 2-26 Let R be an equivalence relation on set A. Then, for any $u, v \in A$,

$$\text{If } [u] \cap [v] \neq \varnothing, \text{ then } [u] = [v].$$

Proof. A graph illustrating this lemma is in Figure 2.7. The equivalence classes of u and v are represented by the rectangles in this figure. Suppose that the intersection is nonempty. Then there is at least one element, I shall call it s, in this intersection, so $s \in [u] \cap [v]$, so $s \in [u]$ and $s \in [v]$.

Now let $x \in [u]$. I will show that this implies that $x \in [v]$, so that $[u] \subseteq [v]$. By assumption, we have $x \in [u]$, and from the previous paragraph, $s \in [u]$. Thus, $x, s \in [u]$. By the previous lemma, Rxs. Because $s \in [v]$, we have Rvs, so also Rsv by symmetry. From $(Rxs \ \& \ Rsv)$ and transitivity, we obtain Rxv. So, by symmetry, Rvx, so $x \in [v]$. Therefore, $[u] \subseteq [v]$. $[v] \subseteq [u]$ is proved in a similar manner. Hence, $[u] = [v]$, and the two rectangles in Figure 2.7 actually turn out to be identical. *Q.E.D.*

An Important Remark. I said that "$[v] \subseteq [u]$ is proved in a similar manner." The reader should verify that this is correct. It is acceptable practice to write that "such and such can be proved in a similar manner" if this is indeed true and one has verified that it is. In general, the author of a proof should *not* rely on similarity in this way unless the similarity is fairly obvious. In the case of this proof, the similarity is obvious from the symmetry in Figure 2.7. If one moves the x from its current rectangle into the other rectangle, and then connects the dots accordingly, it is clear that a similar argument will prove that $[v] \subseteq [u]$.

If we say that equivalence classes are *overlapping* iff they are not disjoint, then this lemma shows that two overlapping equivalence classes must be identical. This is an important fact for the kind of classification system that we seek. If this fact were not true, then there could be an element s that is a member of two different, but overlapping, equivalence classes. This element would not belong to a unique classifying cell of the partition that we hope to form. In Figure 2.6, there are three disjoint E-equivalence classes, corresponding to the three shapes. There are two disjoint F-equivalence classes corresponding to the two sizes. There are six disjoint $(E \cap F)$-equivalence classes corresponding to the six different combinations of shapes and sizes.

If we apply contraposition to Lemma 2-26, we obtain the following: If two equivalence classes are not identical, then they are disjoint. Although the lemmas are abstract and general, they are easy to understand if one compares them to concrete examples such as that in Figure 2.6. If we summarize the lemmas, we see that Lemma 2-25 says that any two elements of the same equivalence class are related by R. This means that they are *equivalent* according to R, i.e., they have the same color, same shape, same size, etc. Lemma 2-26 implies that no element can be in more than one distinct equivalence class of a given relation R. Yet we still have not shown that *every* element of A is in an equivalence class. This is done in the next theorem, which states how an equivalence relation partitions its domain. We finally make use of the reflexivity of R.

Theorem 2-27   If R is an equivalence relation on A, then A/R is a partition of A.

Proof. By definition, $A/R = \{ [u] \mid u \in A \}$. It is necessary to prove that the three parts of the definition of partition are satisfied.

1. Let $C \in A/R$. Then $C = [u]$, for some $u \in A$. Since R is reflexive, Ruu, so $u \in [u]$, so $u \in C$, so C is nonempty.

2. Let $x \in A$. Since Rxx, $x \in [x]$. Hence, x is in at least one of the equivalence classes in A/R.

3. By the previous lemma, if two sets in A/R are not identical, then they are disjoint.

By 1 to 3 above, and the definition of 'partition', A/R is a partition of A.

$$\textbf{\textit{Q.E.D.}}$$

   This theorem shows that an equivalence relation R *determines* the partition A/R. The graph of an equivalence relation is given in Figure 2.8. The nodes (black dots) are the elements of A, which is represented by the large ellipse. The relation R is represented by a graph, which falls into three separate parts. Any two nodes that are connected by an arc of the graph are related by R, and therefore they form an ordered pair of the relation R.

   Because R is symmetric, this is an undirected graph (no arrowheads on the arcs), so any two connected nodes actually correspond to two ordered pairs, going in both directions. Because R is reflexive, each node has a little loop arc back to itself. These loops also have no arrowheads. Because R is transitive, whenever nodes satisfy Rxy & Ryz, they also satisfy Rxz. This can be seen most clearly in the case of the two arcs that cross each other in the middle of the central square. Transitivity is also exemplified by other arcs.

   There are three disjoint equivalence classes (or cells). The left moon-shaped segment contains three equivalent elements of A. One of these elements is labeled a. It is equivalent to the other two elements in this segment of A. The three elements in this segment constitute $[a]$, which is one cell of the partition of A. These elements in $[a]$ are each related to a, to one another, and to themselves (since R is reflexive). Compare this situation with Lemma 2-25. Suppose we choose some element in $[a]$ other than a. Let us denote this element by a_1. Then Raa_1. We then have $a_1 \in [a]$ and (of course) $a_1 \in [a_1]$. Hence, $[a] \cap [a_1] \neq \varnothing$. By Lemma 2-26, $[a] = [a_1]$. Thus, if we choose any of the elements in $[a]$ and

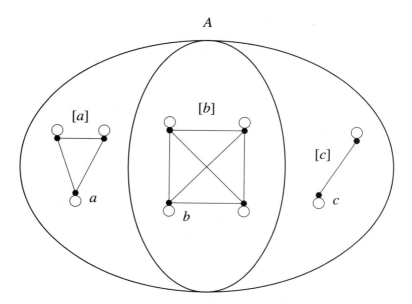

Figure 2.8. *Graph of an Equivalence Relation*

form its equivalence class, we end up with the same three elements in $[a]$. The four elements in the central, smaller ellipse form another equivalence class. The third equivalence class is made up of the two elements in the right moon-shaped segment. These three equivalence classes are the cells of a partition of A.

Now suppose we are given a partition of A. Does this partition determine an equivalence relation? The answer is given in the next theorem, but a simple example suggests the answer. Imagine there is a large, crowded prison in which each prisoner is assigned to one and only one prison cell. Also, every prison cell has at least one prisoner assigned to it, and some of these cells have more than one. Moreover, the prison cells have solid walls and are locked, so no prisoner can get from one prison cell to another. Let P be the population of all prisoners in this prison, and C, D denote sets of prisoners in prison cells, and let Π be the set of all such sets of prisoners. From the description of the prison,

1. Every set C in Π is nonempty.

2. For any x in P, there is some $C \in \Pi$ such that $x \in C$.

3. If C, $D \in \Pi$ and $C \neq D$, then $C \cap D = \varnothing$.

Therefore, by Definition 2-23, Π is a partition of P, and the set of prisoners in a prison cell is a set theoretical cell of the partition Π. Now, for prisoners x, y, define

Rxy iff (x and y are assigned to the same prison cell), i.e.,

Rxy iff there exists a cell $C \in \Pi$ such that ($x \in C$ & $y \in C$).

Then

- R is reflexive: Every prisoner x is assigned to the same cell as himself, so we have Rxx.

- R is symmetric: If Rxy, then x and y are in the same prison cell, so are also in the same cell of Π. But then y and x are also in this same cell, so Ryx.

- R is transitive: If we have Rxy and Ryz, then x is in the same cell as y, and y is in the same cell as z. But no prisoner is in two different cells, so y is in one and only one cell. The same is true for x and z, so x is in the same cell as z, so Rxz.

Therefore, R is an equivalence relation on P. The proof of the next theorem is an abstract generalization of this argument.

Theorem 2-28   Let Π be a partition of set A. Define R on A by

Rxy iff there exists a cell $C \in \Pi$ such that ($x \in C$ & $y \in C$).

Then R is an equivalence relation on A.

Proof. In order to prove that R is an equivalence relation, we must prove that it is reflexive, symmetric, and transitive on A. Let x, y, $z \in A$.

1. Since Π is a partition, any $x \in A$ is in some cell of Π, so we may suppose that $x \in C$, where $C \in \Pi$. Then ($x \in C$ & $x \in C$), so Rxx. Hence, R is reflexive.

2. Suppose that Rxy. Then there is some cell C such that ($x \in C$ & $y \in C$). So, ($y \in C$ & $x \in C$), so Ryx. Thus, R is symmetric.

3. Suppose that Rxy & Ryz. Then there are cells, C, D, such that

$$(x \in C \text{ & } y \in C)$$

and

$$(y \in D \text{ & } z \in D).$$

But then $y \in C$ & $y \in D$, so $C \cap D \neq \varnothing$. By the definition of partition, $C = D$. Hence, $x \in C$ & $z \in C$, so Rxz. And so, R is transitive.

By 1 to 3 above, R is an equivalence relation. **Q.E.D.**

Example.

Problem Statement. Let the domain be Re and define E by: For any x, y in Re,

$$Exy \text{ iff } (y = x \text{ or } y = -x).$$

1. Determine whether E is an equivalence relation on Re and justify your answer.

2. Let r be any real number. If E is an equivalence relation, characterize $[r]$.

Discussion. E is an equivalence relation on Re iff E is reflexive, symmetric, and transitive on Re. Because this problem is about a particular domain, and about a particular relation defined on this domain, we may (and probably must) use special features of the domain and the definition. If x is anything, then $x = x$, so E is obviously reflexive. Since $(y = x \text{ or } y = -x)$ iff $(x = y \text{ or } x = -y)$, E is also symmetric. By now it should be clear that E comprises pairs of the following forms and no others:

$$< x, x >, \quad < x, -x >, \quad < -x, x >, \quad < -x, -x > .$$

Consider any two of these pairs, such as $< x, -x >$, $< -x, x >$, in which the inner two elements match. Then, by inspection (i.e., checking all of the cases), we see that the first and fourth elements, in this case x and x, form an ordered pair, in this case $< x, x >$, which is also in the set of four pairs. Thus, E is transitive. The write-up should be more precise, and it can now be stated as a theorem.

Written Solution. *Theorem.* Let the domain be Re and define E by: For any x, y in Re,

$$Exy \text{ iff } (y = x \text{ or } y = -x).$$

Then

1. E is an equivalence relation on Re.

2. If $r \in Re$, $[r] = \{ r, -r \}$.

Proof.

1. Let x, y, z be any elements of Re.

[Reflexivity] Since $x = x$, by Add, ($x = x$ or $x = -x$), so Exx, so E is reflexive.

[Symmetry] Suppose that Exy. Then ($y = x$ or $y = -x$), so ($x = y$ or $x = -y$), so Eyx, so E is symmetric.

[Transitivity] Suppose that Exy & Eyz. Then we have

$$(y = x \text{ or } y = -x) \ \& \ (z = y \text{ or } z = -y).$$

We now consider the possible cases.

Suppose that $y = x$ together with ($z = y$ or $z = -y$). Then, either $z = x$ or $z = -x$.

Suppose that $y = -x$ together with ($z = y$ or $z = -y$). Again, $z = x$ or $z = -x$.

Therefore, in all four possible cases, Exz, so E is transitive.

2. By definition, $[r] = \{ z \mid Erz \}$; so

$$[r] = \{ z \mid z = r \text{ or } z = -r\} = \{ r, -r \}. \ \textbf{\textit{Q.E.D.}} \ \blacksquare$$

Equivalence relations and partitions are useful. Several additional examples are given in the Exercises. We now turn to a general way of describing multiple relations on a common domain. This method of description is also useful when one is working with ordering relations.

Definition 2-29 Let D be a nonempty set. A *relational system* with domain D is an ordered $(n + 1)$-tuple of the form

$$\mathcal{S} = \langle D, R_1, R_2, \cdots, R_n \rangle,$$

where each R_i is either a subset of D or a relation on D. We consider a subset to be a relation of arity 1. Each R_i may have any arity $\geqslant 1$.

The idea of a relational system is introduced in order to provide a concise and systematic way of talking about a structured set, i.e., a set D together with certain of its subsets and relations on it. Denoting a relational system by an ordered $(n + 1)$-tuple inside large angle brackets is just a useful convention (which will be generalized later). This notation provides a convenient way to refer to the domain of the relational system and to its relations, in a specified order. Some relational systems are familiar. Let Re be the real numbers; then

$$\langle Re, =, < \rangle$$

is the relational system with the set of reals as its domain, together with the relations $=$ and $<$ on this domain. Other relational systems are not so familiar. Let P be the set of all people who have ever lived on the island of Madagascar. Use the predicates

Bx: x is born on the island,

Fxy: y is the father of x,

Mxy: y is the mother of x.

Then $\mathcal{M} = \langle P,\ B,\ F,\ M \rangle$ is a rather complex relational system. Without detailed historical knowledge, there is not much that one can be sure about this relational system. One assertion that seems correct is this: For any $x \in B$, there exists $y \in P$ such that Mxy.

We can also represent the Omega Corporation as a relational system,

$$\langle O,\ M,\ R,\ S,\ P,\ D \rangle,$$

where O, M, etc., have the meanings previously described for OmegaCorp. For practice, one should write out the members of O, M, etc.

There is an infinite variety of relational systems. Among these are some of special interest. We have already considered systems of the form $\langle D,\ E \rangle$, where E is an equivalence relation. In any such system, E partitions D. Other interesting and useful systems are those in which a relation orders the elements of the domain. It is difficult to say exactly how one should define *order* in general, but there are certain types of relations that have traditionally been considered ordering relations. We shall briefly look at a few of these.

Definition 2-30 $\langle A,\ R \rangle$ is a *simple order* iff R is transitive, antisymmetric, and strongly connected on A.

A familiar simple order is $\langle Re,\ \leqslant \rangle$, where (as usual) Re is the set of real numbers. It would be a big digression to prove this here, so we simply assume that one is familiar with the properties of $\leqslant$ on the reals. In the next chapter, it will be proved that $\leqslant$ is a simple order on the set of natural numbers, Nat.

By the preceding definition, any simple order R on A is strongly connected. By the definition of strongly connected, it follows that, for any x in A, (Rxx or Rxx). Hence, by Taut, for any x, Rxx, so R is also reflexive on A. There is a closely related order that is irreflexive.

Definition 2-31 $\langle A,\ R \rangle$ is a *strict simple order* iff R is transitive, asymmetric, and connected on A.

A strict simple order is asymmetric on A, so, if $x \in A$, we have if Rxx, then not Rxx. Thus, by Clav, not Rxx. So, R is also irreflexive on A. A familiar example of a strict simple order is $\langle Re, < \rangle$.

Simple orders are often called *linear orders*, and strict simple orders are often called *strict linear orders*. The term *linear* is used because these orders "line up" the elements of their domains. The notion of lining up elements is only intuitive, but it can be understood by examining some simple graphs. We will look at strict simple orders, but the examination of simple orders is similar. First, consider Figure 2.9. Could any strict simple order R have three distinct (mutually nonidentical) elements a, b, c such that Rab and Rac, but neither Rbc nor Rcb? The answer is no, because this would be a violation of the connectedness condition. Then, the situation portrayed in Figure 2.9 cannot happen with a strict simple order. Instead, either Rbc or Rcb must hold and, in either case, the three points line up. Also, because connectedness requires that any two distinct elements be related, no graph of a strict simple order can have two or more separate subgraphs. Could a strict simple order have cycles as illustrated in Figure 2.10? Again, the answer is no. If we have Rab, Rbc, and Rca, then by transitivity, we would also have Rac. But then we would have both Rca and Rac, which violates asymmetry. So it follows that R may have no such cycle of three elements, and it certainly can have no cycle of two elements. Using methods to be introduced in Chapter 3, it can be proved that cycles with k nodes are excluded for all natural numbers $k \geqslant 2$.

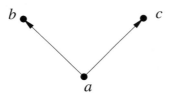

Figure 2.9.

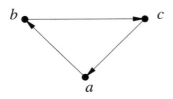

Figure 2.10.

So far it has been argued that any element a in the domain of R has a unique "next" element, if it has any "next elements" at all. Also, this next element cannot be a or any element b such that Rba. Could we have three distinct elements such that Rac and Rbc? Then, by connectedness, we would again have Rab or Rba. In either case, three elements again line up. Although this is not a proof, it should be clear that a strict simple order "lines up" its domain into a linear graph. From this, one might get the impression that all "orders" line up their domains. By convention, however, other relations that lack this feature are also considered to be orders. Here is one of the best known types.

Definition 2-32 $\langle A, R \rangle$ is a *partial order* iff R is reflexive, antisymmetric, and transitive on A.

Here is something to think about: An equivalence relation is reflexive, symmetric, and transitive, and partitions its domain. We do not consider this partition to be an ordering. Yet a partial order is reflexive, antisymmetric, and transitive, and it is considered to be an order. Why? The next example is an illustration. Before studying this example, it may be helpful to review the descriptions of the number systems in the Introduction, Section 0.3.1.

Example.

Let Re be the set of real numbers, Rat be the set of rational numbers, Int be the set of integers, Rat^+ be the positive rationals, Int^+ be the positive integers, A be the set of all animals, C be the set of all cats, L be the set of all lions, I be the set of all iguanas, and P be the set of all planets in the solar system. Now let

$$D = \{\, A,\ C,\ L,\ I,\ P,\ Re,\ Rat,\ Rat^+,\ Int,\ Int^+ \,\}.$$

Notice that the elements of D are sets; they will be related by a relation between sets. Consider $\mathcal{U} = \langle D, \subseteq \rangle$. From theorems in the previous section, we know that $\subseteq$ is reflexive, antisymmetric, and transitive, so this relational system is a partial order. Its graph, ignoring the arcs corresponding to reflexivity (little loops from a node back to itself), is given in Figure 2.11. ■

Because it lacks symmetry, a partial order does not partition the domain like an equivalence relation, but it allows that some elements may be "higher" or "greater" than others. In so far as it is not strongly connected, a partial order does not force the domain elements to "line up" as does a simple order. A partial order may even leave some pairs of elements incomparable, such as the pair Rat^+ and Int, and the pair Rat and A. It can even have isolated elements like the set P in this example.

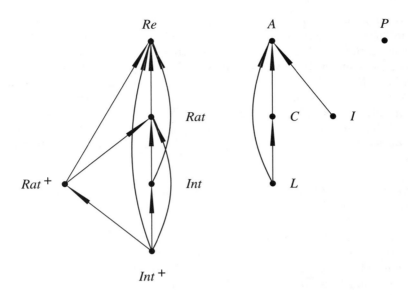

Figure 2.11. *A Partial Order (reflexive loops omitted)*

    Partial orders are common in real-world situations, for instance, when weak preferences exist between alternatives. Suppose that a café has ten entrées on its menu and that we have some (perhaps partial) information about Raquel's tastes in food. Let Lxy mean that Raquel likes entrée y at least as much as she likes entrée x. Then, for any x, Lxx, so L is reflexive. Also, suppose that, if Lxy and Lyz, then Lxz, so L is transitive. A person's preferences are not always transitive, especially where large numbers of choices are involved. In this simple example, it is reasonable to assume that Raquel's preferences (to the extent that they are revealed in the data available to us) are transitive. Finally, suppose that there are not two different entrées x, y such that Lxy & Lyx. In other words, there are no pairs of choices such that she is indifferent between them. Since only ten entrées are available, it is not impossible that her preferences would work out in this manner (again, as revealed in our data). But from this last assumption, L is (vacuously) antisymmetric because Lxy & Lyx is always false. Thus, L is a partial order on the set of entrees, and a graph of it would not necessarily be linear. This example is contrived, but there are more realistic and important ones that are similar to it. We will consider some other examples of important ordering relations in the Exercises and in the next chapter. In particular, E 2-76, at the end of the next exercise set, introduces another kind of ordering relation called a *quasi order*, which is a generalization of the concept of partial order.

EXERCISES

General Instructions

Throughout this book, if an exercise is simply the statement of a theorem or a metatheorem, then the task is to prove this theorem or metatheorem. In your proof, you may use the definitions given in the text. Unless stated to the contrary, you may also use theorems proved in the text.

Exercises for which answers are provided at the back of the book are marked with "Answer Provided."

In this chapter, unless instructed otherwise, you may assume elementary facts of arithmetic and algebra in these exercises.

Finally, please note that exercises that introduce important principles, useful applications, are especially difficult, or are special in some way, are indicated by bold print. In some cases, parenthetical comments describe the special features of the problem. If such comments are missing, the exercise is nonetheless of special value in introducing a new concept, principle, or technique.

E 2-35 Let the domain of discourse be the real numbers, Re. Four relations are defined as follows, where x, $y \in Re$,

Pxy iff $y = x^2 - 2$,

Lxy iff $y = 2x + 1$,

Rxy iff $y \geqslant x^2 - 2$,

Sxy iff $y \leqslant 2x + 1$.

1. Sketch a plot of each of these relations on the Cartesian coordinate system. Use different shadings or colors to indicate regions of the plane.

2. What are the elements (ordered pairs) of the relation $P \cup L$?

3. What are the elements of the relation $P \cap L$? (This requires solving a simple quadratic equation.)

4. Using set abstraction notation, write a definition for the relation $R \cap S$. On your plot, represent $R \cap S$ with special shading or coloration.

E 2-36 The domain of discourse, D, is a set of people, and you are given the following predicates that apply to this domain:

$father(x, y)$: y is the biological father of x,

$mother(x, y)$: y is the biological mother of x,

$sister(x, y)$: y is a biological sister of x,

$brother(x, y)$: y is a biological brother of x.

We assume that some people in D are related by these predicates and others are not. Also, let a denote a particular person in D. Using this notation, write definitions by abstraction of the following sets. Consider only biologically related people.

1. P = the set of ordered pairs $< x, y >$ in which y is a parent of x.

2. P_a = the set of parents of a.

3. Sis_a = the set of sisters of a.

4. Sib = the set of ordered pairs $< x, y >$ in which y is a sibling of x.

5. Sib_a = the set of siblings of a.

E 2-37 (Use of relations in data storage and processing) Suppose we have information about some cars that are denoted by their license plate numbers. There are two sets of data represented by binary relations as follows:

$$C =$$
{ <223ABC, red>, <105XJQ, blue>, <804EWV,green>,

<831EOZ, blue>, <834BTJ, red>, <979XPT, yellow>,

<333NJL, green>, <113NPO, black>, <660AAA,white> }

$$S =$$
{ <223ABC, compact>, <105XJQ, midsize>,

<804EWV, compact>, <834BTJ, large>, <054HHH, midsize>,

<979XPT, large>, <333NJL, compact >, <113NPO, compact> }

A new 3-ary relation J is defined as follows: For any license number n, color x, and size y, $Jnxy \Leftrightarrow (Cnx \ \& \ Sny)$.

1. Using the above data, find the set of all triples in J . Exhibit these triples in a relation table with column headings, *license number, color, size*.

2. A new relation R is defined by $Ryxn \ \Leftrightarrow \ Jnxy$. Make a table that exhibits all triples of R of the form, $< compact, x, n >$.

3. Find the license numbers of all cars (if any) that are compact and either red or green.

E 2-38 Let $P = \{a, b, c, d, e\}$ be a set of people named $a, b, \cdots, S = \{f, m\}$ be the set of sexes (*female, male*), and $V = \{l, r, n\}$ be a set of possible voting results, based on a fictitious exit poll, in a recent election in Northern Antarctica (l = vote for the *left party*, r = vote for the *right party*, n = *no vote*). Table 2.3 has some data from the election in the relational form, *voted(x, y, z)*.

voted(x, y, z)		
x	*y*	*z*
a	m	l
b	f	n
c	f	l
d	m	r
e	f	r

Table 2.3. *Sample Voting Results*

Using the table data, list the elements in each of the following sets:

1. $A = \{x \mid voted(x, f, l)\}$.

2. $M = \{< x, y > \mid voted(x, y, r)\}$.

3. $N = \{< x, z > \mid \text{ there is } y \text{ such that } voted(x, y, z) \ \& \ z \neq n\}$.

4. $R = \{< u, v, s > \mid \text{ there is } z_1 \text{ such that } voted(u, s, z_1)$
   $\& \text{ there is } z_2 \text{ such that } voted(v, s, z_2) \ \& \ u \neq v\}$.

5. Express the meaning of $Ruvs$ in (fairly) ordinary English.

E 2-39 The Omega Corporation is described earlier in the text. Using the information in the text and in Figure 2.4, list all of the elements of the sets M, R, S, P. Display the elements of D in a relation table.

E 2-40 There are three sets, $B_1 = \{a, b, c\}$, $B_2 = \{d\}$, $B_3 = \{e, f, g\}$. Some of the elements are connected by directed lines (with arrowheads) as shown in Figure 2.12. Let Cxy denote the connection relation, i.e., $x \rightarrow y$. For example, we have Cae and Cdb. Using this information, including all of the connections in the figure, answer the following questions. Be sure to distinguish the individual elements in $B_1 \cup B_2 \cup B_3$ from ordered pairs of elements.

1. List the elements in $B_1 \times B_2$.

2. List the elements in $(B_2 \times B_3)^{-1}$.

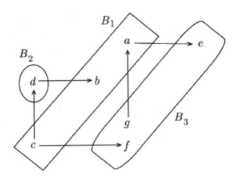

Figure 2.12. *Exercise 2-40*

3. List the elements in the *domain of the relation* C.

4. List the elements in $\{ <x,\, y> \mid <x,\, y> \in B_1 \times B_3 \ \ \& \ \ Cxy \}$.

5. Using definition by abstraction, write a *general specification* of all elements $x \in B_1$ that are such that there is some $y \in B_1 \cup (B_2 \cup B_3)$ such that Cxy or Cyx.

E 2-41 *Theorem.* Let $A \neq \varnothing$ be a set, and let $I = \{ <x, x> \mid x \in A \}$. Then $I \subseteq \mathcal{P}(\mathcal{P}(A))$. (Hint. Use Definition 2-15 of an ordered pair.)

E 2-42 (Answer Provided) *Theorem.* Let A, B, C be nonempty sets. Then

$$A \times (B \cap C) = (A \times B) \cap (A \times C).$$

E 2-43 *Theorem.* Let A, B, C be nonempty sets. If $A \subseteq B$, then

$$(A \times C) \subseteq (B \times C).$$

E 2-44 *Theorem.* Let A, B, C be nonempty sets. Then

$$A \times (B \cup C) = (A \times B) \cup (A \times C).$$

E 2-45 *Theorem.* Let A, B, C be nonempty sets. Then

$$A \times (B - C) = (A \times B) \cap (A \times C').$$

E 2-46 *Theorem.* Let $R, S \subseteq (A \times B)$ be nonempty binary relations. Then

$$(R \cup S)^{-1} = (R^{-1} \cup S^{-1})$$

and

$$(R \cap S)^{-1} = (R^{-1} \cap S^{-1}).$$

E 2-47 Let the domain be the set of real numbers, Re. For $x, y \in Re$, let xy be the product of x with y. Define binary relations on Re as follows: For any $x, y \in Re$,

$$Ixy \Leftrightarrow x = y \ (\text{i.e., } I = \{ <x, y> \mid x = y \}),$$

$$Hxy \Leftrightarrow xy = 4 \ (\text{i.e., } H = \{ <x, y> \mid xy = 4 \}).$$

1. What is $M = I \cap H$? Determine exactly what pairs are in it. Show your work.

2. Is M symmetric on Re? Give a proof or a counterexample.

3. Is $U = I \cup H$ symmetric on Re? Give a proof or a counterexample.

E 2-48 Give an example of a binary relation that is connected on a set A, but not strongly connected on A.

E 2-49 (Answer Provided for 1) *Theorem.* Let R be a binary relation on nonempty set A, and let R^{-1} be the converse of R. Then

1. If R is asymmetric on A, then R^{-1} is asymmetric on A.

2. If R is irreflexive on A, then R^{-1} is irreflexive on A.

3. If R is transitive on A, then R^{-1} is transitive on A.

4. If R is symmetric on A, then R^{-1} is symmetric on A.

E 2-50 (A typical, simple theorem about relations) *Theorem.* Let R be a binary relation on $A \neq \emptyset$. If R is transitive and irreflexive on A, then R is asymmetric on A.

E 2-51 For each of the following assertions, give either a proof that it is *true* or a counterexample that shows it to be *false*.

1. Define R by: For any $x, y \in Re$,

$$R = \{ <x, y> \mid y < x - 1 \}.$$

Then R is symmetric on Re.

2. Define R by: For any $x, y \in Re$, $Rxy \Leftrightarrow x^2 = y^2$. Then R is connected on Re.

3. Define R by: For any $x, y \in Re$, $Rxy \Leftrightarrow x^2 - 2xy + y^2 = 25$. Then R is symmetric on Re.

4. Let $x \in Re$ and $|x|$ be the absolute value of x, i.e.,

$$|x| = x \text{ if } x \geqslant 0, \text{ and } |x| = -x \text{ if } x < 0.$$

Define the relation R on Re by: Rxy iff $|x - y| \leqslant 1$. Then R is an equivalence relation on Re.

5. If E is an equivalence relation on set A, then, for any a, b, $c \in A$,

$$(Rbc \ \& \ Rab) \ \Rightarrow \ Rca.$$

6. Let R be a nonempty symmetric and transitive relation on A. Suppose that, for every $x \in A$, there exists $y \in A$, such that Rxy. Then R is also connected on A.

7. For any $x, y \in Re$, define Rxy iff $x^2 = y^2$. Then R is an equivalence relation on Re.

E 2-52 *Theorem.* Let R be a strongly connected relation on $A \neq \varnothing$. Then

1. R is reflexive on A.

2. R^{-1} is strongly connected on A.

3. R is connected on A.

E 2-53 Let R be a binary relation on the nonempty set A.

1. *Theorem.* If R is asymmetric on A, then R is irreflexive on A.

2. *Theorem.* If R is asymmetric on A, then R is antisymmetric on A.

3. Give an example of a relation that is antisymmetric but not asymmetric.

4. *Theorem.* If R is transitive and irreflexive on A, then R is asymmetric on A.

E 2-54 Let S, A, B be nonempty sets with $A \subseteq S$ and $B \subseteq S$. Define R on S by

$$Rxy \ \Leftrightarrow \ ((x \in A \ \Leftrightarrow \ y \in A) \ \& \ (x \in B \ \Leftrightarrow \ y \in B)).$$

Prove that R is an equivalence relation on S.

E 2-55 *Theorem.* Let A, B be any nonempty sets with $B \subseteq A$, and let E be an equivalence relation on A. If, for any x, $y \in A$,

$$Fxy \ \Leftrightarrow \ (x, y \in B \ \& \ Exy),$$

then F is an equivalence relation on B.

E 2-56 (Answer Provided) Let Γ be the set of *all* $\mathcal{SC}$ sentences, and for any ϕ, $\psi \in \Gamma$, define

$$T\phi\psi \quad \text{iff} \quad (\phi \text{ and } \psi \text{ are tautologically equivalent}).$$

Prove that T is an equivalence relation on Γ.

E 2-57 Let x, y, u, $v \in Nat^+$. Define R by

$$R < x, y >< u, v > \Leftrightarrow xv = yu,$$

where xv and yu are numerical products. Prove that R is an equivalence relation on $Nat^+ \times Nat^+$.

E 2-58 *Theorem.* If R and S are equivalence relations on set $A \neq \varnothing$, then $R \cap S$ is an equivalence relation on A.

E 2-59 Give an example of a binary relation R on a set A, such that R is strongly connected on A and is also transitive on A, but R is not an equivalence relation on A. Show that your example relation is not an equivalence relation on A.

E 2-60 *Definition.* Let R be a binary relation on A. R is *countertransitive* iff for every x, y, z in A, $(Rxy \ \& \ Ryz) \Rightarrow Rzx$.

Theorem. R is an equivalence relation on A iff R is reflexive and countertransitive on A.

E 2-61 Prove or give a counterexample: Let $\langle A, S \rangle$ be any relational system in which S is asymmetric and transitive. Then S is connected on A.

E 2-62 Let $\langle A, R, S \rangle$ be a relational system in which R is a binary relation on A, and $S = R \cup R^{-1}$.

1. Prove that S is symmetric on A.

2. Give a counterexample to show that S need not be reflexive on A.

3. Let I be the identity relation on A, $I = \{ < x, x > \mid x \in A \}$. Prove that $S \cup I$ is reflexive and symmetric on A.

E 2-63 By definition, a binary relation E is an equivalence relation on its domain iff E is (i) reflexive, (ii) symmetric, and (iii) transitive on its domain. Give counterexamples that show that no two of the conditions (i), (ii), (iii) imply the remaining condition.

E 2-64 Let $B = \{0, 1\}$ and $S = B \times B \times B$. Thus, S is the set of all ordered triples whose elements are 0 or 1. For example, $< 0, 1, 1 >$, $< 0, 1, 0 >$, and $< 0, 0, 0 >$ are some of the elements of S. Let R be a relation on S defined as follows: For any $x, y \in S$,

$$Rxy \text{ iff } x \text{ and } y \text{ contain exactly the same number of 0s.}$$

For example,

$$R < 0, 1, 1 >< 1, 0, 1 > .$$

1. Prove that R is an equivalence relation on S.

2. Draw the graph of R.

3. Determine all of the equivalence classes of R.

E 2-65 Let S be the set of all nonempty sets containing any of the three numbers, 1, 2, 3. Thus, S contains elements like $\{1\}$, $\{1, 3\}$, $\{1, 2, 3\}$, etc. Define a relation $\mathcal{R}$ on S as follows: For any $A, B \in S$,

$$\mathcal{R}AB \text{ iff } A \cap B \neq \varnothing.$$

State whether each of the following assertions is true or false, and justify each answer with a proof or a counterexample:

1. $\mathcal{R}$ is reflexive on S.

2. $\mathcal{R}$ is symmetric on S.

3. $\mathcal{R}$ is transitive on S.

4. $\mathcal{R}$ is an equivalence relation on S.

E 2-66 Suppose one has a domain D of toy building blocks with the following properties: There are exactly two disjoint sizes, *big*, B, and *little*, L, and exactly three disjoint colors, *purple*, P, *red*, R, and *green*, G. Define two relations on D by

$$Sxy \text{ iff } ((Bx \ \& \ By) \text{ or } (Lx \ \& \ Ly)),$$

$$Cxy \text{ iff } ((Px \ \& \ Py) \text{ or } (Rx \ \& \ Ry) \text{ or } (Gx \ \& \ Gy)).$$

Using the definitions and given assumptions

1. Briefly describe the meanings of S and C.

2. Prove that S and C are equivalence relations on D.

3. Prove that $S \cap C$ is an equivalence relation on D.

4. Briefly describe the kinds of partition cells that are generated by $S \cap C$. What is the maximum number of cells that might occur? Why?

E 2-67 Let S be a set and P be a nonempty set of nonempty subsets of S. Define E_P on S as follows: For any x, y in S,

$$E_P xy \Leftrightarrow (\text{for every } A \in P \ (x \in A \Leftrightarrow y \in A)).$$

1. Describe an example in which S is a set of at least ten uniformly colored marbles and in which P has nonempty sets for red, green, blue, large, and small marbles. Large and small are disjoint, and any two different colors are disjoint. Describe in words the meaning of E_P for your example. Briefly explain why E_P is an equivalence relation on S, and describe the cells of the partition of S that it determines.

2. *Theorem* Let S be a set and P be a nonempty set of nonempty subsets of S. Let E_P be the relation defined above. Then E_P is an equivalence relation on S.

   (Remark. This theorem yields a very general method for *classifying* the elements of S according to the P-subsets they have in common. Two elements of S are of the same *type* or *kind* iff they fall into exactly the same P-subsets of S. See Causey (1977), Chapter 3, for generalizations and applications.)

E 2-68 *Theorem.* If $\langle A, R \rangle$ is a strict simple order, then $\langle A, R^{-1} \rangle$ is also a strict simple order.

E 2-69 *Theorem.* If $\langle A, R \rangle$ is a partial order, then $\langle A, R^{-1} \rangle$ is a partial order.

E 2-70 Define R on $Nat \times Nat$ as follows: For any x, y, u, v in Nat,

$$R < x, y > < u, v > \text{ iff } (x \leqslant u \ \& \ y \leqslant v).$$

R is a binary relation between pairs. It might help to think of R as a relation between points on the Cartesian coordinate system; then you can draw a sketch of how it appears.

1. Prove that R is a partial order on $Nat \times Nat$.

2. Give a counterexample to the assertion that R is an equivalence relation on $Nat \times Nat$.

3. Give a counterexample to the assertion that R is a simple order on $Nat \times Nat$.

E 2-71 (Answer Provided) *Theorem.* Let $\langle A, L \rangle$ be a strict simple order. Then

1. L is irreflexive on A.

2. For any $x, y \in A$, one and only one of the following holds: $x = y$, Lxy, Lyx.

(Remark. The second statement is called the *trichotomy law.* The fact that L is asymmetric, and hence irreflexive, together with connectedness, will help to prove the trichotomy law.)

E 2-72 *Theorem.* Let $\langle A, L \rangle$ and $\langle A, M \rangle$ be strict simple orders with $L \subseteq M$. Then $L = M$.

E 2-73 Let $A = \{a, b, c, \cdots, x, y, z\}$ be the set of letters in the English alphabet. Let P (*precedes*) be the usual alphabetical ordering of these letters, so that Pab, Pbe, Peg, Pdq, $\cdots$, etc. Notice that $\langle A, P \rangle$ is a strict simple order. Now let Greek letters, α, β, etc., stand for letters in A, and let

$$W = \{\alpha\beta \mid \alpha, \beta \in A\},$$

where $\alpha\beta$ denotes the result of writing the letter α followed by the letter β. Thus, W is the set of all two-letter "words" (strings), such as 'ab', 'ee', 'zx', etc.

Now define L on W as follows:

If $word_1, word_2 \in W$, and $word_1 = \alpha\beta$ and $word_2 = \gamma\delta$, then

$Lword_1word_2$ iff $(P\alpha\gamma$ or $(\alpha = \gamma$ & $P\beta\delta))$.

Use the L relation to sort the following list of two-letter "words" into alphabetical ordering:

lm, ma, an, bk, zz, ab, bx, jj, mn, it, to, aa, ij, xb, ik, ji, zy, ac, mm.

(Remark. After doing this problem, you may want to look at the next one.)

E 2-74 Let A, P, W, and L be the same as in the preceding exercise. Prove that $< W, L >$ is a strict simple order. (Remark. The proof requires considering several cases, so please have patience with it. L is called the *lexicographic ordering* of W. It can be generalized to words with n letters. This ordering is widely used in mathematics and computing theory. It is also used to order the words in dictionaries, phone directories, etc.)

E 2-75 *Theorem.* Let $\langle A, <_A \rangle$ and $\langle B, <_B \rangle$ be strict simple orders. Define L on $A \times B$ as follows: For any $< x, y >$ and $< u, v >$ in $A \times B$,

$$L < x, y >< u, v > \quad \text{iff} \quad (x <_A u \text{ or } (x = u \And y <_B v)).$$

Then $\langle A \times B, L \rangle$ is a strict simple order. (Remark. This theorem is an abstract generalization of the previous exercise. For example, suppose that we have two ordered sequences (or arrays) of numbers. This theorem provides one method for ordering their Cartesian product.)

E 2-76 (Application example) Let $\langle D, Q \rangle$ be a relational system in which Q is reflexive and transitive on D. Such a system is called a *quasi order*. Every partial order is a quasi order but, in addition, partial orders are antisymmetric. Since they are not required to be antisymmetric, quasi orders are more convenient for representing some kinds of data than are partial orders. In psychological and economic theories, a quasi order is sometimes used to represent a person's preferences among a set of choices.

Given a quasi order, we can define, for x, y in D,

$$Ixy \Leftrightarrow (Qxy \And Qyx).$$

The relation I is called the *indifference relation* of Q.

1. Use this definition of I together with the reflexivity and transitivity of Q to prove that I is an equivalence relation on D.

2. Suppose a restaurant has nine items on its menu, $d_1, \cdots, d_9$. Let Pxy represent that Bob prefers menu item y to item x. Bob expresses the following preferences:

$$Pd_1d_2, \; Pd_2d_4, \; Pd_4d_2, \; Pd_2d_3, \; Pd_3d_5, \; Pd_5d_7,$$

$$Pd_7d_5, \; Pd_6d_7, \; Pd_6d_8, \; Pd_8d_6, \; Pd_6d_9.$$

In addition to this data, the only other information we have is that P is a quasi order.

2a. Sketch the graph of Bob's preferences, and indicate the equivalence classes on this graph.

2b. What are the equivalence classes corresponding to I?

2c. Which, if any, is true: Pd_4d_6, Id_4d_6, or Pd_6d_4?

2d. Which, if any, is true: Pd_1d_7, Id_1d_7, Pd_7d_1?

Give reasons for your answers.

2.3 Functions

2.3.1 Basic Ideas

Many people are familiar with functions that are specified by procedures for calculating some numerical result. We also want to consider functions that operate on strings of characters, lists of objects, sets of sentences of a language, and sets of elements of any kind. In set theory, and in mathematics more generally, the following has become accepted as the standard definition:

Definition 2-33 Let A and B be sets. A *function* f from A to B, denoted by $f : A \longrightarrow B$, is a relation between A and B (a subset of $A \times B$) such that

(i) For any $x \in A$, there exists $y \in B$, such that $< x, y > \in f$, i.e., fxy,

and

(ii) If $x \in A$ and $y, z \in B$, and fxy and fxz, then $y = z$.

A function f is often called a *mapping* or a *map* of A to B. A is called the *domain* of f and B is the *codomain* of f. If $x \in A$, $y \in B$, and fxy, then x is said to *map* to y, and y is the *image* of x under f. Also, x is called a *preimage* of y under f.

The arrow in $f : A \longrightarrow B$ is very different in meaning from the arrow used in sentential calculus, as in $(P \to Q)$. Please do not confuse them. Saying that B is *the* codomain of f is misleading, since codomains are not unique. This will be explained in detail shortly. First, observe that condition (i) is an existence condition. It guarantees that, for any x in A, there exists at least one relative y in B; in other words, that fxy holds. For this reason, it is often said that the function f is *defined* on A. Condition (ii) is a uniqueness condition. It guarantees that for any x in A, there is at most one relative in B. This is expressed by saying that if fxy and fxz are both true, then y and z must be the same element of B. Thus, x has a unique relative in B.

Consider the relation Bxy, which abbreviates 'y is a brother of x'. We may have $B \, bob \, jim$ and also $B \, bob \, bill$, so B is not a function, since *bob* has more than one brother. Contrast this with Mxy, which abbreviates 'y is the (biological) mother of x'. Then $M \, bob \, y$ holds for one and *only one* y. Since there is exactly one mother of Bob, we may denote her by a special term, such as $mother(bob)$ or even $M(bob)$. In general, if f is a function, instead of writing fxy, we may write

$$y = f(x).$$

It is important to understand that this *functional notation*, $f(x)$, is justified only because there is a unique f-relative of x when f is a function.

How does one prove that a given relation, f, is a function? One proves that it satisfies (i) and (ii) of Definition 2-33. Unfortunately, there is no one way to do this, because the proof will depend on the particular relation with which one is working. In most cases, proving (i) is easy; proving (ii) may require some effort or ingenuity.

Examples.

1. Let $A = Re = B$, and consider the relation

$$fxy \Leftrightarrow y = x^2.$$

Let x, $y \in Re$ and suppose fxy & fxz. Then we have

$$y = x^2 \quad \& \quad z = x^2,$$

so $y = z$. Hence, f is a function from Re to Re.

2. Let $A = Re = B$, and consider the converse relation

$$f^{-1}xy \Leftrightarrow fyx \Leftrightarrow x = y^2.$$

We know that the square of a real number is nonnegative, so $x \geqslant 0$. Thus, the domain of f^{-1} is the set of nonnegative reals. So f^{-1} cannot be a function defined on Re. Is it a function on the nonnegative reals?

Since $4 = 2^2 = (-2)^2$, $f^{-1}42$ and $f^{-1}4 -2$ are both true, i.e., $< 4, 2 >\in$ f^{-1} and $< 4, -2 >\in f^{-1}$. So, 4 does not have a unique f^{-1}-relative, so f^{-1} is not a function from the nonnegative reals to Re.

3. Let W be the set of words in an English-language dictionary. For any word, w, let $g(w)$ denote the number of characters (including hyphens) in w. Then g is a function from W to $Nat = \{0, 1, 2, \cdots\}$, since each word contains a definite number of characters. However, many words map into the same number, e.g., $g(cat) = g(dog) = g(the) = g(rat) = 3$. Consequently, the converse of g is not a function. Also, notice that some numbers in the codomain have no words mapped to them. For example, fortunately, there are no English words with 1,000,000 characters. Thus, the function never takes the value 1,000,000. ■

Let f map Rat^+ to Rat^+, where $f(x) = x/2$. Clearly,

$$f \subseteq Rat^+ \times Rat^+,$$

so by Definition 2-33, "the" codomain is Rat^+. Yet $Rat^+ \subseteq Re^+$, so $f \subseteq Rat^+ \times Re^+$. We can also consider f to be a function from Rat^+ to Re^+, in which case "the" codomain is Re^+. As a result, a function does not have a unique codomain. For this reason, some logicians do not like to use this terminology. Nevertheless, 'codomain' is widely used in the mathematical literature, so I will use it also. Once we have described f as a function from A to B, by convention we will call B the codomain, even though other sets, of which B is a subset, could have been used.

Definition 2-34 If f is a function from A to B, the *image of A under f* is

$$Img(f, A) = \{\, y \mid \text{there exists } x \in A \text{ such that } y = f(x) \,\}.$$

If $C \subseteq A$, we also define

$$Img(f, C) = \{\, y \mid \text{there exists } x \in C \text{ such that } y = f(x) \,\}.$$

The equation defining $Img(f, A)$ is often written in the simpler form,

$$Img(f, A) = \{\, f(x) \mid x \in A \,\}.$$

This use of the term *image* is different from that in the previous definition.* If y is an element of the codomain, then $y \in Img(f, A)$ iff there is some x in the domain such that f maps x to y. In this case, from the preceding definition, it is also true that y is the image of x under f, and x is a preimage of y. There may be more than one such x; in the previous dictionary example, *cat, dog, the, rat* all map to 3. It is often said that 3 can be *pulled back* to *cat, dog,* etc., which means that *cat, dog,* etc., are all preimages of 3. For any function f, the image $Img(f, A)$ is always a subset of the codomain. It often does not equal the entire codomain. In the dictionary example, $Img(g, W)$ is only a relatively small proper subset of *Nat*. A typical domain A, codomain B, and image $Img(f, A)$ are sketched in Figure 2.13. In this diagram, the image is a proper subset of the codomain.

Definition 2-35 A function $f : A \longrightarrow B$ is *onto B* (*maps onto B*) iff for every $y \in B$, there exists $x \in A$, such that $y = f(x)$. A function f that maps onto its codomain B is often called an *onto mapping*, and such a mapping may be denoted by

$$f : A \xrightarrow{\;onto\;} B.$$

---

*Mathematical writings often use $f(A)$ for what I write as $Img(f, A)$. But f could be defined on a set of sets S, and we might have $A \in S$, then $f(A)$ would be the value returned when A is the argument to f. Ambiguity results if we then also use $f(A)$ to denote the image of the set A under a function defined on A, so it is better not to use $f(A)$ for the image of the set A under f. Some other notations are used, but they can be difficult to remember, or easily confused with other symbolism. Also note that the term *range of f* is often used for $Img(f, A)$.

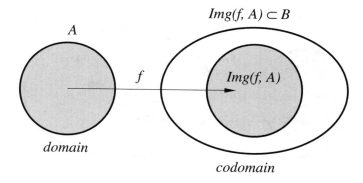

Figure 2.13. *Image of a Function*

If f maps A onto B, then every $y \in B$ can be pulled back to one or more elements of A. Thus, $Img(f, A) = B$, and we can say that $Img(f, A)$ *covers* B. A picture of a mapping onto the codomain B is given in Figure 2.14. The fact that this mapping is onto B does not prevent more than one element of A from being mapped to the same element of B.

Definition 2-36 A function $f : A \longrightarrow B$ is *one–one* iff for every x_1, $x_2 \in A$, if $f(x_1) = f(x_2)$, then $x_1 = x_2$. Such a function may also be called a one–one (or 1–1) mapping and be denoted by

$$f : A \xrightarrow{one-one} B$$

or by

$$f : A \xrightarrow{1-1} B.$$

The contrapositive of 'if $f(x_1) = f(x_2)$, then $x_1 = x_2$' is the statement, 'if $x_1 \neq x_2$, then $f(x_1) \neq f(x_2)$'. Therefore, if f is one–one, distinct elements of A are mapped to distinct elements of B. This also implies that if $z \in Img(f, A)$, then z can be pulled back to exactly one element, say a, in A, because no element of A other than a is mapped to z. Figure 2.15 schematically represents a one–one function that does not map onto B.

Examples

1. Let $A = Nat$ and

$$B = \{\, y \mid \text{there exists } k \in A \text{ such that } y = 2k \,\}.$$

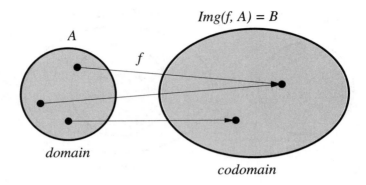

Figure 2.14. *f Maps onto the Codomain*

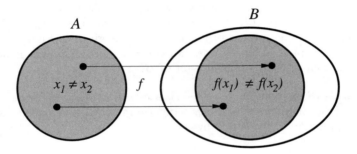

Figure 2.15. *A One–One Function*

B is the set of all even natural numbers, $\{0, 2, 4, \cdots\}$. Define f on A as follows: For every $x \in A$, $f(x) = 2x$. Thus, $f(0) = 0$, $f(1) = 2$, $f(2) = 4$, etc.

Suppose that $y \in B$. Then, by the definition of B, there exists some $k \in A$ such that $y = 2k$. Hence, y can be pulled back to the preimage k, i.e., $y = 2k = f(k)$. Therefore, f maps A onto B and $B = Img(f, A)$.

Now let $u, v \in A$ and suppose that $f(u) = f(v)$. Then, from the definition of f, we have $2u = 2v$. Whence, by elementary algebra, we obtain $u = v$. Therefore, if $f(u) = f(v)$, then $u = v$, so f is one–one from A to B. Since f is both 1–1 and onto, we can write

$$f : A \xrightarrow[onto]{1-1} B.$$

This simple function maps A both one–one and onto B. Such a function establishes a one-to-one correspondence between its domain and its codomain, i.e., each element of the domain can be matched with exactly one element of the codomain, and vice versa. This example shows that it is possible to have a one-to-one correspondence between the elements of a set A and the elements of a proper subset B of A. This is possible because both A and B are infinite sets; systematic study of the properties of infinite sets is beyond the scope of this book.

2. Let $f(x) = 2x$ as before, but now consider f as a function from *Nat* to *Nat*, so that *Nat* is now both the domain and the codomain. Then f is one–one as before. Does f map onto *Nat*? To show that it does not, we must show that there is at least one element of the codomain, *Nat*, that cannot be pulled back to an element of the domain, which is also *Nat*. In other words, we must show that the image is a proper subset of the codomain. The number 3 is in the codomain, but $3 \notin Img(f, Nat)$, since there is no number $k \in Nat$ such that $2k = 3$. Thus, f does not map *Nat* onto *Nat*.

3. Define h on *Nat* by

$$h(x) = \begin{cases} 1 \text{ if } x \text{ is odd,} \\ 0 \text{ if } x \text{ is even.} \end{cases}$$

This elaborate expression means that the value of $h(x)$ is 1 if x is odd, and the value of $h(x)$ is 0 if x is even. With the understanding that $x \in Nat$, x must be even if it is not odd, so we can restate this definition in the following, more compact format:

$$h(x) = [\text{ if } x \text{ is odd then 1 else 0 }].$$

I will say that such an equation is expressed in the *if-then-else format*. The expression on the right-hand side of the identity sign says that $h(x)$ has, or *returns*, the value 1 if x is odd, and it returns the value 0 otherwise. The purpose of the square brackets is to group together the right-hand side of the defining equation. They have no special meaning; they are just used to make the equation a little easier to read. Both forms of definition mean the same thing; they are just different notations. Notice that h maps onto $\{0, 1\}$, but is not one–one.

4. Let the domain and codomain both be *Re*, and consider the absolute value function, $|x|$, that satisfies

$$|x| = [\text{if } x \geqslant 0 \text{ then } x \text{ else } -x].$$

Then $|5| = |-5|$, so $|x|$ is not one–one. Also, $|x|$ never returns a negative value, so it does not map onto *Re*. ■

To prove that a given relation is a function, one must prove that it satisfies the definition of a function. Similarly, to prove that a given function is onto, or that it is one–one, it is necessary to prove that it satisfies the corresponding definition. There is no one way to accomplish these goals because the details of the proofs will usually depend heavily on the characteristics of the domains and functions with which one is working. Sometimes it is possible to prove a general result that applies to an entire class (i.e., set) of functions. Here is a typical theorem of this kind. To prove the result, we assume the elementary algebra of the real number system. We will say that a *linear function f* on the reals is one satisfying $f(x) = ax + b$, where a, b are particular (constant) real numbers and $a \neq 0$.

Theorem 2-37 Any linear function $f : Re \to Re$ is one–one and maps onto *Re*.

Proof.

1. I will first prove that it is one–one. Let $f(x) = ax + b$, for all $x \in Re$, where $a \neq 0$, b are particular real numbers. Let u, $v \in Re$ and suppose that $f(u) = f(v)$. Then

$$f(u) = au + b,$$

and

$$f(v) = av + b,$$

so

$$au + b = av + b.$$

Hence,

$$au = av.$$

But $a \neq 0$, so we can divide both sides of this equation by a. Alternately, since $a \neq 0$, $1/a$ exists, and we can multiply both sides of the equation by $1/a$. These two procedures give the same result, namely,

$$u = v,$$

so f is one–one.

2. Now let $u \in Re$. To prove that f maps onto *Re*, we must show that there exists $x \in Re$ such that $f(x) = u$. In other words, there must be a solution to

$$f(x) = ax + b = u.$$

But, since $a \neq 0$, we know from elementary algebra that a solution exists, namely,

$$x = \frac{u - b}{a},$$

so f maps onto Re. **Q.E.D.**

The Cartesian coordinate system plot of $ax + b$ is a straight line with slope a. The condition $a \neq 0$ is necessary in order to complete the previous proof. This is no accident. If we allow $a = 0$, then the function reduces to $f(x) = b$, which corresponds to a straight line of zero slope. This function always returns the value b, so is definitely not one–one!

This is a good place to observe that functions play an important role in the algebraic manipulation of equations. Suppose that $f : A \longrightarrow B$. Let $u, v \in A$ and also suppose that $u = v$. In other words, u and v are two terms for the same element of A. Then we have, for some $y, w \in B$,

$$fuy \text{ and } fvw.$$

But, because $u = v$, from the definition of *function*, it follows that $y = w$. In other words: If f is a function and is defined for u, v, and if $u = v$, then we may infer that $f(u) = f(v)$. The result of applying the function to both sides of the equation is another equation.

There is another way to get this result: Again suppose that $u = v$. As f is a function, the notation $f(u)$ denotes a unique object. Everything is identical to itself, so $f(u) = f(u)$. Since $u = v$, we can substitute an occurrence of v for an occurrence of u in an algebraic expression. Substituting v for u in the right side of $f(u) = f(u)$ yields $f(u) = f(v)$. It may be useful to compare this discussion with the related presentation of the laws of identity in the Introduction, Section 0.3.3.

2.3.2 Compositions and Inverses

On the set of positive real numbers, Re^+, let $f(x) = 5x + 1$. Then

$$Img(f, Re^+) =$$

$$\{ y \mid \text{there exists } x \in Re^+ \ \& \ y = 5x + 1 \} =$$

$$\{ 5x + 1 \mid x \in Re^+ \}.$$

If $u \in Img(f, Re^+)$, then $1 \leqslant u$. On Re, let $g(x) = x^3$, so the domain of g is Re. Thus, $Img(f, Re^+)$ is a subset of the domain of g. So, g returns a value for every element in the image of Re^+ under f, and we can define a new function h: for any $x \in Re^+$,

$$h(x) = g(f(x)) = (5x + 1)^3.$$

We have applied one function, g, to the values returned by another function, f. This is often useful and motivates the next definition.

Definition 2-38 Let

$$f : A \longrightarrow B$$

and

$$g : C \longrightarrow D,$$

with

$$Img(f, A) \subseteq C.$$

The *composition of g with f* is denoted by $g \circ f$ and is defined by: For any $x \in A$,

$$g \circ f(x) = g(f(x)).$$

This definition must be *justified*, for it would not make any sense to treat $g \circ f$ as a function if it is not a function. But it is easy to see that the composition really is a function from A to D. Let $x \in A$. Then

$$f(x) \in Img(f, A) \subseteq C,$$

so $f(x) \in C$. Therefore, $g(f(x))$ is defined and is an element of D. This shows that the first part of the definition of a function is satisfied. Now suppose that $u, v \in A$ and that $u = v$. Then, in so far as f is a function, $f(u) = f(v)$. Then, since g is a function, $g(f(u)) = g(f(v))$, so $g \circ f$ satisfies the second part of the definition of a function. The composition, $g \circ f$, of two functions is illustrated schematically in Figure 2.16.

Suppose that we have three functions,

$$f : A \longrightarrow B,$$

$$g : C \longrightarrow D,$$

$$h : E \longrightarrow F,$$

and that

$$Img(f, A) \subseteq C,$$

$$Img(g, C) \subseteq E.$$

For any $x \in A$, we have

$$(h \circ (g \circ f))(x) = h((g \circ f)(x)) = h(g(f(x))),$$

and

$$((h \circ g) \circ f)(x) = (h \circ g)(f(x)) = h(g(f(x))).$$

Hence, we may write an identity statement relating the functions

$$h \circ (g \circ f) = (h \circ g) \circ f.$$

This shows that composition satisfies an associative law. But what exactly is composition? Suppose that we have four sets, A, B, C, D, as in Definition 2-38. Let S_1 be the set of all functions from A to B with images of A contained in C, and let S_2 be the set of all functions from C to D. If $f \in S_1$ and $g \in S_2$, then $g \circ f$ is a function from A to D. In other words, the *composition operator*, $\circ$, denotes a function that maps a pair of functions, f, g to a new function, $g \circ f$. The composition function accepts two functions as inputs, and returns a function as its output. Since any function is a relation, and any relation is a set, the composition operator is just another, rather complicated, kind of set.

There are some computer programming languages, for example, LISP, in which the programs consist largely of definitions of functions and the application of these functions to various kinds of data. In such languages, it is standard practice to use some previously defined functions in the definitions of new functions, and such definitions often use composition of functions. In fact, in LISP one can define a function, let us call it "compose," that behaves just like our $\circ$. LISP uses prefix notation, so in LISP, (compose g f) would be the same as $g \circ f$, where f, g are any functions with domains and images such that their composition is defined.

We now turn to the important topic of inverse functions. It has already been seen that, in general, the converse relation of a function is not a function. For instance, the converse of $f(x) = x^2$, on the domain Re, is not a function. In special cases, the converse is a function, which is called the *inverse function*. Consider $h(x) = 5x$ on Re. Then, for x in the image, we have $h^{-1}xy$ iff hyx, so $h^{-1}xy$ iff $x = 5y$. Suppose that we have $h^{-1}xu$ and $h^{-1}xv$. Then $x = 5u$ and $x = 5v$, so $5u = 5v$, whence, $u = v$, so h^{-1} is a function. The key difference between f and h is the fact that f is not one–one, whereas h is. The following definition, lemmas, and theorem state basic facts about inverse functions.

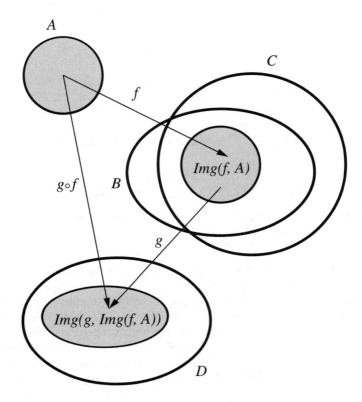

Figure 2.16. *Composition of Two Functions*

Definition 2-39 Let f be a function from A to B. A function h from

$$Img(f,\ A) \subseteq B$$

to A is an *inverse* of f iff for every $x \in A$,

$$h \circ f(x) = x.$$

According to this definition, an inverse of a function f need not be defined on the entire codomain of f. Of course, if f also maps its domain onto its codomain, then an inverse will be defined everywhere on the codomain. Suppose that $y \in Img(f,\ A)$. Then there exists $x \in A$ such that $y = f(x)$. Hence,

$$f(h(y)) = f(h(f(x))).$$

By the definition of inverse, we have $h(f(x)) = x$, so we obtain

$$f(h(y)) = f(x) = y.$$

Thus, we have *both*: $h \circ f(x) = x$ and $f \circ h(y) = y$. It also turns out that if f has an inverse, then it has only one. Before proving this fact, we need a lemma about identity of functions.

Lemma 2-40 Let $g : A \longrightarrow B$ and $h : A \longrightarrow C$. Then

$$g = h \iff \text{ for every } x \in A, \; g(x) = h(x).$$

Proof. The functions g and h are each sets of ordered pairs; $g \subseteq A \times B$ and $h \subseteq A \times C$. So, $g = h$ iff the set of pairs that is g is the same as the set of pairs that is h.

1. Suppose first that $g = h$, and let $x \in A$. If $y = g(x)$, then gxy, so also hxy, so $y = h(x)$. Therefore, $g(x) = h(x)$.

2. Conversely, suppose that $g(x) = h(x)$, for all $x \in A$. Then

$$Img(g, A) = \{ \, g(x) \mid x \in A \, \} = \{ \, h(x) \mid x \in A \, \},$$

so the image of A under g is the same as the image of A under h. Let I be this image. For any $x \in A$, $y \in I$,

$$hxy \iff y = h(x).$$

Since

$$g(x) = h(x),$$

also

$$hxy \iff y = g(x),$$

so

$$hxy \iff gxy.$$

Therefore, $h = g$. **Q.E.D.**

Special cases of this lemma should be familiar from elementary algebra. For example, on the real numbers the function represented by

$$(x - 1)(x - 2)$$

is identical to the function represented by

$$x^2 - 3x + 2,$$

since they each return the same output value for any input value of x.

Lemma 2-41 Let $f : A \longrightarrow B$. If there exists an inverse function of f, then this inverse is unique.

Proof. As usual, to prove uniqueness we assume that there are two such things, then prove that they must be the same (identical). Suppose that both g and h are inverses of f. We need to prove that they are the same function. Let

$$y \in Img(f, A),$$

so it can be pulled back to some element in A, i.e., there is some $x \in A$ such that $y = f(x)$. Then, from the definition of inverse function, we have

$$g(y) = g(f(x)) = g \circ f(x) = x,$$

and

$$h(y) = h(f(x)) = h \circ f(x) = x.$$

Hence, for any $y \in Img(f, A)$,

$$g(y) = h(y),$$

so $g = h$ by the previous lemma. ***Q.E.D.***

Because an inverse of f is unique, we may now call it *the* inverse of f. The next lemma characterizes this inverse.

Lemma 2-42 Let $f : A \longrightarrow B$. If there exists an inverse function of f, then this inverse is a one–one function and maps $Img(f, A)$ onto A.

Proof.

1. [onto] Let g be an inverse of f, and $x \in A$. Then $f(x) \in Img(f, A)$, so $g \circ f(x) = x$. Hence, any element $x \in A$ can be pulled back to an element $f(x) \in Img(f, A)$ such that g maps $f(x)$ to x, so g maps onto A.

2. [one–one] Now let $u, v \in Img(f, A)$ and suppose that $g(u) = g(v)$. We need to show that $u = v$. Since g is the inverse of f, $g(u), g(v) \in A$. Since f is a function on A, we may infer that $f(g(u)) = f(g(v))$. But since g is the inverse of f, we obtain $u = v$ (by the discussion following Definition 2-39), so g is one–one on $Img(f, A)$. ***Q.E.D.***

We now come to the theorem relating the one–one property to the existence of an inverse function.

Theorem 2-43 Let $f : A \longrightarrow B$. Then f has an inverse function iff f is one–one. Moreover, if the inverse exists, then it is identical to the converse relation f^{-1}.

Proof.

1. [Necessity] It is easy to see that the one–one condition is necessary for the existence of an inverse. Suppose that f has the inverse function g. Let x, $y \in A$ and assume that $f(x) = f(y)$; we need to show that $x = y$. Since $f(x) = f(y)$ and g is a function, we also have $g(f(x)) = g(f(y))$. But g is the inverse of f, so we also have

$$x = g(f(x)) = g(f(y)) = y,$$

so f is one–one.

(*Remark.* This theorem is a good example of an *existence theorem* in mathematics. The preceding argument establishes a necessary condition for the existence of something, in this case, an inverse function. Typically, proving such a necessary condition is relatively easy. Proving that a condition is sufficient for the existence of something usually requires more work. Given the sufficient condition, we must somehow construct the entity in question (or at least prove by some argument that it exists) and show that it has the requisite features. Fortunately, for the current problem, the previous examples suggest that the converse relation, f^{-1}, is the inverse function of f when f is one–one.)

2. [Sufficiency] Let f be one–one. I claim that this is sufficient for the existence of an *inverse function*, f^{-1}. To avoid confusion, I will temporarily denote the *converse relation* of f by f^c. I now claim that f^c is the desired inverse function f^{-1} (which is unique by Lemma 2-41). In order to prove this, we must show that f^c is a function and that it satisfies the definition of an inverse of f. Let $v \in A$ and $u \in Img(f, A)$.

   a. [f^c is a function] From the definition of converse relation, we have

$$f^c uv \iff fvu.$$

Since f is a function, we can rewrite this as

$$f^c uv \iff u = f(v).$$

Now, if u is any element in $Img(f, A)$, it can be pulled back to some $v \in A$ such that $u = f(v)$, so also $f^c uv$. Therefore, f^c satisfies part (i) of the definition of a function from $Img(f, A)$ to A.

Now I show that f^c also satisfies part (ii) of the definition of a function, i.e., that f^c relates any element in $Img(f, A)$ to a *unique* element of A. As usual, we prove uniqueness of an object by assuming that there could be two such objects and then proving that these two possible objects are actually identical.

Let v_1, $v_2 \in A$, and suppose that $f^c u v_1$ and $f^c u v_2$. By the definition of converse relation and the fact that f is a function, we have that $u = f(v_1)$ and $u = f(v_2)$, so $f(v_1) = f(v_2)$. To show that f^c is a function, we still have to show that $v_1 = v_2$. But since $f(v_1) = f(v_2)$, and f is one–one on A, $v_1 = v_2$. Thus, $u \in Img(f, A)$ has exactly one relative in A, so f^c is a function from $Img(f, A)$ to A.

b. [f^c is the inverse of f] There could be many functions from $Img(f, A)$ to A, so we still need to show that f^c is the inverse of f. Since it has been proved that f^c is a function, we may now use functional, as well as relational, notation for it. Let x be any element of A and suppose that $y = f(x) \in Img(f, A)$. Then we have $f x y$, so also $f^c y x$, and hence $x = f^c(y)$. Substituting $f(x)$ for y, we obtain $x = f^c(f(x))$. So, for every $x \in A$,

$$f^c \circ f(x) = x.$$

Consequently, f^c satisfies Definition 2-39 of an inverse function, and since the inverse of a function is unique, it is *the inverse* of f. Therefore, the converse relation f^c is the inverse function f^{-1}.   **Q.E.D.**

Here are two more useful definitions relating to functions. First, an *identity function* maps each element of a set to itself:

Definition 2-44 Let A be a nonempty set. The *identity function* on A is $i_A : A \longrightarrow A$ given by $i_A(x) = x$ for all $x \in A$.

Notice that i_A is one–one and maps A onto A. Suppose that

$$f : A \xrightarrow[onto]{1-1} B.$$

If $x \in A$, then

$$i_A(x) = x = f^{-1} \circ f(x).$$

If $y \in B$, then

$$i_B(y) = y = f \circ f^{-1}(y).$$

These observations can be expressed by simple *functional equations*,

$$f^{-1} \circ f = i_A,$$

and

$$f \circ f^{-1} = i_B,$$

that state identity relations between symbols denoting functions.

The next definition is very useful in the study of various mathematical systems, including abstract algebraic systems.

Definition 2-45 Let A be a nonempty set, and let A^k be the set of all ordered k-tuples of elements of A, i.e., the kth–dimensional Cartesian product of A. Let $f : A^k \longrightarrow A$ and let $B \subseteq A$. B is said to be *closed under* f iff: For any k-tuple, $\mathbf{b} = <b_1, \cdots, b_k> \in B^k$, $f(\mathbf{b}) \in B$.

For example, let A be Re, and $p : Re \times Re \longrightarrow Re$, where $p(x, y)$ is the product of x and y. Let $B = \{ x \mid -5 \leqslant x \leqslant 5 \}$. Then $p(-4, -3) = 12 \notin B$, so B is not closed under p. On the other hand, the integers, $Int \subseteq Re$, and if $i, j \in Int$, then $p(i, j) \in Int$. Hence, Int is closed under p. Of course, we are drawing these conclusions from our basic knowledge of arithmetic. The closure properties of a set under various functional mappings can become a complicated issue.

Examples.

Problem Statement. For each of the following descriptions of relations,

$$f \subseteq A \times B,$$

state whether the relation is a function. If it is, state whether it is one–one and whether it maps onto its codomain. Justify your answers. In each case, $x \in A$ and $y \in B$.

1. A is the set of books and documents in the U.S. Library of Congress. Assume that there is at least one pair of duplicate books, i.e., two books that are identical in printing, paper, binding, etc. B is Re^+. $fxy : y$ is the numerical value of the weight of x measured in grams.

2. A and B are each sets of uniformly colored marbles. A and B each have 20 marbles. There are 5 different colors among the marbles in A and 6 different colors among the marbles in B. A and B each have at least two blue marbles. $fxy : y$ has the same color as x.

3. $A = B = \{ a, b, c, d \}$, the set of nodes in Figure 2.17. If x, y are nodes, fxy iff there is one arc starting at x and ending at y.

4. A, B, and f are the same as in the previous question, except that the graph is changed to Figure 2.18.

5. Now let $A = B = \{a, b, c\}$, and change the graph of f to Figure 2.19.

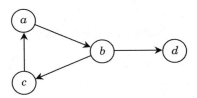

Figure 2.17.

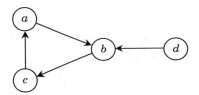

Figure 2.18.

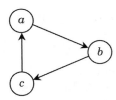

Figure 2.19.

Discussion. These questions simply require applying definitions to the given information. The discussions are included in the written solutions.

Written Solutions.

1. Every book or document has some unique weight, so f is a function. In order for f to be one–one, for every x, y in A, $f(x) = f(y) \Rightarrow x = y$. Since we assume that there is a pair of duplicate books, each book in this pair should have the same weight. Say that these two duplicate books are a and b. Then,

$f(a) = f(b)$ & $a \neq b$, so f is not one–one. The function f also does not map onto B, since

$$10^{1,000,000,000,000} \in Re,$$

and we know that no book in the Library of Congress weighs that much, even in grams. Clearly, we are making use here of some nonmathematical, factual information, so this is a bit of applied mathematics.

2. Let a_1, a_2 be two blue marbles in A and b_1, b_2 be two blue marbles in B. Using relational notation, i.e., writing f as a predicate denoting a relation, we have fa_1b_1 and fa_1b_2. Therefore, $a_1 \in A$ does not have a unique relative in B, so f is not a function. Notice that only one blue marble in A is enough to prevent f from being a function.

x	$f(x)$
a	b
b	c
c	a
d	b

Table 2.4. *Relation Table for Figure 2.18*

3. The node d has no relative, so even if f were a function, it would not be defined *on* A, i.e., A could not be the domain of f since f does not satisfy part (i) of Definition 2-33. In addition, node b has two relatives, so part (ii) of Definition 2-33 is also violated. This graph does not represent a function.

4. The graph has been changed to Figure 2.18. Let us display its data in a relation table, Table 2.4. Now f returns an output value for each input value in A, so condition (i) is satisfied. (In the graph, every node has some arc originating at that node.) Also, each input value has a unique output value, so condition (ii) is satisfied. (In the graph, no node has two or more arrows originating at the node.) Therefore, f is a function on A. But $Img(f, A) = \{a, b, c\} \neq A$, so the image is only a proper subset of the codomain, A. (One can see in the graph that node d has no arc entering it.) Thus, f does not map onto A. Also, since $f(a) = f(d)$, f is not one–one.

5. This case is very simple: f maps A one–one and onto A. In fact, f is a *permutation* function on A; see E 2-90. ■

Example.

Problem Statement.

1. *Theorem.* Let f be any one–one function from Re^+ onto Re^+. For any $x \in Re^+$, define

$$h(x) = \frac{a}{f(x)},$$

where the constant $a \in Re^+$, i.e., a is a fixed positive real number. Then

$$h : Re^+ \xrightarrow[\text{onto}]{1-1} Re^+.$$

2. For $u \in Re^+$, find an explicit algebraic formula for an element $x \in Re^+$ such that $h(x) = u$, in other words, an explicit algebraic formula for $x = h^{-1}(u)$. You may use the symbol, f^{-1}, for the inverse of f in your formula.

Discussion.

1. First, it is easy to see that h maps Re^+ to Re^+.

To prove that h is one–one, we let $x, y \in Re^+$ and assume that $h(x) = h(y)$. We then need to prove that $x = y$. This is very simple algebra.

To prove that h maps onto Re^+, let u be any element of Re^+. We then need to prove that u can be "pulled back" to some preimage, say $v \in Re^+$, such that

$$u = h(v).$$

In other words, we need to show that, given any $u \in Re^+$, this equation can be solved for some $v \in Re^+$. Suppose there is such a $v \in Re^+$. By the definition of h, we would have

$$u = h(v) = \frac{a}{f(v)},$$

so we would have

$$f(v) = \frac{a}{u}.$$

But clearly, $a/u \in Re^+$. Since f maps onto Re^+, the equation $f(v) = a/u$ has a solution $v \in Re^+$, so the desired $v \in Re^+$ does exist. It will be easy to write out the details.

2. The previous discussion makes it clear that we just need to do a little more algebra to solve Part 2.

Written Solution.

1. *Theorem.* Let f be any one–one function from Re^+ onto Re^+. For any $x \in Re^+$, define

$$h(x) = \frac{a}{f(x)},$$

where the constant $a \in Re^+$, i.e., a is a fixed positive real number. Then

$$h : Re^+ \xrightarrow[onto]{1-1} Re^+.$$

Proof.

(i) $[h : Re^+ \longrightarrow Re^+]$ Since f maps Re^+ to Re^+, for any $x \in Re^+$, $f(x) \in Re^+$. Also, $a \in Re^+$. Since the quotient of two positive real numbers is a unique positive real number,

$$h(x) = \frac{a}{f(x)} \in Re^+$$

returns a unique output in Re^+ for every input value in Re^+. Thus, h is a function from Re^+ to Re^+.

(ii) $[h \text{ is } 1\text{--}1]$ Let $x, y \in Re^+$ and assume that $h(x) = h(y)$. By the definition of h,

$$h(x) = \frac{a}{f(x)} = h(y) = \frac{a}{f(y)},$$

from which it follows algebraically that $f(x) = f(y)$. Since f is 1–1, $x = y$, which proves that h is 1–1.

(iii) $[h \text{ maps onto } Re^+]$ Let $u \in Re^+$. Then

$$\frac{a}{u} \in Re^+.$$

But it is given that f maps onto Re^+, so there exists some $v \in Re^+$ such that

$$f(v) = \frac{a}{u},$$

so

$$u = \frac{a}{f(v)} = h(v).$$

Therefore, u has a preimage v, so h maps onto Re^+. **Q.E.D.**

2. By the theorem just proved,

$$h : Re^+ \xrightarrow[onto]{1-1} Re^+.$$

Hence, by Lemmas 2-41, 2-42, and Theorem 2-43, it has a unique inverse function

$$h^{-1} : Re^+ \xrightarrow[onto]{1-1} Re^+.$$

For any $u \in Re^+$, let $x = h^{-1}(u)$; we want an explicit formula for $h^{-1}(u)$. Clearly, we have

$$u = h(x) = \frac{a}{f(x)},$$

so

$$f(x) = \frac{a}{u}.$$

Since $u > 0$, the quotient $a/u \in Re^+$. But also, f has an inverse, so we can now use its inverse to solve for x:

$$f^{-1} \circ f(x) = x = f^{-1}\left(\frac{a}{u}\right).$$

We now verify that x is the preimage with respect to h of u,

$$h(x) = h \circ f^{-1}\left(\frac{a}{u}\right) = h\left(f^{-1}\left(\frac{a}{u}\right)\right) = \frac{a}{f\left(f^{-1}\left(\frac{a}{u}\right)\right)} = \frac{a}{\left(\frac{a}{u}\right)} = u.$$

Therefore, by the definition of inverse function, the unique inverse of h is

$$h^{-1}(u) = f^{-1}\left(\frac{a}{u}\right). \ \blacksquare$$

Systematic development of the main number systems is beyond the scope of this book, although the basic foundations of Nat are presented in the next chapter. So far, we have assumed familiar properties of Nat, Rat, etc., as briefly described in the Introduction. For instance, the square root of any negative number is not even in Re. On the other hand, there are various procedures for at least estimating square roots of positive reals, and these procedures often lead to numbers with long decimal fractions. Recall that Rat^+ is the set of all positive numbers that can be expressed in the form a/b, where a and b are integers with $b \neq 0$. It is therefore natural to wonder whether Rat^+, the set of positive rational numbers, is closed under the square root function. If we can find one element of Rat^+ that does not have a rational square root, we will know that Rat^+ is not closed under that function. In fact, $\sqrt{2}$ is not rational.

This is a famous mathematical result that was proved by the ancient Greeks several centuries B.C.E. It is presented here as an example of a proof that a particular set is not closed under a particular function. The proof shows that when the function is applied to a particular element of the set, the function does not return a member of that set. Thus, it is also an example of a proof of the nonexistence of a mathematical entity of a certain type. Finally, it is a nice example of an informal RAA proof.

Example.

Problem Statement. Theorem. $\sqrt{2}$ is not a rational number.

Discussion. Basically, we need to prove that there does not exist any rational number that equals $\sqrt{2}$. How can one prove that something does not exist? One way to do this is to assume that the thing does exist and derive a contradiction from this assumption. This is what we shall do here:

Suppose that $\sqrt{2}$ is rational. Then there exists a rational number i/j such that $\sqrt{2} = i/j$. We also assume, from elementary arithmetic, that any such fraction can be reduced to lowest terms. Therefore, we may assume without loss of generality that i and j have no common factor greater than 1. For example, suppose that we start with a fraction such as 24/36, which is not in lowest terms. The numbers 24 and 36 have 2 as a common factor; also, 3 is a common factor, and there are others. The highest common factor is 12. If we divide both numerator and denominator by 12, we get 2/3, which is in lowest terms, i.e., 2 and 3 have no common factor greater than 1. We shall assume that all such common fractions can be reduced in this manner to lowest terms. This is well known from elementary arithmetic, and it will be proved in Chapter 3.

The proof that $\sqrt{2}$ is not rational is simple, but is not immediately obvious. We assume, for RAA, that $\sqrt{2} = i/j$, where i and j are integers, and the fraction is assumed without loss of generality to be reduced to lowest terms. Then we have $2 = i^2/j^2$, so $i^2 = 2j^2$. Therefore, i^2 is an even integer since it is a multiple of 2. But any odd integer has the form $2k + 1$ for some integer k. Thus, the square of an odd integer has the form,

$$(2k + 1)^2 = 4k^2 + 4k + 1 = 4(k^2 + k) + 1,$$

which is odd. So, the square of an odd integer is odd. Since i^2 is even, i must consequently be even. Hence, there exists some natural number, say, m, such that $i = 2m$. Therefore, $i^2 = 4m^2$, from which $j^2 = 2m^2$. The rest is easy. Please try to finish it before reading the solution that follows.

Written Solution. Theorem. $\sqrt{2}$ is irrational.

Proof. For RAA, suppose that $\sqrt{2}$ is a rational number. Then there exist integers i and j such that $\sqrt{2} = i/j$. We can assume without loss of generality that i/j is reduced to lowest terms, so that i and j have no common factor greater than 1.

We now have $2 = i^2/j^2$, so $i^2 = 2j^2$. Since i^2 is even and the square of an odd number is odd, i must be even. Thus, there is some number m such that $i = 2m$, and then $i^2 = 4m^2$. Therefore, we have

$$i^2 = 2j^2 = 4m^2,$$

so $j^2 = 2m^2$. Hence, j^2 is even. By the same kind of reasoning that was applied to i, it follows that j is also even. It follows that, both i and j are even, so they have the common factor 2. This contradicts the assumption that i/j is reduced to lowest terms. ***Q.E.D.*** ■

EXERCISES

General Instructions

Throughout this book, if an exercise is simply the statement of a theorem or a metatheorem, then the task is to prove this theorem or metatheorem. In your proof, you may use the definitions given in the text. Unless stated to the contrary, you may also use theorems proved in the text.

Exercises for which answers are provided at the back of the book are marked with "Answer Provided."

In this chapter, unless instructed otherwise, you may assume elementary facts of arithmetic and algebra in these exercises.

Finally, please note that exercises that introduce important principles, useful applications, are especially difficult, or are special in some way, are indicated by bold print. In some cases, parenthetical comments describe the special features of the problem. If such comments are missing, the exercise is nonetheless of special value in introducing a new concept, principle, or technique.

E 2-77 (Answer Provided) Consider the relational system $\langle Re^+, f \rangle$, where

$$f = \{ \langle x, y \rangle \mid xy = 7 \},$$

and xy is the product of x and y.

1. Prove that f is a function.

2. Now introduce functional notation by

$$y = f(x) \Leftrightarrow <x, y> \in f.$$

Prove that f is one–one. Write an expression for the inverse of f; show that it is the inverse.

E 2-78 For $x, y \in Re$, let $m(x, y) = m(<x, y>) = xy$ be the product of x and y, i.e., x times y. Define the following functions on Re: $f(x) = x + 1$, $g(x) = x - 1$, $h(x) = m(f(x), g(x))$. Prove or disprove each of the following:

f is one–one on Re. f is onto Re.

g is one–one on Re. g is onto Re.

h is one–one on Re. h is onto Re.

E 2-79 Briefly state what the error is in this *alleged* "proof." Assume the elementary properties of the integers.

Let f be the function from the integers to the integers satisfying, for any integer x, $f(x) = x^2 - 16$. Then

$$f(4) = f(-4) = 0.$$

Let f^{-1} be the inverse of f. Then

$$f^{-1} \circ f(4) = f^{-1} \circ f(-4).$$

Hence,

$$4 = -4.$$

E 2-80 Let f be a one–one function from Re^+ onto Re^+, i.e.,

$$f : Re^+ \xrightarrow[onto]{1-1} Re^+.$$

Define

$$g : Re^+ \longrightarrow Re^+$$

using the following: For any $x \in Re^+$, $g(x) = 1/f(x)$. Prove that g is also one–one and that it maps onto Re^+.

E 2-81 Let f be any one–one function from Re^+ onto Re^+. Let the constants $a, b \in Re^+$. For any $x \in Re^+$, let

$$h(x) = \frac{a}{b + f(x)}.$$

1. Prove that h is a one–one function from Re^+ to Re^+.

2. Prove that, for all possible values of a, $b \in Re^+$, h does not map Re^+ onto Re^+.

E 2-82 Let A be a nonempty set and let $\Pi = \{ C_1, \cdots, C_k \}$ be a partition of A with the cells $C_1, \cdots, C_k$. Define a relation P between A and Π by: For any $x \in A$, and any $C \in \Pi$,

$$P x C \iff x \in C.$$

1. Prove that P is a function from A to Π.

2. Prove that $P : A \xrightarrow{onto} \Pi$.

3. Describe a specific example of an A and a corresponding Π for which the function P is not one–one on A.

E 2-83 *Theorem.* Let E be an equivalence relation on nonempty set A. Define the relation g as follows: For any x, y in A, $gx[y] \iff Exy$. Then

1. g is a function, $g : A \longrightarrow A/E$.

2. For any $x \in A$, $g(x) = [x]$.

3. $g(x) = g(y) \iff [x] = [y]$.

4. $g(x) = g(y) \iff Exy$.

E 2-84 Let A, B, C be nonempty sets with functions

$$f : A \longrightarrow B,$$

and

$$g : B \longrightarrow C.$$

Prove or give a counterexample to

1. If f maps A onto B, and g maps B onto C, then $g \circ f$ maps A onto C.

2. If $g \circ f$ maps A onto C, then f maps A onto B and g maps B onto C.

E 2-85 Let R be a nonempty binary relation on $A \neq \varnothing$, and let R^{-1} be the converse relation of R. Let $f : R \longrightarrow R^{-1}$, where $f(< x, y >) =< y, x >$, for every $< x, y > \in R$. Prove the following:

1. $f : R \xrightarrow{1-1} R^{-1}$.

2. $f : R \xrightarrow{\text{onto}} R^{-1}$.

3. For every $< u, v > \in R^{-1}$, $f^{-1}(< u, v >) = < v, u >$, where f^{-1} is the inverse function of f.

E 2-86 For this problem, assume that the possible values of the variable x are real numbers, and let f return output values in Re according to the formula

$$f(x) = \frac{x+1}{x-1}.$$

Let D be the set of all values of x for which f is defined, i.e., D is the set of values of x for which f returns real number outputs. Clearly, D is the domain of f.

1. What are the elements of D?

2. Is f one–one on its domain D? Prove or disprove.

3. Does f map onto Re? Prove or disprove.

4. Does f have an inverse? Justify your answer. As usual, you may use theorems stated earlier in the text.

5. If f has an inverse, write an algebraic formula for this inverse. Justify that your formula is a function and prove that it is f^{-1}.

E 2-87 (Answer Provided) *Theorem.* Let A, B, C be nonempty sets and suppose that

$$f : (A \cup B) \longrightarrow C.$$

Then $Img(f, (A \cup B)) = Img(f, A) \cup Img(f, B)$.

E 2-88 Let A and B be nonempty sets, and suppose that

$$f : (A \cup B) \longrightarrow C.$$

1. Prove that $Img(f, (A \cap B)) \subseteq Img(f, A) \cap Img(f, B)$.

2. Show by an example that it is possible to have

$$Img(f, (A \cap B)) \neq Img(f, A) \cap Img(f, B).$$

3. Prove that, if f is one–one, then

$$Img(f, A) \cap Img(f, B) \subseteq Img(f, (A \cap B)).$$

E 2-89 Define $f(x, y) = f(<x, y>) = 2^x 3^y$, for $x, y \in Nat$. Assume elementary facts of arithmetic and algebra (including properties of exponents) to prove that f maps $Nat \times Nat$ one–one to Nat.

E 2-90 Let $A = \{1, 2, 3, 4\}$. A function that maps A one–one and onto A is called a *permutation* of A. The identity function, i_A, is a permutation, and there are many others, e.g., the function p such that $p(1) = 2$, $p(2) = 1$, and $p(x) = x$, for $x = 3, 4$. Prove that the composition of any two permutations of A is also a permutation of A. (Hint. Let p_1 and p_2 be two permutations of A. Each of them maps A one–one and onto A. You need to prove that $p_2 \circ p_1$ also maps A one–one and onto A. It may help to look at Figure 2.20. Let $x \in A$, and consider that p_1 maps x to $p_1(x)$, and then p_2 maps the latter to $p_2(p_1(x)) = p_2 \circ p_1(x)$. Consider how these functions and their inverses can map elements back and forth around this diagram.)

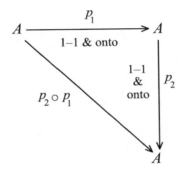

Figure 2.20.

E 2-91 Assume standard facts of elementary arithmetic and algebra. Consider the following subsets of Re: Rat, Rat^+, Int, and the set, N_e, of all negative even integers. For each of these subsets determine whether it is closed under addition, multiplication, and the reciprocal function. Justify your answers.

E 2-92 *Theorem.* Let $S = \{x + y\sqrt{2} \mid x, y \in Rat\} \subseteq Re$. Then S is closed under addition and multiplication. (Remark. You may assume basic facts of arithmetic and algebra, and assume that Rat is closed under addition and multiplication.)

E 2-93 *Theorem.* Let A, B be nonempty sets, and suppose that

$$f : A \xrightarrow{1\text{--}1} B.$$

Let C be a nonempty subset of A. Then

$$C = Img(f^{-1}, Img(f, C)).$$

E 2-94 Let A be a nonempty set with function $f : A \longrightarrow A$. Suppose that, for every $x \in A$, $f \circ f(x) = x$. Prove that f maps A one–one and onto A. Also, what is f^{-1}? Justify your answer.

E 2-95 *Theorem.* Let A be a nonempty set and let f, g be one–one functions from A onto A. Then

1. $g \circ f : A \xrightarrow[\text{onto}]{1-1} A$, and

2. $(g \circ f)^{-1} = f^{-1} \circ g^{-1}$.

E 2-96 (Answer Provided) *Theorem.* Let $\langle A, L \rangle$ be a relational system such that, for any $x, y \in A$, one and only one of the following holds:

$$x = y, \ Lxy, \ Lyx.$$

Let $f : A \longrightarrow A$ such that, for any $x, y \in A$, if Lxy, then $Lf(x)f(y)$. Then f maps A one–one to A.

E 2-97 *Definition.* Let $f : Re \longrightarrow Re$. Re is said to be *strictly increasing* iff for any $x, y \in Re$,

$$x < y \ \Rightarrow \ f(x) < f(y).$$

Assume that $<$ is a strict simple order on Re. Prove the following:

Theorem. If f is a strictly increasing function from Re to Re, then

1. f is one–one on Re, and

2. f^{-1} is strictly increasing on $Img(f, Re)$. (Hint. Look at E 2-71 and E 2-96.)

E 2-98 Let A, L, and f satisfy the antecedent conditions of the theorem in E 2-96. Give a counterexample to show that f need not map onto A. (Hint. Try a simple function on Re.)

E 2-99 Let A be a nonempty set and let $f : A \xrightarrow[\text{onto}]{1-1} A$. Let B be any nonempty subset of A. Define the function f_B by:

$$f_B = \{ < x, y > \ | \ x \in B \ \ \& \ \ y = f(x) \}.$$

f_B is called the *restriction of f to B*. Prove or give a counterexample to each of the following:

1. f_B is a one–one function on its domain B.

2. f_B maps B onto B.

E 2-100 Let A, B be nonempty sets and $f : A \longrightarrow B$. Define R on A by

$$Rxy \iff f(x) = f(y).$$

1. Prove that R is an equivalence relation on A.

2. Let $a \in A$ and $[a]$ be the R-equivalence class of a. Prove that all elements in $[a]$ have the same f value, i.e., f maps every element in $[a]$ into exactly one element of B .

3. Suppose that it is proposed to define a new function,

$$F : A/R \longrightarrow B,$$

using the following: For any $x \in A$,

$$F([x]) = f(x).$$

Prove that F really is a function.

4. In the special case where f is one–one, prove that every cell of A/R has exactly one member.

E 2-101 *Theorem.* Let $\langle A, L \rangle$ be a relational system in which the binary relation L is reflexive and transitive. Define the relation $C \subseteq A \times \mathcal{P}(A)$ as follows: For any $x \in A$ and $s \in \mathcal{P}(A)$,

$$Cxs \iff s = \{u \in A \mid Lux\}.$$

Then

1. C is a function from A to $\mathcal{P}(A)$.

2. For every $x \in A$, $x \in C(x)$.

3. For every x, $y \in A$, $(C(x) \subseteq C(y)) \iff Lxy$.

4. If L is reflexive, transitive, and symmetric, then $Img(C, A)$ is a partition of A.

5. If L is reflexive, transitive, and antisymmetric, then C is one–one.

E 2-102 *Theorem.* Let $\langle A, L \rangle$, $\langle B, R \rangle$ be relational systems in which L and R are binary relations. Let

$$f : A \xrightarrow[onto]{1-1} B.$$

Suppose that, for every x, $y \in A$,

$$Lxy \Leftrightarrow Rf(x)f(y).$$

Then if L is a strict simple order on A, then R is a strict simple order on B.

(Remark. $\langle A, L \rangle$ is said to be *isomorphic to* $\langle B, R \rangle$ when there is a one–one function f mapping A onto B satisfying the above biconditional. Such a one–one function mapping A onto B is said to *preserve* the ordering relations and such a function is called an *isomorphism*. Isomorphisms and other kinds of related mappings play an important role in much of modern mathematics.)

2.4 Relational Systems

In Definition 2-29, a relational system was defined to be an ordered $(n + 1)$-tuple,

$$\langle D, R_1, R_2, \cdots, R_n \rangle,$$

where D is a nonempty domain and the R_i are k-ary relations on D, where $k \geqslant 1$. When $k = 1$, the "relation" is just a subset of D. Since functions are special kinds of relations, some of the R_i may be functions. This motivates a more detailed definition of *relational system*, which will be stated shortly. It will first be convenient to amplify our conventions regarding functional notation. Up to this point we have discussed functions from a set A to a set B without giving detailed consideration to the structure of the domain and codomain. Since these may be any sets, it is possible that A is a set of ordered k-tuples and that B is a set of ordered m-tuples. We will focus attention on A. Suppose that $A = A_1 \times A_2 \times \cdots \times A_k$. Then a function $f : A \longrightarrow B$ is a function,

$$f : A_1 \times A_2 \times \cdots \times A_k \longrightarrow B.$$

In this situation, it is standard practice to write

$$f(x_1, x_2, \cdots, x_k)$$

as an abbreviation of

$$f(< x_1, x_2, \cdots, x_k >).$$

We will follow this practice and will be especially interested in the situations where $B = A$.

Definition 2-46 In the following, let f be a function. If

$$f : A \longrightarrow A,$$

then f is a *unary operation* on A. If

$$f : A \times A \longrightarrow A,$$

then f is a *binary operation* on A. If f is a function from a set of ordered k-tuples of A to A, then f is a *k-ary operation* on A.

The reciprocal function, $1/x$, and the square root function, $\sqrt{x}$, are unary operations on Re^+. They map Re^+ to Re^+. The functions $+$ and $\times$ are binary operations on Re. Binary operations are often written in *infix* form, in which the function symbol is placed between its two arguments. For example, one may write $2 + 3$ instead of $+(2, 3)$, and 3×8 instead of $\times (3, 8)$. Nevertheless, these binary operations are just functions. They map $Re \times Re$ to Re. In general, a k-ary operation on A is a $(k + 1)$-ary relation on A. In mathematics, one often works with a system consisting of a nonempty domain of elements, some special, *distinguished elements* in this domain, some relations on the domain, and some operations on the domain. For this reason, it is convenient to restate the definition of a relational system as follows:

Definition 2-47 Let D be a nonempty set. A *relational system* on D is an ordered $(k + m + n + 1)$-tuple,

$$\langle\, D, c_1, \cdots, c_k, R_1, \cdots, R_m, f_1, \cdots, f_n \,\rangle,$$

where $(k + m + n + 1) > 1$, and where the c_i are distinguished elements of D, the R_i are relations on D, and the f_i are operations on D. D is the *domain* of the relational system.

The "distinguished elements" are just some elements of D that may be of special interest, e.g., zero and one in the domain of real numbers. If there are no distinguished elements, this definition is equivalent to the definition of *relation system* given previously. Allowing the distinguished elements extends the previous definition slightly. A familiar relational system is

$$\langle Re, <, +, \times \rangle.$$

If we wish to introduce special names for distinguished elements (e.g., zero and one), we can write

$$\langle Re, 0, 1, <, +, \times \rangle.$$

In the next chapter, we will examine relational systems with the domain Nat and the distinguished element 0. Relational systems are also useful in characterizing abstract algebraic systems. As an illustration, I will briefly describe the *theory of groups*.

Definition 2-48 A relational system $\langle G, i, * \rangle$, in which $i \in G$ and $*$ is a binary operation on G, is a *group* iff it satisfies the following axioms G1 through G3. The possible values of the variables, x, y, z are elements of G.

G1. For any x, y, z, $x * (y*z) = (x*y)*z$.

G2. For any x, $x*i = x$.

G3. For any x, there exists an element x^{-1}, such that $x*x^{-1} = i$.

The expression $x * y$ is often abbreviated as xy. If we do this, from G1, $x(yz) = (xy)z$. G1 says that the operation $*$ is associative. G2 says that i is an identity element; but the axioms do not explicitly state that there is a *unique* identity element. Later I will prove that there is only one such identity element. In G3, the use of the superscript –1 does not refer to the converse of a relation or inverse of a function. The symbol ' x^{-1} ' denotes an element of G, and such an element is called an *inverse of x*. The axioms do not explicitly state that x^{-1} is unique, so x^{-1} denotes *an inverse element of x*. Later I will prove a theorem that states that it is unique. A group is *any* relational system satisfying Definition 2-48. This is an abstract and general definition, but there are many examples of special groups.

Examples.

1. $\langle Re^+, 1, \times \rangle$, where $\times$ is multiplication of positive real numbers. Multiplication is associative. An identity element is 1, since $x \times 1 = x$. An inverse (element) of x is the reciprocal, $1/x$, since $x \times (1/x) = 1$.

2. $\langle Int, 0, + \rangle$. Addition is associative. An identity element is 0, since $x + 0 = x$. An inverse of x is $-x$, since $x + (-x) = 0$. The relational system $\langle Int^+, + \rangle$ is *not* a group; it has no identity element and no inverse elements.

3. Let

$$G = \{f \mid f : Re \xrightarrow[\text{onto}]{1-1} Re\}.$$

G is the set of all one–one functions from Re onto Re. Since $i_{Re} \in G$, G is nonempty. In fact, G is a huge infinite set.

If $f \in G$, then (since f is one–one), there exists an inverse function,

$$f^{-1} : Img(f, Re) \longrightarrow Re,$$

by Theorem 2-43. Since f also maps onto Re, $Img(f, Re) = Re$. Applying Lemma 2-42, the inverse function f^{-1} maps Re one–one and onto Re. Hence, if $f \in G$, then the inverse function $f^{-1} \in G$.

Now consider the system, $\langle G, i_{Re}, \circ \rangle$, where $\circ$ is composition of functions. It is easy to check that the composition of any two functions in G is also a function in G. It was shown in Section 2.3.2 that composition is associative. G contains an identity element, i_{Re}, in that, for any $f \in G$, $f \circ i_{Re} = f$. From the previous paragraph, for any $f \in G$, the inverse function $f^{-1} \in G$. Also, $f \circ f^{-1} = i_{Re}$. Therefore,

$$\langle G, i_{Re}, \circ \rangle$$

is a group. The elements of this group are functions, the group operation is composition of these functions, an identity element is the identity function on Re. In this example, since the elements of the group are functions, an inverse of an element is just the inverse of this function. This is a special feature that does not hold for groups in general.

In the first example, the group operation is multiplication, which is commutative. The addition operation, in the second example, is also commutative. The definition of *group* does not require that the operation be commutative. Two functions in the G of the third example are $f(x) = x + 1$ and $g(x) = 2x$. It is easy to check that both of these map Re one–one and onto Re (indeed, this follows from Theorem 2-37). Consider the compositions

$$f(g(x)) = f(2x) = 2x + 1,$$

and

$$g(f(x)) = g(x + 1) = 2(x + 1) = 2x + 2.$$

In particular, $f(g(0)) = 1$ and $g(f(0)) = 2$, so $f \circ g \neq g \circ f$, and the group operation is not commutative.

4. Let S be any nonempty set, and

$$H = \{f \mid f : S \xrightarrow[\text{onto}]{1-1} S\}.$$

It is easy to generalize the previous example and verify that $\langle H, i_S, \circ \rangle$ is a group. Suppose that $S = \{1, 2, 3, 4\}$. A function f that maps S one–one onto S is called a *permutation* of S. If we consider the elements of S in some particular order, say,

$$< 1, 2, 3, 4 >,$$

then (in general) a permutation puts them into a new order. For example, suppose that $f(1) = 3$, $f(2) = 1$, $f(3) = 4$, and $f(4) = 2$. Then we get the new order

$$< 3, 1, 4, 2 > .$$

The identity permutation is an exception; it does not change order. The set of all permutations on a finite set is a group under functional composition. ∎

   There is an astonishingly large variety of groups; some have a finite number of elements and some are infinite. The elements of a group may be physical operations that transform some object or set of objects. For example, consider a cube that has no special features on its sides, edges, or corners. It can be rotated in space in such a way that it looks the same as it was before the rotation. There are many ways to do this. Also, it can be reflected in a mirror and still look the same. These types of operations are called *symmetry operations*. The symmetry operations of a cube can be composed, they have inverses, and there is an identity operation (that does nothing at all). It can be shown that the symmetry operations of a figure, like a cube, form a group. Groups of symmetry operations are widely used in chemistry and physics. Symmetry groups also apply to many patterns found in nature, in manmade structures, and in various kinds of artifacts; see Weyl (1982). Symmetries play an important role in the art of M. C. Escher; see Escher (1971). More generally, many kinds of groups are useful in pure and applied mathematics.

   If there are so many different kinds of groups, what is the point of the very abstract definition of a group? This is a good question that has an important answer: By using the abstract definition, we are able to prove theorems that apply to *every group*. Even though special groups have special properties, all groups share some important properties. If we prove a theorem using only logic, set theory, and the abstract definition of group, then we know that this theorem holds for every group. This kind of abstraction yields powerful results and is widely used in mathematics and other fields.

   The study of the properties of groups is a special branch of mathematics, *the theory of groups*, which fills many books. A few typical results about groups are given in the next theorem. They are only elementary illustrations of abstract group theorems. Before proceeding with the theorem, we should note an important feature of operations. Suppose that $x, y, z \in G$ and that $x = y$. Since $*$ is a binary operation (a function) on G, we have

$$x * z = *(x, z) = *(y, z) = y * z.$$

The equality holds because we are just applying the $*$ function to the identical pairs, $< x, z > = < y, z >$. Hence,

$$\text{if } x = y, \text{ then } x * z = y * z.$$

   Suppose that a group element x is operated on its right side by, say, z to form $x * z$. We shall call this a *right multiplication*. If z is applied to the left side of x to form $z * x$, then we have a *left multiplication*. The previous equation shows that we can apply right multiplication to both sides of a group equation and the result is a new equation (in the above, the result is $x * z = y * z$). Similarly,

$$\text{if } x = y, \text{ then } z * x = z * y.$$

In algebraic jargon, we can say this: If a group equation is transformed by a left multiplication operation on both sides of the equation, or by a right group multiplication on both sides of the equation, the result is another group equation. On the other hand, since $*$ may not be commutative, we cannot (in general) do a left multiplication on one side of the equation and a right multiplication on the other side. These observations apply to any binary operations, not just group operations. They will be used in the following proofs, in which the group operation, $x * y$, is abbreviated as xy.

Theorem 2-49   Let $\langle G, i, * \rangle$ be a group with an identity element i. Let $x, y, z \in G$. Then

1. If $xz = yz$, then $x = y$ (right cancellation).

2. $ix = x$.

3. $i = x^{-1}x$, where x^{-1} is an inverse of x.

4. If $zx = zy$, then $x = y$ (left cancellation).

5. G has a unique identity element.

6. x has a unique inverse.

Proofs.

1. Suppose that $xz = yz$. Then, since z has an inverse z^{-1},
   $$x = xi = x(zz^{-1}) = (xz)z^{-1} = (yz)z^{-1},$$
   so
   $$x = (yz)z^{-1} = y(zz^{-1}) = yi = y.$$
   This is a right cancellation law.

2. We have
   $$xx^{-1} = i = ii = i(xx^{-1}) = (ix)x^{-1}.$$
   Applying the right cancellation law,
   $$x = ix.$$
   Also, since $xi = x$ (by G2, Definition 2-48), we have $ix = xi$. Thus, i commutes with any x, although arbitrary group elements x, y may not commute with each other.

3. From the previous result, we have $ix^{-1} = x^{-1}i$, so

$$ix^{-1} = x^{-1}i = x^{-1}(xx^{-1}) = (x^{-1}x)x^{-1}.$$

Applying the right cancellation law,

$i = x^{-1}x$.

Since, by G3, $xx^{-1} = i$, we have $xx^{-1} = x^{-1}x$. Therefore, any element x commutes with its inverse, x^{-1}.

4. Suppose that $zx = zy$. Then

$z^{-1}(zx) = z^{-1}(zy)$.

$(z^{-1}z)x = (z^{-1}z)y$.

Applying the previous result,

$ix = iy$,

so by 2 above,

$x = y$.

This is the left cancellation law.

5. In addition to i, suppose there is an identity element e. Then, for any x, $xe = x$. But also $xi = x$. So $xi = xe$, and by left cancellation, $i = e$.

6. In addition to x^{-1}, suppose that y is also an inverse of x. Then $xy = i = xx^{-1}$. By left cancellation, $y = x^{-1}$. **_Q.E.D._**

These are simple theorems, but they hold for all groups. For example, we now know that 1 is *the unique identity element in* $\langle\ Re^+,\ 1,\ \times\ \rangle$. We know that $-x$ is *the unique inverse* of x in $\langle\ Int,\ 0,\ +\ \rangle$ and so on. This demonstrates the power of abstraction. Many beautiful and nontrivial theorems about groups have been proved. Similar techniques apply to other abstract algebraic systems; see Warner (1990).

Chapter 3

Recursion and Mathematical Induction

3.1 The Natural Number System

3.1.1 Introduction

The natural numbers are 0, 1, 2, $\cdots$. In elementary school, one learns how to add, multiply, and perform other operations on these numbers. In more advanced mathematics, one may study many beautiful theorems about the natural numbers. In the previous chapter, I have already used the symbol *Nat* to denote the set of natural numbers and made use of some of the basic features of this system of numbers. In this chapter, we will study the natural numbers in more detail, beginning with a set of axioms or postulates for them. There are several reasons for doing this, including the following: First, by starting with simple postulates, we will be able to see more clearly how to justify proofs of theorems about the natural numbers. This is an important part of the *foundations of mathematics*, the study of the justification and development of mathematical systems. Second, we will see how to use the method of *definition by recursion*, which has many applications in other contexts, including writing computer programs. Third, the *proof techniques* that we will use are applicable in other contexts, including proofs of the correctness of programs. More generally, we will look at some similarities between features of the natural numbers and features of other things, such as strings and lists of characters. Finally, it is amazing and intellectually enlightening to see how intriguing properties of the natural number system, and of list data structures, arise from a few apparently very simple axioms.

Before starting our systematic development, it may help to look at a simple example of proof by *mathematical induction*. In elementary algebra, one usually learns formulas for summing series of numbers. Let the function, $sum(k)$, denote the *sum* of the natural numbers, 0 through k, inclusively. This is often written as

$$sum(k) = 0 + 1 + 2 + \cdots + k,$$

where it is assumed that we are already familiar with the basic arithmetical operations, and where it is understood that the three dots (ellipsis) represent the intermediate numbers between 2 and k. Although the meaning of the ellipsis may be clear in such a simple example, in the case of a more complex series, the use of three dots could be ambiguous because there might be more than one "obvious" way to fill in the intermediate steps. It would be nice to have a definite algorithm or procedure that specifies exactly how all intermediate steps of the calculation are to be performed. One way to accomplish this is by a *recursive definition*. For any $k \in Nat$, let

$$sum(k) = 0 \text{ if } k = 0$$

and

$$sum(k+1) = sum(k) + (k+1).$$

From the definition, we also have

$$sum(k+1) = (k+1) + sum(k).$$

At first glance, this form of definition may appear circular and possibly defective, since the term *sum* occurs on both sides of the second equation. Yet we can calculate the following:

$$sum(0) = 0,$$

$$sum(1) = sum(0+1) = (0+1) + sum(0) = 1 + 0 = 1,$$

$$sum(2) = sum(1+1) = (1+1) + sum(1) = 2 + 1 = 3,$$

and so on. The definition at least *seems* to provide a computational procedure that produces correct results. Yet we want to define the function *sum* for all natural numbers; it is not obvious that the definition accomplishes this. One purpose of this chapter is to *justify* definitions of this general type and to study how they can be generalized in form and applied in various contexts. For now, let us just assume that we have an adequate definition of *sum*.

Although our recursive definition provides a computational procedure, there may be a shorter, more convenient way to compute values of $sum(k)$ for various values of k. In fact, after computing the values of *sum* for several numbers, one may eventually conjecture that, for all $n \in Nat$, $sum(n) = n(n+1)/2$. To prove this, one uses the method of *mathematical induction*. The proof of the following theorem illustrates this method.

Theorem 3-1 Let $sum(k)$, as previously defined, be the sum of the natural numbers 0 through k. Then, for any $n \in Nat$,

$$sum(n) = n(n+1)/2.$$

Proof. For any $k \in Nat$, let $\phi(k)$ abbreviate the predicate,

$$sum(k) = k(k+1)/2.$$

Observe that $\phi(k)$ really is a predicate that applies to *Nat*. For any number $k \in Nat$, $\phi(k)$ is true iff this number satisfies this equation. If we can prove that, for any $n \in Nat$, $\phi(n)$, then we will have proved our theorem.

Part 1. We first prove that $\phi(k)$ is true when $k = 0$. By the definition of *sum*, if $k = 0$, then $sum(k) = 0$. Also, by *direct calculation*, if $k = 0$,

$$k(k+1)/2 = 0(0+1)/2 = 0.$$

Thus, if $k = 0$, then

$$sum(k) = k(k+1)/2,$$

so $\phi(k)$ is true when $k = 0$, i.e., $\phi(0)$ is true.

Part 2. Now let k be any number in *Nat* and *assume* that $\phi(k)$ is true. I claim that this assumption implies that $\phi(k+1)$ must also be true. First, by the definition of *sum*,

$$sum(k+1) = (k+1) + sum(k) = sum(k) + (k+1).$$

By the assumption that $\phi(k)$ is true, we have

$$sum(k) = k(k+1)/2.$$

Substituting the last equation into the previous one yields

$$sum(k+1) = \frac{k(k+1)}{2} + (k+1)$$

$$= (k+1)\left(\frac{k}{2} + 1\right)$$

$$= (k+1)(k+2)/2$$

$$= (k+1)((k+1)+1)/2.$$

Hence,

$$sum(k+1) = (k+1)((k+1)+1)/2,$$

which is the statement of $\phi(k+1)$.

We now know that $\phi(0)$ is true, and also that, for any k,

$$\text{if } \phi(k), \text{ then } \phi(k+1).$$

Intuitively, if the predicate ϕ is satisfied by any natural number k, then it is also satisfied by the next natural number, $k + 1$. Since it is satisfied by 0, it is therefore satisfied by 1, and it is also satisfied by 2, etc. We conclude that it is satisfied by *all* natural numbers, i.e., for all $n \in Nat$, $sum(n) = n(n + 1)/2$.

$$\textbf{\textit{Q.E.D.}}$$

This proof begs for some discussion. It is clear that we have assumed some basic facts about arithmetic and algebra. It will be seen later in this chapter how these assumptions can be justified. Many of them ultimately depend on the method of mathematical induction, which was applied in this proof. The basic idea of an induction proof is this: We prove that some predicate $\phi(k)$ is true when $k = 0$. We also prove that, for any $k \in Nat$, if $\phi(k)$, then $\phi(k + 1)$. From these two results, we conclude, as was done at the end of the proof, that $\phi(n)$ is true for every natural number n. It is a fundamental axiom about the natural numbers that this type of proof is acceptable. This axiom is presented in the next section and is used extensively in this chapter.

Notice that, after $\phi(k)$ is stated, a proof of this type naturally divides into two main sections, Parts 1 and 2 as shown. In Part 1, $\phi(0)$ is proved. This is called the *base case* of the proof, because it requires proving that the first number in question (0 in this proof) satisfies ϕ. It is very important to understand exactly what was proved here. Some people, when they first see this proof, get the mistaken impression that the equation $sum(k) = k(k + 1)/2$ is just assumed, and that 0 is substituted into it. This is not what is done. The equation is merely the *definition* of the predicate $\phi(k)$, and defining a predicate does not guarantee that it is satisfied by anything. In Part 1, we must prove that $\phi(0)$ is true. We must prove that the equation really is satisfied by 0. To do this, we independently compute $sum(0)$ and $0(0 + 1)/2$. We see that they both equal 0, so they are equal to each other.

In Part 2 of the proof, we let k stand for any (arbitrary) natural number. This means that k is not any number in particular. We prove that, if $\phi(k)$ is true, then $\phi(k + 1)$ must also be true. This part of the proof is called the *induction step*. The assumption that $\phi(k)$ is true is called the *induction hypothesis*.

The entire proof is therefore divided into two parts: the base case and the induction step. When presenting such proofs, it is first necessary to state a clear definition of the predicate $\phi(k)$. Then one proceeds to state the base case, and prove it. One then states the induction hypothesis, $\phi(k)$, and proves the induction step, namely, that $\phi(k)$ implies $\phi(k+1)$.

It may be helpful to consider mathematical induction proofs from a slightly different point of view. The above proof used the predicate $\phi(k)$, which is just $sum(k) = k(k+1)/2$. Let *Nat* be the domain of discourse. Using the Axiom of Separation, we can define the set

$$S = \{k \mid \phi(k)\} = \{k \mid sum(k) = k(k+1)/2\}.$$

Using this definition of S, one can see the following:

$$0 \in S \iff \phi(0)$$

and

$$(k \in S \implies k+1 \in S) \iff (\phi(k) \implies \phi(k+1)).$$

S is the set of all numbers that satisfy ϕ. Hence, proving the base case amounts to proving that $0 \in S$. Proving the induction step amounts to proving that, if $k \in S$, then $k+1 \in S$. The entire induction proof amounts to showing that ϕ is satisfied by all natural numbers, in other words, that $S = Nat$. We are now at a good place to begin the systematic development of the natural number system.

3.1.2 Peano's Axioms and the Induction Principle

There are several approaches one can take in stating axioms for *Nat*. If a formal system is desired, one can use the language of predicate calculus, which is the subject of the next chapter. One method is to formulate the axioms for *Nat* directly in predicate calculus. However, as will be seen in Chapter 4, there are different versions of predicate calculus with different representational powers. It turns out that various axiom systems for the natural numbers can be formulated in different versions of predicate calculus, and the results can be significantly different characterizations of what are "numbers." Another method is to axiomatize set theory in predicate calculus and then define the natural numbers as special kinds of sets that arise in this axiomatic set theory. Historically, a less formal approach was used: One assumes that the set *Nat* exists, one states axioms about some of the properties of this set within informal set theory, and then one proves theorems, using informal set theory together with the axioms about *Nat*. Although I will not reconstruct the actual history, this is essentially the approach that will be taken here.

In this chapter, I will often use i, j, k, m, n as variables ranging over natural numbers. Some other variables, such as p, q, r, u, v, x, y, z, $\cdots$, may also be used. What are now called "Peano's Axioms" for *Nat* were independently

invented by R. Dedekind and G. Peano in the 1880s. A somewhat similar set
of axioms was also introduced in Peirce (1881).* Peano's Axioms can be presented in the following manner:

Peano's Axioms. There is a relational system, $\langle$ *Nat*, 0, *s* $\rangle$, which
satisfies the following axioms:

1. There is a special element of *Nat* denoted by 0.

2. *s* is a function from *Nat* to *Nat*, i.e., $s : Nat \longrightarrow Nat$. This
   function *s* is called the *successor function*.

3. *s* is one–one.

4. There does not exist $k \in Nat$ such that $s(k) = 0$.

5. (*Induction Axiom*) Let S be any subset of *Nat*.
   IF:

      (i) $0 \in S$,
        and
      (ii) for any $k \in Nat$, $k \in S \Rightarrow s(k) \in S$,

   THEN: $S = Nat$.

According to the *Modus Operandi* of Chapter 2, we might try to define *Nat*
in terms of the elements of some domain of discourse. Peano's Axioms, as just
stated, do not do this. This is not surprising, since the natural numbers are
such simple and basic mathematical objects. If we invented some domain of
discourse, D, and then defined *Nat* as a subset of D, assuming that D exists
would be at least as questionable as assuming the existence of *Nat*. This seems
to be an inevitable result of the informal approach used here. In order to try
to avoid this difficulty, one might resort to using a formalized axiomatic system, but we are not using that approach here. On the other hand, the *Modus
Operandi* is intended to be applied to some set of objects whose existence is
assumed. Because natural numbers are such familiar and basic mathematical
objects, we will assume they exist. At least provisionally, we will also assume
that Peano's Axioms characterize the natural numbers. We will proceed to
develop some of the logical consequences of the axioms. These consequences
are called *Peano arithmetic*, and they include many of the theorems in the
branch of mathematics called *number theory*. Thus, we will see how some of
the basic theorems of number theory follow from the axioms, and these results

---

*Peirce (1881) is reprinted in Hartshorne and Weiss (1960), pp. 158–170. Beth (1966),
Chapter 6, describes the ideas of Dedekind and Peano, and provides specific references. Some
detailed historical discussion is also in Wang (1957).

will support our assumption that Peano's Axioms do characterize the natural numbers. We begin with a definition and with a fundamental theorem that will be used in the proofs of many other theorems. *Throughout this chapter, whenever lowercase s is used as a function on Nat, s is the successor function.*

Definition 3-2 Let j, $k \in Nat$; k is the *successor* of j iff $k = s(j)$. Also, j is a *predecessor* of k iff $k = s(j)$.

Theorem 3-3   Let k be any element of *Nat*. Then $k = 0$ or there exists a unique $j \in Nat$ such that $k = s(j)$; in other words, $k = 0$ or k has a unique predecessor.

Proof. Let $S = \{\, 0 \,\} \cup Img(s, Nat)$. With the help of the Induction Axiom, I will first prove that $S = Nat$.

Case 0. By the definition of S, $0 \in S$.

Case $s(k)$. We make the hypothesis that $k \in S$. In order to apply the Induction Axiom, I will prove that this hypothesis implies that $s(k) \in S$. Now if $k \in S$, then either $k = 0$ or $k \in Img(s, Nat)$. I now consider two subcases:

1. If $k = 0$, then $k = 0 \in Nat$ by Axiom 1. It follows from Axiom 2 that $s(k) = s(0) \in Img(s, Nat) \subseteq S$. Hence, if $k = 0$, then $s(k) \in S$.

2. Suppose that $k \in Img(s, Nat)$. It also follows from Axiom 2 that

$$Img(s, Nat) \subseteq Nat,$$

   so if $k \in Img(s, Nat)$, then $k \in Nat$. Hence, $s(k) \in Img(s, Nat)$, so $s(k) \in S$.

In both subcases, if $k \in S$, then $s(k) \in S$. Applying Axiom 5 (the Induction Axiom) to Case 0 and Case $s(k)$, $S = Nat$. This concludes the first part of the proof.

Now let k be any element of *Nat*. Then $k \in S$, so

$$k = 0, \text{ or } k \in Img(s, Nat).$$

In the latter case, there exists $j \in Nat$ such that $k = s(j)$. So, any element k of *Nat* is either 0 or else k has a predecessor.

It is now necessary to prove that predecessors are unique. Uniqueness is proved in the usual way: Assume that more than one such thing exists, and then show that they must be the same. So suppose there are predecessors i and j of k, such that $k = s(i) = s(j)$. By Axiom 3, s is one–one, so $i = j$. Thus, when $k = s(j)$, this predecessor j is unique. ***Q.E.D.***

It should be clear that Axiom 5 is a strong statement, for it does not just refer to individual elements of *Nat*; it also refers to subsets of *Nat*. It is called the *Induction Axiom*, and it largely determines the structure of the set of natural numbers, as will be seen shortly. It should also be apparent that the use of Axiom 5 in the preceding proof is very similar to the earlier proof regarding the *sum* function. In fact, the Induction Axiom is often stated in another form which matches the form of proof used for *sum*. In our mathematical English language, which includes the language of informal set theory, let $\phi(k)$ be a predicate that applies to the natural numbers. For instance, suppose that we already have the relation $<$ and the operation $+$, then $\phi(k)$ could be the predicate: '$k < s(k)$', or $\phi(k)$ might be '$k < (k + k)$'. In the previous section, we used '$sum(k) = k(k + 1)/2$' as $\phi(k)$. We will soon define $+$, $<$, and other basic numerical functions and predicates. Using these, it will be possible to define a huge variety of other $\phi(k)$ predicates. For now, we assume that such predicates can be formulated suitably in our informal mathematical English. With this assumption, we can restate the Induction Axiom as follows:

Induction Principle (IP). Let $\phi(k)$ be any predicate that applies to *Nat*.

    IF:

        (1) $\phi(0)$,

        and

        (2) for any $k \in Nat$, $\phi(k) \Rightarrow \phi(s(k))$,

    THEN: For all $n \in Nat$, $\phi(n)$.

It is easy to see the similarity between Axiom 5 and the Induction Principle (IP). In the proof of the theorem about the *sum* function, I used IP and applied it to the $\phi(k)$ that is $sum(k) = k(k+1)/2$. I am assuming here that $k+1 = s(k)$, a fact that will be proved later. It was also pointed out in the discussion that this theorem could have been proved using the set of numbers that satisfy $\phi(k)$ and showing that this set equals all of *Nat*. If the proof had been stated in terms of this set, we would have directly applied Axiom 5 to get the result. Here is an informal justification of the equivalence of Axiom 5 with the Induction Principle.

First assume Peano's Axiom 5, and also assume (1) and (2) in IP above. We want to prove the consequent of IP. In order to use Axiom 5, we need to form some set S, and this set must be suitably related to ϕ. We use the Axiom of Separation. Define $S = \{ x \mid x \in Nat \ \& \ \phi(x) \}$. Then $0 \in S$ since, by (1), $\phi(0)$ is true. Let k be any element of Nat, and assume that $k \in S$. Then $\phi(k)$. By (2), $\phi(s(k))$ is true, so $s(k) \in S$. Hence, if $k \in S$, then $s(k) \in S$. By Axiom 5, $S = Nat$, so for any $n \in Nat$, $n \in S$, so $\phi(n)$. This shows that Axiom 5 implies the Induction Principle (IP).

Conversely, let $S \subseteq Nat$, and assume IP and (i) and (ii) of Peano's Axiom 5. We now want to prove the consequent of Axiom 5, namely, that $S = Nat$. This time we start with the set S and need to form a suitably related predicate to which one can apply IP. In the previous paragraph, we defined the set S in terms of the given predicate. This time we define a predicate, P, in terms of the given set, S, as follows: For any $k \in Nat$,

$$Pk \ \Leftrightarrow \ k \in S.$$

We are assuming that our mathematical language can provide a predicate P for any subset S of the natural numbers. For some formal languages, this assumption is not true, in which case IP becomes more restricted by the particular formal language that is used. However, for this informal argument, we make the assumption that P is available. Then Pk is a particular instance of $\phi(k)$ in IP. From (i) and (ii) of Axiom 5, we have

$$P0,$$

and

$$Pk \ \Rightarrow \ Ps(k).$$

It follows that, by IP, for all $n \in Nat$, Pn. Thus, $Nat \subseteq S$. Since also $S \subseteq Nat$, we have $S = Nat$. This is the consequent of Axiom 5, which was to be proved.

This argument shows that, in the context of our informal treatment of the natural numbers, we may treat Axiom 5 and the Induction Principle as equivalent. Because of this equivalence, from now on they will both be called the 'Induction Principle,' abbreviated as IP. Either one of these formulations may be used in proofs, although the formulation in terms of a predicate $\phi(k)$ will be used more often. In the context of formalized languages, these matters become complicated by the fact that a formalized language imposes restrictions on the types of predicates, $\phi(k)$, that are expressible in the language.

Before moving on, I will present a general schema for induction proofs using IP.

Schema for Proofs Using the Induction Principle (IP).

The goal is to use IP to prove that, for all $n \in Nat$, $\phi(n)$, where ϕ is a predicate applying to natural numbers.

Induction Hypothesis: Precisely state $\phi(k)$. This is just the predicate ϕ, with k in place of n.

Base Case. Prove that $\phi(0)$.

Induction Step. Assume the induction hypothesis, $\phi(k)$, together with special features of the particular problem. From these assumptions, prove that $\phi(s(k))$.

If the base case and the induction step are accomplished, then from IP it follows that $\phi(n)$ for all $n \in Nat$.

One might get the impression that assuming the induction hypothesis is equivalent to assuming the goal statement to be proved. This is not the case. The goal statement is the claim that ϕ is true of *every* natural number. The induction hypothesis is the statement that ϕ is true of an *arbitrary number k*. The induction step involves proving that if $\phi(k)$, then $\phi(s(k))$. This is a conditional with the antecedent $\phi(k)$, and the induction step uses the Conditional Proof (CP) strategy described in Chapter 1. It may help to restate these points in the following format. The goal is to prove:

$$\text{For every } n \in Nat, \phi(n).$$

The induction step statement to be proved is

$$\text{For every } k, \text{ if } \phi(k), \text{ then } \phi(s(k)),$$

which can also be expressed as

$$\text{For every } k, \phi(k) \Rightarrow \phi(s(k)).$$

The goal statement can be called a *categorical statement* that every natural number satisfies ϕ. The induction step statement is a *conditional statement* that if an arbitrary k satisfies ϕ, then $s(k)$ must also satisfy ϕ. The categorical goal statement requires that every natural number satisfy ϕ. On the other hand, the conditional statement would be true if no k satisfies ϕ, because then the antecedent of each conditional, for each k, would be false. This is the reason why the induction step by itself does not imply the goal statement—we also need to prove the base case, $\phi(0)$. (It will be seen later that sometimes the base case may be 1 or some other specific number.) Still, we normally use the CP strategy to help prove the conditional, induction step statement. In using the CP strategy, we assume $\phi(k)$ as an extra premise to help us deduce $\phi(s(k))$. But assuming $\phi(k)$, for an arbitrary k, is not equivalent to the categorical statement that every natural number satisfies ϕ.

The preceding schema is just a general outline for proofs; there is no guarantee that following this outline will always accomplish the goal. The schema says nothing about two important questions: When should one try to use IP, rather than some alternative form of proof? If an induction proof is attempted, how does one create a suitable induction hypothesis $\phi(k)$? The rest of this chapter contains many examples of induction proofs. By emulating these examples, and using additional proof heuristics that will be given later, it will turn out that answers to these two questions are often easy to obtain for particular problems.

3.1.3   Definition by Recursion

We now have the Induction Principle (IP), which is the basis for mathematical induction proofs. For example, IP plays a vital role in the earlier proof about the *sum* function, but we still have not justified the recursive definition of *sum*. As we saw, the definition provided a procedure for calculating the values of the function. More exactly, it specified an output value for the input value 0, and it specified how to compute the output value for the input $k + 1$, *assuming* that we already have the value of the function for the input k. But *sum* is supposed to be a function that maps *Nat* to *Nat*, so it is a set of ordered pairs $< k, x >$ such that for every $k \in Nat$, there exists exactly one $x \in Nat$ such that $< k, x > \in sum$. How can we be sure that we have really defined a set of ordered pairs of this form? The answer is provided by the next theorem, the Recursion Theorem. The proof of this theorem is lengthy, but the details of the proof are not essential for what follows in this book. Our main interest in the theorem is the fact that it justifies recursive definitions. It will be useful to study the proof, since it contains several subproofs illustrating mathematical induction.

Before considering the theorem in detail, it will help to describe what it is about. Recall that Section 3 of Chapter 2 includes examples of existence and uniqueness problems regarding certain types of functions, in particular, the existence and uniqueness of inverse functions. Existence and uniqueness problems are very common in mathematics. For example, consider these elementary questions: Do there exist integers x and y satisfying the equation $x + y = 5$? If such a pair of integers exists, is it unique? Obviously such integers exist, since $1 + 4 = 5$. Moreover, this solution is not unique: For any integer x, the pair x and $y = 5 - x$ is a solution, so there is an infinite set of solutions to the equation. Now consider these questions: Do there exist positive integers x and y satisfying the equation $x + y = 5$? If such a pair of integers exists, is it unique? We now have the following four ordered pairs of solutions for $< x, y >$: $< 1, 4 >$, $< 2, 3 >$, $< 3, 2 >$, $< 4, 1 >$ and, if we ignore the order of x and y, we have two solutions. In this example, we have the existence of solutions, but not uniqueness. For contrast, consider these questions: Do there exist positive integers $x > 3$ and y satisfying the equation $x + y = 5$? If such a pair of integers exists, is it unique? Now there exists the solution pair $< 4, 1 >$, and it is unique. Finally, consider these questions: Do there exist integers x and y satisfying the two equations $x + y = 5$ and $x - y = 2$? If such a pair of integers exists, is it unique? We have $x = 5 - y = 5 - (x - 2) = 7 - x$, so $2x = 7$. For any integer x, $2x$ is an even number, whereas 7 is odd, so they cannot be equal. Actually, x must be $7/2$, which is not an integer. Thus, no integer solution exists (although a rational number solution exists).

These examples are algebra problems in which the possible solutions are certain types of numbers. From Chapter 2 we know that there are analogous problems in which the possible solutions are functions. For example: Is there a function $f : Re \longrightarrow Re$ that satisfies, for any real numbers x and y, the equation $f(x+y) = f(x) + f(y)$? If such a function f exists, is it unique? The "unknown quantity" here is not a number, but instead is a function. A little thought yields the solution $f(x) = x$, the identity function on Re. But this solution is not unique, because a little more thought suggests a set of solutions,

$$S = \{\, f : Re \longrightarrow Re \mid \text{there is nonzero } a \in Re$$

$$\text{such that, for every } x \in Re,\ f(x) = ax \,\}.$$

This is the set of all functions from the reals to the reals that have the form $f(x) = ax$, where a is a nonzero constant real number. For any such function, and all $x, y \in Re$,

$$f(x + y) = a(x + y) = ax + ay = f(x) + f(y),$$

so all such functions are solutions, and we do not have uniqueness. It is an interesting, further question whether any solution of this existence problem is an element of S, but this question is beyond the scope of this book. This example is presented here to emphasize that there are interesting existence and uniqueness questions about functions as well as numerical "unknowns." The Recursion Theorem is about the existence and uniqueness of some very important types of functions.

Suppose that M is any nonempty set and that F is any function mapping M to M. Let $e \in M$. All of this is very general: M might have only the one element e, or M might be a large set such as Re, the set of real numbers. All we assume is that we are given $e \in M$ and $F : M \longrightarrow M$, together with $\langle Nat, 0, s \rangle$. Under these assumptions, the Recursion Theorem states that there is a unique function,

$$r : Nat \longrightarrow M,$$

satisfying

$$r(0) = e,$$

and

$$\text{for every } k \in Nat, \ r(s(k)) = F(r(k)).$$

Because of this theorem, we can use these two conditions as a *recursive definition* of the function r. The conditions also provide a computational procedure for r, if we can compute values of F. From the recursive definition, we obtain

$$r(0) = e,$$

$$r(s(0)) = F(r(0)) = F(e),$$

$$r(s(s(0))) = F(r(s(0))) = F(F(e)),$$

and so on. To compute $r(s(0))$, the definition says: "Compute $r(0)$, then use this value to compute $F(r(0))$." More generally, to compute $r(s(k))$, we must first compute $r(k)$. In computer programming jargon, when we compute the value of the function $r(k)$, we are *calling* the function r for the input k. Since computing $r(s(k))$ requires first computing $r(k)$, it is said that the function r (for the input $s(k)$) *calls itself* for the input k. This is described as a *recursive function call*. Here are a couple of simple examples.

Example.

Consider the trivial case where $M = \{e\}$. Then $F(e) = e$, so

$$r(0) = e = r(s(0)) = r(s(s(0))) = \cdots. \ \blacksquare$$

Example.

Suppose that we are given the real numbers, Re, and $F : Re \longrightarrow Re$, where $F(x) = x^2$ for any $x \in Re$. Assume the standard laws for exponents, and also suppose that $e = 3$. Define r on Nat by

$$r(0) = 3,$$

and

$$r(s(k)) = F(r(k)).$$

Then we have

$$r(0) = 3, \; r(s(0)) = F(r(0)) = F(3) = 3^2,$$

$$r(s(s(0))) = F(r(s(0))) = F(3^2) = (3^2)^2 = 3^4, \; \cdots. \; \blacksquare$$

We cannot directly apply the Recursion Theorem in this form to define our *sum* function. However, a later version of the Recursion Theorem (Theorem 3-18, in Section 3.3.1) will justify the definition of *sum*, as well as the factorial, and many other familiar functions. The current version of the theorem, Theorem 3-4, which states the existence of the unique function r from *Nat* to M, is illustrated in Figure 3.1.

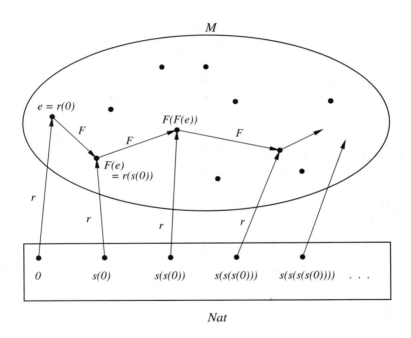

Figure 3.1. *The Recursion Theorem*

Theorem 3-4 **(Recursion Theorem).** Let M be a set, $e \in M$, $F : M \longrightarrow M$, and s be the successor function on *Nat*. Then there exists a unique function

$$r : Nat \longrightarrow M$$

satisfying

$$(1)\ r(0) = e$$

and

$$(2)\ \text{for every } k \in \textit{Nat},\ r(s(k)) = F(r(k)).$$

Proof. (Note. This proof is fairly long and complex. The reader may wish to skip the proof on a first reading, and then return to it later, after studying some of the later proofs in this chapter.)

In order to prove this theorem, we must somehow define r, prove that it really is a function from *Nat* to M, that it satisfies conditions (1) and (2) in the statement of the theorem, and that it is unique. To motivate the definition of r, the proof is developed in a methodical, stepwise fashion. It can be rewritten more concisely.

Since r is to be a function from *Nat* to M, we must have $r \subseteq \textit{Nat} \times M$. There are many such subsets. They are all relations between *Nat* and M, but we are interested only in ones that are functions satisfying (1) and (2). To facilitate the proof, for any R, define

$$\Delta(R)\ \text{iff}$$

$$(R \subseteq \textit{Nat} \times M)\ \&\ R0e\ \&\ (\text{for every } k \in \textit{Nat},\ x \in M,\ Rkx \ \Rightarrow\ Rs(k)F(x)\,).$$

Δ is a predicate that applies to relations, and there are relations that satisfy it. For instance, *Nat* $\times M$ does, since it contains all pairs of elements between *Nat* and M. In general, there will be many relations satisfying Δ, but most of them will not be functions. The rest of the proof is divided into several parts.

Part 1. I will first prove that the domain of any relation satisfying Δ is *Nat*. Suppose that $\Delta(R)$ and define the predicate ϕ by, for any $k \in \textit{Nat}$,

$$\phi(k)\ \text{iff there exists } x \in M \text{ such that } Rkx.$$

Induction will show that all natural numbers satisfy ϕ.

Case 0. Since $R0e$ is true, we have $\phi(0)$.

Case $s(k)$. Let $k \in \textit{Nat}$ and assume the induction hypothesis: $\phi(k)$. Then we have Rkx, for some $x \in M$. Since $\Delta(R)$, we have $Rs(k)F(x)$. Thus, if $\phi(k)$, then also $\phi(s(k))$, so we have proved the induction step. By IP, for all $n \in \textit{Nat}$, $\phi(n)$. This shows that the domain of R is *Nat*. Thus, if R also happens to be a function, then it is defined on all of *Nat*, i.e., it is a function from *Nat* to M.

Part 2. We still need to prove that such a function R exists. For now, suppose that R is a function satisfying Δ. Then we have $R0e$, so $R(0) = e$. Also, if we have $R(k) = x$, then Rkx, so also $Rs(k)F(x)$, so $R(s(k)) = F(x) = F(R(k))$. So, if R is a function, then it satisfies (1) and (2) in the statement of the theorem.

Part 3. Suppose that R_1 and R_2 are functions satisfying Δ. Let $\psi(k)$ be the predicate

$$R_1(k) = R_2(k).$$

We apply IP again.

Case 0. By Part 2, we have $R_1(0) = e = R_2(0)$.

Case $s(k)$. Assume the induction hypothesis $\psi(k)$, so $R_1(k) = R_2(k)$. By Part 2,

$$R_1(s(k)) = F(R_1(k)) \text{ and } R_2(s(k)) = F(R_2(k)).$$

Therefore, $R_1(s(k)) = R_2(s(k))$, which is $\psi(s(k))$. Hence, for all $n \in Nat$, $R_1(n) = R_2(n)$, so R_1 and R_2 are the same function.

Part 4. If we put together the results of Parts 1–3, we obtain the following: If there is any function satisfying Δ, then it is defined on Nat, it satisfies (1) and (2), and it is unique. I still have not proved that such a function exists, but suppose that it does, and let r_0 be this unique function. Let R be any relation satisfying Δ. I claim that $r_0 \subseteq R$. Let $\chi(k)$ be the predicate: $Rkr_0(k)$.

Case 0. $r_0(0) = e$. Since $R0e$, we have $R0r_0(0)$.

Case $s(k)$. Suppose that $\chi(k)$. We have $r_0(s(k)) = F(r_0(k))$. Since $\chi(k)$, $Rkr_0(k)$. Since $\Delta(R)$, $Rs(k)F(r_0(k))$, which is $Rs(k)r_0(s(k))$, which is $\chi(s(k))$.

Using IP, the two cases yield, for all $n \in Nat$, $Rnr_0(n)$. Let $u \in r_0$. Then, for some n, $u =< n, r_0(n) >$. But then $Rnr_0(n)$, so $u \in R$. Hence, $r_0 \subseteq R$.

We now know that if the unique function r_0 exists, it must be a subset of any R such that $\Delta(R)$. In other words, it must be the *smallest* such R. This is an important clue, suggesting that we define

$$T = \{\, R \mid \Delta(R) \,\}$$

and

$$r = \bigcap T.$$

From the definition of $\bigcap$, if $u \in r$, then $u \in R$, for all $R \in T$. Thus, $r \subseteq R$, for all R in T. Moreover, it is easy to check that, since $r = \bigcap T$, $\Delta(r)$ is true; so also $r \in T$. Thus, $r \in T$ and $r \subseteq R$ for all $R \in T$. Suppose there is another relation, say t, and that also $t \in T$ and $t \subseteq R$ for all $R \in T$. Then we would have $r \subseteq t$ and $t \subseteq r$, so $r = t$. This shows that $r = \bigcap T$ is the unique, smallest set in T, i.e., the unique, smallest relation satisfying Δ. Clearly, if r is a function, then it must be the unique function described in the statement of the theorem.

Part 5. All that remains is to prove that r really is a function. Again we use induction, this time on the predicate, ρ, given by

$$\rho(k) \text{ iff for every } x, y, (rkx \ \& \ rky) \Rightarrow x = y.$$

A number k satisfies ρ iff r relates k to a unique object in M. If ρ is satisfied by all $n \in Nat$, then r is a function. The induction proof uses *reductio ad absurdum* (RAA). It will be shown that if $\rho(k)$ is false for some k, then we can define a new relation that satisfies Δ and is smaller than r. This contradicts the fact that r is the (unique) smallest relation satisfying Δ. This same method of deriving a contradiction is used for both the base case and the induction step.

> Case 0. Suppose, for RAA, that $\rho(0)$ is false. Since $r0e$ is true, there must be some $e_1 \in M$, such that $r0e_1$ is also true and $e_1 \neq e$. Define $r_1 = r - \{ <0, e_1> \}$. Then r_1 is a proper subset of r, so r_1 is smaller than r. But r_10e is true. Also, if r_1kx, then rkx, so $rs(k)F(x)$. Since 0 has no predecessor, $s(k) \neq 0$, so the pair
>
> $$< s(k), F(x) > \neq < 0, e_1 > .$$
>
> Thus, $r_1s(k)F(x)$. Therefore, $\Delta(r_1)$. But r is the smallest relation satisfying Δ, so this is the contradiction we sought. Hence, $\rho(0)$.
>
> Case $s(k)$. Assume the induction hypothesis: $\rho(k)$. We want to prove that $\rho(s(k))$. The domain of r is Nat, so rkx holds for some $x \in M$. By $\rho(k)$, x is unique. From rkx, we have $rs(k)F(x)$. Now suppose, for RAA, that $\rho(s(k))$ is false. Then there must be some $u \in M$, $u \neq F(x)$, such that also $rs(k)u$. Define $r_2 = r - \{< s(k), u >\}$. Since 0 has no predecessor,
>
> $$< s(k), u > \neq < 0, e >,$$
>
> so r_20e is true.
>
> Let $j \in Nat$, $v \in M$, and suppose that r_2jv. We need to prove that $r_2s(j)F(v)$. There are two subcases: $j \neq k$ or $j = k$. Keep in mind that r and r_2 are the same, except that r_2 lacks the pair $< s(k), u >$.

Subcase $j \neq k$. From r_2jv, we have rjv, and therefore $rs(j)F(v)$. But s is one–one, so $s(j) \neq s(k)$. Therefore,

$$< s(j),\ F(v) > \neq < s(k), u >,$$

so

$$< s(j),\ F(v) > \in r_2,$$

or $r_2s(j)F(v)$. Thus, when $j \neq k$, if r_2jv, then $r_2s(j)F(v)$.

Subcase $j = k$. From r_2jv, we again have rjv. From $j = k$, we obtain rkv. But we also have rkx, where x is unique by the induction hypothesis. From the uniqueness, $v = x$, so $F(v) = F(x)$. Because we have rkx, we also have $rs(k)F(x)$, which is $rs(k)F(v)$. From the RAA hypothesis, $u \neq F(x)$. So, $u \neq F(v)$. Thus,

$$< s(k),\ F(v) > \neq < s(k), u >,$$

so we obtain $r_2s(k)F(v)$. And so, when $j = k$, if r_2jv, then $r_2s(j)F(v)$.

Putting the two subcases together shows that, whenever r_2jv, then

$$r_2s(j)F(v).$$

Since we also have r_20e, $\Delta(r_2)$. But this is the second contradiction we have sought, as r_2 is a proper subset of r. So the induction step has also been proved. Therefore, by IP, r is a function. **Q.E.D.**

The rest of this chapter develops these and related ideas in considerable detail, and it illustrates and discusses many different kinds of applications of recursive definition and mathematical induction. First, here is an example, followed by some exercises.

Example.

Problem Statement. Consider this sequence of rational numbers

$$1/1,\ 1/3,\ 1/5,\ 1/7,\ 1/9,\ 1/11,\ \ldots .$$

Assuming that the simple pattern of this sequence continues, write a recursive definition of a function r from *Nat* to *Rat*$^+$ whose output values are the numbers in this sequence. Show how your definition fits the Recursion Theorem. Compute a table of values that r returns for the inputs 0, $s(0)$, $s \circ s(0)$, $s \circ s \circ s(0)$.

Discussion. Since r is defined on *Nat*, we must begin with $r(0)$. The set M in this problem is Rat^+, and $1/1$ is obviously the constant e in the Recursion Theorem. It is also clear that the sequence consists of reciprocals of odd natural numbers. We need to determine the function F that maps Rat^+ to Rat^+. Consider the term $1/5$. How is it related to the next term $1/7$? The reciprocal of $1/5$ is 5. Compute $5 + 2 = 7$. This is the denominator of $1/7$. In other words, we have

$$1/7 = \frac{1}{\frac{1}{(1/5)} + 2} \; .$$

This suggests the following *recurrence* relationship: Let a_k be the kth term of the sequence of values. Then

$$a_{s(k)} = \frac{1}{(1/a_k) + 2} \; .$$

In other words, this recurrence relation states how one term is related to the next term. This relation is still just a guess, but the written solution will show that it fits the sequence.

k	$r(k)$
0	$1/1$
$s(0)$	$\dfrac{1}{1/r(0) + 2} = \dfrac{1}{1 + 2} = 1/3$
$s \circ s(0)$	$\dfrac{1}{1/r(s(0)) + 2} = \dfrac{1}{1/(1/3) + 2} = 1/5$
$s \circ s \circ s(0)$	$\dfrac{1}{1/r(s \circ s(0)) + 2} = \dfrac{1}{1/(1/5) + 2} = 1/7$

Table 3.1.

Written Solution. Let $M = Rat^+$, $e = 1/1$, and $F : Rat^+ \longrightarrow Rat^+$, where

$$F(x) = \frac{1}{(1/x) + 2}$$

for all x in Rat^+. Define $r : Nat \longrightarrow Rat^+$ by

(1) $r(0) = e$, and (2) for every $k \in Nat$, $r(s(k)) = F(r(k))$.

Using the given values, we have

(1) $r(0) = 1/1$, and (2) for every $k \in Nat$, $r(s(k)) = \dfrac{1}{(1/r(k)) + 2}$.

The first few values are given in Table 3.1. ■

Exercises

General Instructions

Throughout this chapter, whenever lowercase s is used as a function on Nat, s is the successor function on Nat. You may use the concepts and developments in Section 3.1.2 and whatever additional information is given in a specific problem. If a problem introduces a set or function pertaining to elements that are not natural numbers, you may use elementary facts about the set or function.

Exercises for which answers are provided at the back of the book are marked with "Answer Provided."

E 3-1 (Answer Provided) Let $M = \{a, b, c\}$, where a, b, c are alphabetic characters. Let $F : M \longrightarrow M$, with the values, $F(a) = b$, $F(b) = c$, $F(c) = a$. Write the recursive definition of a function, $r : Nat \longrightarrow M$, which satisfies $r(0) = a$ and uses M and F as in the statement of the Recursion Theorem. Also, what are the values of $r(k)$ for

$$k = s(0),\ s(s(0)),\ s(s(s(0))),\ s(s(s(s(0)))),\ s(s(s(s(s(0)))))?$$

E 3-2 Assume that we are given $M = Re^+$ and the function on the positive reals given by $F(x) = x/2$. Use F to write the recursive definition of a function $r : Nat \longrightarrow Re^+$ satisfying $r(0) = 1$. Also, compute the values of $r(k)$ for

$$k = s(0),\ s(s(0)),\ s(s(s(0))),\ s(s(s(s(0)))),\ s(s(s(s(s(0)))))).$$

E 3-3 Assume that we are given $M = Re$ and the function on the reals given by $F(x) = x + 1$. Use F to write the recursive definition of a function $r : Nat \longrightarrow Re$ satisfying $r(0) = 0$. Also, compute the values of $r(k)$ for

$$k = s(0),\ s(s(0)),\ s(s(s(0))),\ s(s(s(s(0)))),\ s(s(s(s(s(0)))))).$$

E 3-4 Let $M = Re^+$, $e = \sqrt{2}$, and $F(x) = \sqrt{x+1}$, for any x in Re^+. Use this data to write a recursive definition of a function r from *Nat* to Re^+ that satisfies the conditions of the Recursion Theorem. Also, compute a table of values of r for the inputs 0, $s(0)$, $s \circ s(0)$, $s \circ s \circ s(0)$, $s \circ s \circ s \circ s(0)$. It is not necessary to evaluate the various square roots; just write them with the radical sign $\sqrt{\ }$.

E 3-5 Consider this sequence of positive integers

$$3, 6, 9, 12, 15, \cdots .$$

Assuming that the simple pattern of this sequence continues, write a recursive definition of a function r from *Nat* to Rat^+ whose output values are the numbers in this sequence. Show how your definition fits the Recursion Theorem. Compute a table of values that r returns for the inputs 0, $s(0)$, $s \circ s(0)$, $s \circ s \circ s(0)$, $s \circ s \circ s \circ s(0)$.

E 3-6 Consider this sequence of rational numbers,

$$1/1, -1/2, 1/4, -1/8, 1/16, \cdots .$$

Assuming that the simple pattern of this sequence continues, write a recursive definition of a function r from *Nat* to *Rat* whose output values are the numbers in this sequence. Show how your definition fits the Recursion Theorem. Compute a table of values that r returns for the inputs 0, $s(0)$, $s \circ s(0)$, $s \circ s \circ s(0)$, $s \circ s \circ s \circ s(0)$.

E 3-7 Let k be any element of *Nat*. Prove that $k \neq s(k)$. (Hint. $0 \neq s(0)$. Why? Now finish the proof by using induction. Please do not use $+$, addition, in your proof. The proof should use only the concepts and developments in Section 3.1.2.)

E 3-8 Let M be a set, $e \in M$, and $F : M \to M$. Define r on *Nat* by

$$r(0) = e, \text{ and, for every } k \in Nat, r(s(k)) = F(r(k)).$$

Assume that F maps M one-one and onto M. Define a function F for which r does not map onto M. Also, define a function F for which r is not a one-one function on *Nat*. These may be two independent examples. Justify your work. (Hint: This will require some guesswork and testing. You might consider a one-one mapping of a small finite set onto itself. You might also consider a simple F on Rat^+, or Re^+.)

3.2 Basic Arithmetic

3.2.1 Some Simple Functions

This section examines some of the basic properties of the natural numbers and of simple functions and relations defined on *Nat*. These developments amount to a brief introduction to that part of number theory that is covered by Peano arithmetic. It is a very interesting part of pure mathematics. However, my main goal is to provide useful examples of recursive definitions and mathematical induction proofs. The techniques that are illustrated have many useful applications.

We begin with a more detailed characterization of the natural numbers. The set M in the Recursion Theorem may be any set. When developing Peano arithmetic, one begins by letting $M = Nat$. Often it is convenient to let the constant $e = 0$. But what functions, F, mapping Nat to Nat can one use? Peano's Axioms provide us with only the successor function, s, so it is the function with which we begin. It turns out that a huge variety of other functions, including addition and multiplication, can be defined by starting with s. We can also use recursive definitions that begin with numbers other than 0.

From now on, I will use 1 as an abbreviation for $s(0)$, 2 as an abbreviation for $s(1)$, etc. Suppose we want to use a recursive definition beginning with 1. Assume that we are given $F : M \longrightarrow M$, and $e \in M$. By the Recursion Theorem, there is a unique function r_1 satisfying

$$r_1(0) = e$$

and

$$r_1(s(k)) = F(r_1(k)).$$

Let $Nat^+ = Nat - \{0\}$. For any $k \in Nat$, define

$$r(s(k)) = r_1(k).$$

Then clearly r is a function. Also, by Theorem 3-3, any element of Nat^+ has the form $s(k)$ for some unique $k \in Nat$, so $r : Nat^+ \longrightarrow M$. Finally,

$$r(1) = r_1(0) = e$$

and, for $k \neq 0$,

$$r(s(k)) = r_1(k) = r_1(s(j)),$$

where $k = s(j)$. Then

$$r_1(s(j)) = F(r_1(j)) = F(r(s(j))) = F(r(k)),$$

so

$$r(s(k)) = F(r(k)).$$

Hence, r satisfies a recursive definition starting with 1. Now suppose that there exists a function q such that

$$q(1) = e$$

and, for $k \neq 0$,

$$q(s(k)) = F(q(k)).$$

Then we could define a new function q_1 such that, for any $k \in Nat$,

$$q_1(k) = q(s(k)).$$

Then $q_1(0)=e$ and $q_1(s(k)) = F(q_1(k))$. These are the same recursion equations satisfied by r_1. By the Recursion Theorem, there is a unique function satisfying these equations, so $r_1 = q_1$. Hence, for any $k \in Nat$,

$$r(s(k)) = r_1(k) = q_1(k) = q(s(k)).$$

If $i \in Nat^+$, then i has a unique predecessor k, such that $i = s(k)$. Thus,

$$r(i) = r(s(k)) = q(s(k)) = q(i).$$

Consequently, r is the unique function satisfying the recursive definition that begins with 1 instead of 0. This shows that we can apply the Recursion Theorem to definitions beginning with 1. Similarly, it can be applied to definitions beginning with 2, or 3, etc. Here is an example starting with 1.

Definition 3-5 Let $k \in Nat^+$. The *predecessor function* is denoted by *prv* and satisfies

$$prv(1) = 0$$

and

$$prv(s(k)) = s(prv(k)).$$

Observe that

$$prv(1) = 0,$$

$$prv(2) = prv(s(1)) = s(prv(1)) = s(0) = 1,$$

$$prv(3) = prv(s(2)) = s(prv(2)) = s(1) = 2.$$

It is clear that *prv* is the function that returns the unique predecessor of k, for any $k \neq 0$. This can be proved by a simple induction argument.

It is often convenient to use *prv* in recursive definitions. Instead of

$$r(0) = e$$

and

$$r(s(k)) = F(r(k)),$$

we can write

$$r(0) = e$$

and, for $k \neq 0$,

$$r(k) = F(r(prv(k))).$$

This definition can be written in a more compact form: Let $k \in Nat$ and define

$$r(k) = [\text{ if } k = 0 \text{ then } e \text{ else } F(r(prv(k)))] .$$

This equation is written in the *if-then-else format* that was introduced in the previous chapter. The right-hand side of the equation means: If $k = 0$, then r "returns" the value e (i.e., $r(0) = e$); otherwise (i.e., if $k \neq 0$), then $r(k) = F(r(prv(k)))$. We will often define functions using this format.

We know that *Nat* contains 0, $s(0)$, $s(s(0)) = s \circ s(0)$, etc., which we now may write in the forms, 0, 1, 2, etc. We also know that any natural number is either 0 or the successor of some natural number. One might hypothesize that any nonzero natural number can be obtained by starting with 0 and repeatedly forming successors for some "number" of times. In other words, one might ask whether any nonzero natural number can be *generated* from zero by repeatedly applying the successor function. Here is a way to show this.

Let $M = \mathcal{P}(Nat) - \{\varnothing\}$, so that the elements of M are all of the nonempty subsets of *Nat*. Let $e = \{ 0 \}$, and for any $S \in M$, let

$$F(S) = S \cup \{ y \mid \text{there exists } x \in S \text{ such that } y = s(x) \} = S \cup Img(s, S).$$

Then $F : M \longrightarrow M$. For example,

$$F(\{1, 7\}) = \{1, 7\} \cup \{2, 8\} = \{1, 7, 2, 8\}.$$

Now use the Recursion Theorem to define

$$r(k) = [\text{ if } k = 0 \text{ then } \{0\} \text{ else } F(r(prv(k)))].$$

Then

$$r(0) = \{0\},$$

$$r(1) = F(r(0)) = F(\{0\}) = \{0\} \cup \{1\} = \{0, 1\},$$

$$r(2) = F(r(1)) = F(\{0, 1\}) = \{0, 1, 2\}.$$

Since r maps to M, for every $k \in Nat$, $r(k)$ is a set of natural numbers. Intuitively, for each k, $r(k)$ is just the set of all natural numbers between 0 and k, inclusively, and the nonzero elements of this set are generated by repeatedly applying the successor function to 0. Now let

$$A = \bigcup \{\, r(k) \mid k \in Nat \,\},$$

so that every nonzero element in A is generated from 0. If $i \in A$, then $i \in r(i)$, so $i \in Nat$. Thus, $A \subseteq Nat$. Notice that $0 \in A$. Also, if $i \in A$, then $i \in r(i)$, so $s(i) \in r(s(i))$. Hence, if $i \in A$, then $s(i) \in A$. By the Induction Axiom, it follows that $A = Nat$. So, any natural number is either 0 or is generated from 0 by repeatedly applying the successor function. It is important to remember that this result depends on our use of informal set theory as well as Peano's Axioms. The exact assumptions that we have used are subject to more refined analysis in careful studies of the foundations of mathematics.

It was shown above that the Recursion Theorem can be applied where the base case (starting number) is 1, or 2, etc. The Recursion Theorem can also be generalized in other ways. To keep things as simple as possible and to save space, I will not state most of these generalizations as explicit theorems. Instead, they will be introduced in a more casual style, as needed. One useful generalization is obtained very easily as follows:

Suppose we are given a set M and two functions,

$$F : M \longrightarrow M$$

and

$$G : Nat \longrightarrow M.$$

Then, for each $i \in Nat$, $G(i)$ is an element of M. Then by the Recursion Theorem, there exists a unique function $r_i : Nat \longrightarrow M$ satisfying

$$r_i(0) = G(i)$$

and

$$r_i(s(k)) = F(r_i(k)).$$

Thus, there exists an infinite set of functions, r_i, where $i \in Nat$. Now define

$$r(i, k) = r_i(k)$$

so that $r : Nat \times Nat \longrightarrow M$. Then r satisfies

$$r(i, 0) = G(i),$$

and

$$r(i, s(k)) = F(r(i, k)),$$

so we can use these two equations to give a recursive definition of $r(i, k)$. As an illustration, I will now define addition on Nat.

Definition 3-6 Let $i, k \in Nat$. The function

$$plus : Nat \times Nat \longrightarrow Nat$$

is defined by

$$plus(i, 0) = i,$$

and

$$plus(i, s(k)) = s(plus(i, k)).$$

In this definition, $M = Nat$, $G(i) = i$, and $F(i) = s(i)$. It can be seen that $plus(5, 0) = 5$, and

$$plus(6, 7) = s(plus(6, 6)) = s \circ s \circ s \circ s \circ s \circ s \circ s(plus(6, 0)) = 13.$$

Using the alternative format, the definition can be stated as

$$plus(i, k) = [\text{ if } k = 0 \text{ then } i \text{ else } s(plus(i, prv(k)))].$$

Multiplication will be defined next, but we will first look at another useful format for recursive definitions. Suppose that we are given a set, M, and two functions, $F : Nat \times M \longrightarrow M$ and $G : Nat \longrightarrow M$. Let $x \in M$. For each $i \in Nat$, define $F_i(x) = F(i, x)$. Then we obtain an infinite set of functions, F_i, and each of these maps M to M. By the Recursion Theorem, for each i there exists a unique function $r_i : Nat \longrightarrow M$ satisfying

$$r_i(0) = G(i)$$

and

$$r_i(s(k)) = F_i(r_i(k)).$$

Now define $r(i, k) = r_i(k)$. Then, by some substitutions, it follows that

$$r(i, 0) = G(i)$$

and

$$r(i, s(k)) = F(i, r(i, k)),$$

so we can use these two equations to give a recursive definition of r. Using the more compact format, we can write

$$r(i, k) = \quad [\ \text{if } k = 0 \text{ then } G(i) \text{ else } F(i, r(i, prv(k)))\ ]\ .$$

This method of generalization can be extended further to define functions of the form $r(i, j, k)$, $r(i, j, m, k)$, etc., in which the recursion is performed on k while the other variables take arbitrary values.

Consider the function $plus : Nat \times Nat \longrightarrow Nat$. If we again let $M = Nat$, we obtain $plus : Nat \times M \longrightarrow M$. Let $G(i) = 0$, for all i. Now the new format can be used to define multiplication.

Definition 3-7 Let $i, k \in Nat$. The function

$$times : Nat \times Nat \longrightarrow Nat$$

is defined by

$$times(i, k) = \quad [\ \text{if } k = 0 \text{ then } 0 \text{ else } plus(i, times(i, prv(k)))\ ]\ .$$

For example, $times(5, 0) = 0$, and

$$times(3,\ 2) = plus(3,\ times(3,\ 1)) =$$

$$plus(3,\ plus(3,\ times(3,\ 0))) = 6.$$

If one has a programming language that includes (or permits the definition of) the successor function and also supports recursive definitions of functions, then the preceding definitions can be used to define addition and multiplication. However, the addition and multiplication functions built into the programming language will usually be more efficient. Programmers may find it instructive to write little programs for these functions and for others that occur later in this chapter.

3.2.2   Additional Arithmetical Definitions

In the following, I will often write $plus(i, k)$ in the more conventional form, $i + k$. Notice that, for any k,

$$k + 1 = plus(k, 1) = plus(k, s(0)) = s(plus(k, 0)) = s(k), \qquad (3.1)$$

which one would expect. Equation (3.1) results simply from the definition of $plus$ and the notational change between $plus$ and $+$. Along with use of the $+$ sign, the conventional notation, ik, will usually be used instead of $times(i, k)$. It is not difficult to define many other arithmetical functions, e.g., i^k. Using Peano's Axioms, one can also prove elementary theorems of arithmetic, such as commutativity of addition,

$$i + j = j + i.$$

It is not obvious from the axioms and preceding definitions that this is true. I will not develop the theory of arithmetic systematically, but the most basic parts of this development will be presented. This will show the remarkable power of Peano's Axioms as a foundation for the natural number system. It will also provide many examples of induction proofs and the heuristic rules guiding the construction of such proofs.

Theorem 3-8   For any $k \in Nat$:

$$0 + k = k, \ 0k = 0, \ 1 + k = s(k), \ 1k = k.$$

Proof. (First equation)

Let $\phi(k)$ be: $0 + k = k$, and use induction on k.

Case 0. By the definition of addition, $0 + 0 = 0$, so the base case holds.

Case $k + 1$. Assume $\phi(k)$. Then we obtain the following:

By notational rewriting,

$$0 + (k + 1) = plus(0, (k + 1)).$$

By Equation (3.1),

$$plus(0, (k + 1)) = plus(0, s(k)).$$

By the definition of $plus$,

$$plus(0, s(k)) = s(plus(0, k)).$$

By notational rewriting,

$$s(plus(0, k)) = s(0 + k),$$

so, by linking these equations, we have,

$$0 + (k + 1) = s(0 + k).$$

Now, using the induction hypothesis, $\phi(k)$, and Equation (3.1), we obtain

$$0 + (k + 1) = s(0 + k) = s(k) = k + 1,$$

which is the statement of $\phi(k + 1)$.

The proofs of the remaining cases are left as exercises. **Q.E.D.**

Note that this proof proceeds until a stage is reached where the induction hypothesis (I.H.) can be applied. After applying the I.H., the proof continues to produce the statement of $\phi(k + 1)$. It is a conditional proof that shows that $\phi(k)$ implies $\phi(k + 1)$, as is explained in the discussion following the Schema for Proofs Using the Induction Principle.

The next theorem involves more than one variable. In this kind of situation, it is important to decide which variable should be used for the mathematical inductive reasoning. Induction proofs typically make use of recursive definitions, and the induction step of the proof usually corresponds to the recursive step of a definition. Therefore, by looking at the form of a recursive definition, one can often get a useful clue regarding the variable to be used for the inductive argument.

Theorem 3-9   For any i, j, $k \in Nat$,

$$i + (j + k) = (i + j) + k \quad \text{(association)},$$

and

$$i + k = k + i \quad \text{(commutation)}.$$

(*Discussion of the Proof.* In the definition of *plus*, the recursion was done on the second of the two arguments in the function. In the equation

$$i + (j + k) = (i + j) + k,$$

i occurs as the first argument to $+$ on both sides, and j occurs as the first argument on the left side. Only k occurs as the second argument to $+$ on both sides of the equation. Thus, we should try induction on k. This is done in Part 1 of the proof.)

Proof.

Part 1. Let $\phi(k)$ be

$$i + (j + k) = (i + j) + k.$$

We allow i, j to be any elements of *Nat* and consider the induction proof cases for k.

Case 0. Using the definition of addition,

$$i + (j + 0) = i + j = (i + j) + 0.$$

Case $k + 1$. Assume $\phi(k)$.

By Equation (3.1) and the definition of *plus*,

$$i + (j + (k + 1)) = i + (j + s(k)) = i + s(j + k) = s(i + (j + k)).$$

Applying the induction hypothesis, $\phi(k)$, we obtain

$$i + (j + (k + 1)) = s(i + (j + k)) = s((i + j) + k).$$

Hence, by the definition of *plus* and Equation (3.1),

$$i + (j + (k + 1)) = s((i + j) + k) = (i + j) + s(k) = (i + j) + (k + 1),$$

which is $\phi(k + 1)$.

(*Remark.* As in the proof of Theorem 3-8, this proof proceeds until a stage is reached where the induction hypothesis (I.H.) can be applied. After applying the I.H., the proof continues until the statement of $\phi(k + 1)$ is obtained. It is a conditional proof that shows that $\phi(k)$ implies $\phi(k + 1)$, as is explained in the discussion following the Schema for Proofs Using the Induction Principle. This kind of proof structure is also used in Part 2 and in other IP proofs.)

Part 2.

(*Discussion.* In the equation $i + k = k + i$, neither variable occurs only as the second argument. Fortunately, the equation has a kind of symmetry. If i and k are swapped with each other, we get $k + i = i + k$, which is equivalent to the original equation. Thus, if we can accomplish an induction proof at all, we would expect to be able to use either variable. I will use k.)

Let $\phi(k)$ be $i + k = k + i$.

Case 0. $i + 0 = i$ by the definition of addition. Also, by the previous theorem, $0 + i = i$. Thus, $i + 0 = 0 + i$, which is the base case.

Case $k + 1$. Assume $\phi(k)$. From Equation (3.1) and the definition of addition, we have

$$i + (k + 1) = i + s(k) = s(i + k).$$

Substituting the $\phi(k)$ equation, and again using the definition of addition,

$$s(i + k) = s(k + i) = k + s(i).$$

By Theorem 3-8,

$$k + s(i) = k + (1 + i).$$

After linking the preceding equations, the result follows by using the associativity of addition:

$$i + (k + 1) = k + (1 + i) = (k + 1) + i,$$

which is $\phi(k + 1)$. **Q.E.D.**

   As expected from the symmetry just mentioned, the same kind of proof can be achieved by induction on i.[*] Other familiar principles of arithmetic can be proved. A few more are given in the following lemma and theorems. Notice that Equation (3.1) and the definition of *times* yield

$$is(k) = i(k + 1) = i + ik.$$

Applying commutativity of $+$ yields

$$i(k + 1) = ik + i.$$

We should also expect that $(k + 1)i = ki + i$. This is proved in the next lemma.

Lemma 3-10   If $i, k \in Nat$, then $(k + 1)i = ki + i$.

Proof.

In the definitions of both *plus* and *times*, the recursion is done on the second argument in the functions. The variable i occurs as the second argument to

---

[*]Boyer and Moore (1979), (1998) describe a system of computational logic that has been implemented as a distinguished computer program. The authors discuss heuristics for proofs, including proofs by induction.

times on both sides of this equation and also occurs as the second argument to *plus*. Thus, I try induction on i. Let

$$\phi(i) \text{ be } (k+1)i = ki + i.$$

Case 0. From the definitions of addition and multiplication, we have $(k+1)0 = 0$ and $k0 + 0 = 0 + 0 = 0$, so $(k+1)0 = k0 + 0$.

Case $i+1$. Assume $\phi(i)$. By the definition of multiplication, and the commutativity of addition,

$$(k+1)(i+1) = (k+1) + (k+1)i = (k+1)i + (k+1).$$

From the induction hypothesis, $\phi(i)$, we obtain

$$(k+1)(i+1) = (ki + i) + (k+1).$$

By repeated use of association and commutation,

$$(k+1)(i+1) = ki + (i + (k+1)),$$
$$= ki + ((i+k) + 1),$$
$$= ki + ((k+i) + 1),$$
$$= ki + (k + (i+1)),$$
$$= (ki + k) + (i+1),$$
$$= (k + ki) + (i+1).$$

By the definition of multiplication,

$$(k+1)(i+1) = k(i+1) + (i+1),$$

which is $\phi(i+1)$. **Q.E.D.**

The next theorem states commutativity and associativity of multiplication, and also the principle that multiplication distributes over addition. In ordinary, intuitive arithmetic, these principles are taken for granted, along with others, such as the associative and commutative properties of addition (proved in Theorem 3-9).* If we believe in the correctness of ordinary arithmetic and want to use Peano's Axioms as its foundation, then we have an obligation to prove that the axioms imply these familiar principles.

---

*Imagine the confusion there would be in supermarket checkouts if addition were not both associative and commutative. The total bill could depend on the way the purchased items were grouped or ordered.

The previous lemma is used in the proof of commutativity; the other proofs are omitted. The proofs are enjoyable exercises using mathematical induction. It is illuminating to contemplate how the Induction Principle is related to the truth of the principles that are proved.

Theorem 3-11   For any i, j, $k \in Nat$,

$$ik = ki \; (commutation),$$

$$i(j + k) = ij + ik \; (distribution),$$

$$i(jk) = (ij)k \; (association).$$

Proof.

[Commutativity] Letting $\phi(k)$ be $ik = ki$, we use induction on k.

Case 0. $i0 = 0$ by definition and $0i = 0$ by Theorem 3-8, so $i0 = 0i$.

Case $k + 1$. By the definition of multiplication and the commutativity of addition, we have

$$i(k + 1) = i + ik = ik + i.$$

Using $\phi(k)$,

$$i(k + 1) = ki + i.$$

Applying the previous lemma yields

$$i(k + 1) = (k + 1)i,$$

which is $\phi(k + 1)$. **Q.E.D.**

Proofs of the distribution and association principles are left as exercises.

The properties of addition and multiplication enable us to simplify some proofs by rewriting expressions in new forms. The following theorem states two "cancellation laws" for equations between natural numbers. It also tells us that if the product of two numbers is 0, then at least one of these numbers must be 0.

Theorem 3-12   For any i, j, $k \in Nat$,

$$(i + k = j + k) \;\Rightarrow\; i = j, \tag{3.2}$$

$$ij = 0 \;\Leftrightarrow\; (i = 0 \text{ or } j = 0), \tag{3.3}$$

$$(k \neq 0 \;\&\; ik = jk) \;\Rightarrow\; i = j. \tag{3.4}$$

Proof.

Part 1. Let

$$\phi(k) \text{ be if } i + k = j + k, \text{ then } i = j.$$

I use induction on k, because it is the second argument to the addition function on both sides of the equation. Since $\phi(k)$ is a conditional, we assume its antecedent and prove its consequent.

Case 0. Suppose that $i + 0 = j + 0$. Then, from the definition of $+$, we immediately obtain $i = j$.

Case $k + 1$. Assume $\phi(k)$ and also suppose that

$$i + (k + 1) = j + (k + 1).$$

From associativity,

$$(i + k) + 1 = (j + k) + 1,$$

so

$$s(i + k) = s(j + k),$$

but s is one–one, so

$$i + k = j + k.$$

Now applying $\phi(k)$, we obtain $i = j$. Therefore, if

$$i + (k+1) = j + (k+1), \text{ then } i = j.$$

This conditional is $\phi(k+1)$, which proves the first assertion.

Part 2. If either $i = 0$ or $j = 0$, then from commutativity and the definition of multiplication, we have $ij = 0$.

For the converse, suppose that $ij = 0$. If $i \neq 0$, then by Theorem 3-3, there is some i_1 such that $i = i_1 + 1$. Hence, by commutation and distribution,

$$ij = (i_1 + 1)j = j(i_1 + 1) = ji_1 + j.$$

Now if $j \neq 0$, then, similarly, $j = j_1 + 1$, for some j_1. Then we would have

$$0 = ij = ji_1 + (j_1 + 1) = (ji_1 + j_1) + 1,$$

which is impossible, since 0 has no predecessor. Therefore, if $i \neq 0$, then $j = 0$. In other words, either $i = 0$ or $j = 0$, which proves the second proposition.

Part 3. Now let the I.H., $\psi(j)$, be as follows: For any $i, k \in Nat$,

$$(\text{if } k \neq 0 \ \& \ ik = jk), \text{ then } i = j.$$

We use induction on j.

Case 0. Suppose $k \neq 0$ and $ik = jk$. Since $j = 0$, $jk = 0$. Thus, $jk = 0 = ik$. Since $k \neq 0$, from Part 2, we have $i = 0$. Hence, $i = j$.

Case $j + 1$. Assume $\psi(j)$. Suppose that $k \neq 0$ and $ik = (j+1)k$. Now k and $j+1$ are both nonzero, so by Part 2, $(j+1)k$ is also nonzero. Hence, $ik \neq 0$. Again by Part 2, $i \neq 0$. Therefore, i has a predecessor: There is some i_1 such that $i = i_1 + 1$. We now have

$$ik = (j+1)k = k(j+1) = kj + k = jk + k$$

and

$$ik = (i_1 + 1)k = k(i_1 + 1) = ki_1 + k = i_1 k + k.$$

Hence,

$$jk + k = i_1 k + k.$$

By Part 1 (cancellation law for addition), we obtain

$$jk = i_1 k.$$

Hence,

$$i_1 k = jk,$$

so by $\psi(j)$

$$i_1 = j.$$

But then $i_1 + 1 = j + 1$, so $i = j + 1$. Thus, $\psi(j)$ implies $\psi(j+1)$. **Q.E.D.**

Starting with Peano's Axioms and elementary set theory, we have now derived the basic principles about equations involving addition and multiplication. Using the preceding theorems, it is not difficult to prove other standard facts about such equations. We now consider the ordering of the natural numbers.

It will be recalled from the previous chapter that a simple order on a domain A is a binary relation that is transitive, antisymmetric, and strongly connected on A. In that chapter, I indicated that $\langle$ *Nat*, $\leqslant$ $\rangle$ is a simple order system. We are now in a position to prove that this relational system is indeed a simple order. However, in order to do this, we first need to state a "reasonable and natural definition" of $\leqslant$ on *Nat*. This is usually done in one of two ways. Suppose that $x \leqslant y$. Then, intuitively, either $x = y$ or else applying the successor function to x a finite number of times should yield y. This approach is fine, but is more complicated than necessary for our purposes. If $x \leqslant y$, then there is some number z (possibly zero) such that $x + z = y$. I will use this second approach.

Definition 3-13 Let $i, j \in Nat$, then

$$i \leqslant j \text{ iff there is } k \in Nat \text{ such that } j = i + k$$

and

$$i < j \text{ iff } (i \leqslant j \,\&\, i \neq j).$$

As is customary, $i \leqslant j$ is read as 'i is less than or equal to j', and $i < j$ is read as 'i is less than j'.

Theorem 3-14 $\langle Nat, \leqslant \rangle$ is a simple order (transitive, antisymmetric, and strongly connected).

Proof. [Transitivity] Let i, j, k, m, $n \in Nat$. If $i \leqslant j$ and $j \leqslant k$, then there exist m, n such that

$$j = i + m$$

and

$$k = j + n.$$

Hence, $k = (i + m) + n$, so $k = i + (m + n)$. Therefore, $i \leqslant k$, so $\leqslant$ is *transitive*.

[Antisymmetry] Now suppose that $i \leqslant j$ and $j \leqslant i$. Then there are m, n such that $j = i + m$ and $i = j + n$. Since $0 + j = j$, we have

$$0 + j = j = i + m = (j + n) + m = j + (n + m) = (n + m) + j.$$

Applying Equation (3.2), the cancellation law for addition,

$$0 = n + m.$$

Now suppose, for RAA, that $n \neq 0$. Then there is some $n_1 \in Nat$ such that $n = s(n_1)$, so

$$0 = s(n_1) + m = m + s(n_1) = s(m + n_1),$$

which is impossible, since 0 has no predecessor. Hence, $n = 0$, so also $m = 0$. Therefore, $i = j$ and $\leqslant$ is *antisymmetric*.

[Strongly connected] The two preceding subproofs used straightforward algebraic transformations. We now need to prove, for any i, $j \in Nat$, that ($i \leqslant j$ or $j \leqslant i$). This requires showing that there exists a number that, added to i, gives j, or else there is a number that added to j gives i. There does not appear to be any obvious way to attack this existence problem by simple algebraic transformations. Instead, we will try induction. Fortunately, the disjunction is symmetrical in i, j, so it should not matter which variable is used for the inductive proof. I shall use j.

Let $\phi(j)$ be ($i \leqslant j$ or $j \leqslant i$).

Case 0. For any i, $i = 0 + i$, by Theorem 3.8. Since $j = 0$, we have $i = j + i$. Thus, by the definition of $\leqslant$, $j \leqslant i$. Hence, $\phi(0)$.

Case $j + 1$. Assume that $\phi(j)$ is true. Since this is a disjunction, we break the proof into the subcases $i \leqslant j$ and $j \leqslant i$.

If $i \leqslant j$, then $i \leqslant j + 1$. Hence, $(i \leqslant j + 1$ or $j + 1 \leqslant i)$, which is $\phi(j + 1)$.

If $j \leqslant i$, then there is m such that $i = j + m$. We now consider two subsubcases:

If $m = 0$, then $i = j$. Thus, $i \leqslant j + 1$, from which $\phi(j + 1)$ again follows.

If $m \neq 0$, then there is some m_1 such that $m = s(m_1)$. Then

$$i = j + m = j + s(m_1) = j + (m_1 + 1) = (j + 1) + m_1.$$

Hence, $j + 1 \leqslant i$, whence $\phi(j + 1)$ again follows.

This completes the induction proof that $\leqslant$ is *strongly connected*. **Q.E.D.**

Using this theorem and the definition of $<$, it is easy to prove that $<$ is a strict simple order.

Theorem 3-15 $\langle Nat,\ < \rangle$ is a strict simple order (transitive, asymmetric, and connected).

Proof. Left as an exercise.

It is also not difficult to prove the usual laws about inequalities. For example, suppose that $i \leqslant j$. Then, for some $x \in Nat$, $i + x = j$. Then, applying previously proved results,

$$j + k = (i + x) + k = i + (x + k) = i + (k + x) = (i + k) + x,$$

so $i + k \leqslant j + k$.

At this point it is convenient to introduce subtraction for natural numbers. Of course, we may not subtract a larger number from a smaller one, so this subtraction function will be restricted.

Definition 3-16 Let $A = \{\ <i, j>\ |\ i, j \in Nat\ \&\ i \geqslant j\ \}$. The *subtraction function* maps A to Nat and is denoted by '$-$'. It is defined as follows:

Let $<i, j> \in A$ and $d \in Nat$. Then

$$i - j = d \Leftrightarrow i = j + d.$$

This definition must be *justified*, i.e., we must prove that it really defines a function. This is easy. We can certainly define the *relation* $S \subseteq A \times Nat$ by

$$S < i, j > d \Leftrightarrow i = j + d.$$

To prove that S is a function, we need to prove that S relates any pair, $< i, j >$ to a unique d. Uniqueness is proved in the usual manner. Suppose that we have $S < i, j > d_1$ and $S < i, j > d_2$. Then $i = j + d_1$ and $i = j + d_2$. Hence, $j + d_1 = j + d_2$, whence $d_1 = d_2$. Therefore, subtraction is a function defined on A.

It is easy to prove by induction that $prv(s(i)) = i$ for all $i \in Nat$. From this and the definition of prv, for $i \geqslant 1$, one gets $s(prv(i)) = i$. Thus, $prv(i) + 1 = i$. Therefore, for $i \geqslant 1$, we have $i - 1 = prv(i)$. From now on, I will often write $i - 1$ instead of $prv(i)$.

We can also show that multiplication distributes over subtraction. Suppose that $i - j = d$, so that also $i = j + d$. From the first of these equations, we obtain

$$k(i - j) = kd,$$

and from the second we obtain

$$ki = k(j + d) = kj + kd.$$

Thus,

$$ki - kj = kd = k(i - j).$$

This concludes our look at the foundations of arithmetic, although we will examine a few theorems of number theory in the next section. *Unless mentioned to the contrary, from now on we will assume the standard properties of addition, subtraction, multiplication, $<$ and $\leqslant$ on Nat. Since we want to consider some applications to the integers, rational, and real numbers, we will also assume their basic features.* In fact, these number systems can all be constructed within set theory, beginning with the natural numbers. Suitable constructions can be found in many sources, including Chapter 6 of Suppes (1972). The next examples may be helpful before trying the following exercises.

Example.

Problem Statement. Let $M = Rat^+$, $e = 1$, and

$$F(x) = \frac{1}{1 + x},$$

for any x in Rat^+. Use this information to write a recursive definition of a function r from Nat^+ to Rat^+ that satisfies the conditions of the Recursion

Theorem. Also, compute a table of values of r for the inputs $s(0)$, $s \circ s(0)$, and $s \circ s \circ s(0)$.

Discussion. This problem is straightforward; we simply write the two functional equations in the Recursion Theorem, starting with $s(0)$ instead of 0.

k	$r(k)$
$s(0)$	1
$s \circ s(0)$	$\dfrac{1}{1 + r(s(0))} = \dfrac{1}{1 + 1} = \dfrac{1}{2}$
$s \circ s \circ s(0)$	$\dfrac{1}{1 + r(s \circ s(0))} = \dfrac{1}{1 + \frac{1}{1+1}} = \dfrac{1}{1 + \frac{1}{2}} = \dfrac{2}{3}$

Table 3.2.

Written Solution. Define r as follows: For any k in Nat^+,

$$r(s(0)) = e = 1 \text{ and } r(s(k)) = F(r(k)) = \frac{1}{1 + r(k)}.$$

The first few values are given in Table 3.2. ■

Example.

Problem Statement. Theorem. Let a, d be natural numbers. Define

$$f(k) = [\text{ if } k = 0 \text{ then } a \text{ else } f(k - 1) + a + d \,].$$

Then for all natural numbers n, $f(n) = (n + 1)a + nd$.

Discussion. Let $\phi(k)$ be the statement that $f(k) = (k + 1)a + kd$. By the definition, $f(0) = a$, so the base case is easy. The induction step is proved by computing $f(k + 1)$.

Written Solution.

Theorem. Let a, d be natural numbers. Define

$$f(k) = [\text{ if } k = 0 \text{ then } a \text{ else } f(k - 1) + a + d \,].$$

Then, for all natural numbers n, $f(n) = (n + 1)a + nd$.

Proof. Let $\phi(k)$ be $f(k) = (k + 1)a + kd$.

Case $k = 0$. By the definition of f, $f(0) = a$. Also,

$$(0 + 1)a + 0(d) = a = f(0),$$

which establishes the base case.

Case $k + 1$. By the definition of f, we have $f(k + 1) = f(k) + a + d$. Assuming the induction hypothesis $\phi(k)$, we also have

$$f(k) = (k + 1)a + kd.$$

Substituting the last equation into the previous one yields

$$
\begin{aligned}
f(k + 1) = f(k) + a + d &= (k + 1)a + kd + a + d \\
&= (k + 2)a + (k + 1)d = ((k + 1) + 1)a + (k + 1)d,
\end{aligned}
$$

which proves $\phi(k + 1)$.   **Q.E.D.**   ∎

Example.

Problem Statement. Theorem. Define $r : Nat^+ \times Nat^+ \times Nat \longrightarrow Rat$: For every i, $j \in Nat^+$ and every $k \in Nat$,

$$r(i, j, k) = \left[\text{ if } k = 0 \text{ then } 1 \text{ else } \frac{ir(i, j, k - 1)}{j} \right].$$

Then

$$r(i, j, k) = \frac{i^k}{j^k}.$$

Discussion. The predicate to be proved true for all k in Nat is $\phi(k)$: For any i, $j \in Nat^+$,

$$r(i, j, k) = \frac{i^k}{j^k}.$$

In the base case, $r(i, j, 0) = 1$ by the definition of r, and

$$\frac{i^0}{j^0} = \frac{1}{1} = 1,$$

by direct calculation. Therefore,

$$r(i, j, 0) = \frac{i^0}{j^0} \, ,$$

so $\phi(0)$ (is true). In the induction step we assume the induction hypothesis $\phi(k)$, and try to prove $\phi(k + 1)$. One way to begin is by calculating

$$r(i, j, k + 1) = \frac{ir(i, j, k)}{j} \, .$$

By the induction hypothesis, $r(i, j, k) = \frac{i^k}{j^k}$; the rest is easy to write up.

Written Solution.

Theorem. Define $r : Nat^+ \times Nat^+ \times Nat \longrightarrow Rat$ as follows:

For every i, $j \in Nat^+$ and every $k \in Nat$,

$$r(i, j, k) = \left[\text{ if } k = 0 \text{ then } 1 \text{ else } \frac{ir(i, j, k - 1)}{j} \right].$$

Then

$$r(i, j, k) = \frac{i^k}{j^k} \, .$$

Proof. Let $\phi(k)$ be as follows: For any i, $j \in Nat^+$,

$$r(i, j, k) = \frac{i^k}{j^k} \, .$$

Case $k = 0$. We have $r(i, j, 0) = 1$ by the definition of r, and

$$\frac{i^0}{j^0} = \frac{1}{1} = 1,$$

by direct calculation. Therefore,

$$r(i, j, 0) = i^0 / j^0,$$

so $\phi(0)$.
Case $k + 1$. Assume $\phi(k)$. We have

$$r(i, j, k + 1) = \frac{ir(i, j, k)}{j} = \left(\frac{i}{j} \right) r(i, j, k).$$

Applying $\phi(k)$ yields

$$r(i, j, k+1) = (i/j)\frac{i^k}{j^k} = \frac{i^{k+1}}{j^{k+1}}.$$

■

Exercises

General Instructions

In this exercise set, unless mentioned to the contrary, you may assume the basic principles of arithmetic and algebra as applied to *Nat*. Several of the problems have more specific instructions; please be sure to follow these instructions where they occur.

Throughout this book, if an exercise is simply the statement of a theorem or a metatheorem, then the task is to prove this theorem or metatheorem. In your proof, you may use the definitions given in the text. Unless stated to the contrary, you may also use theorems proved in the text.

Exercises for which answers are provided at the back of the book are marked with "Answer Provided."

Finally, please note that exercises that introduce important principles or useful applications are especially difficult, or are special in some way, are indicated by bold print. In some cases, parenthetical comments describe the special features of the problem. If such comments are missing, the exercise is nonetheless of special value in introducing a new concept, principle, or technique.

E 3-9 Consider this sequence of positive rational numbers,

$$1/1^2, \ 1/2^2, \ 1/3^2, \ 1/4^2, \ 1/5^2, \ \cdots .$$

Assuming that the simple pattern of this sequence continues, write a recursive definition of a function r from Nat^+ to Rat^+ whose output values are the numbers in this sequence. Show how your definition fits the Recursion Theorem. Compute a table of values that r returns for the inputs $s(0)$, $s \circ s(0)$, $s \circ s \circ s(0)$, $s \circ s \circ s \circ s(0)$.

E 3-10 (Answer Provided) Complete the proof of Theorem 3-8 using results in the text that precede it.

E 3-11 In the text, the commutativity of $+$, $i + k = k + i$, was proved by induction on k (see Theorem 3-9). Prove this again by using induction on i and by using results that precede this theorem.

E 3-12 Complete the proof of Theorem 3-11 using results that precede it.

E 3-13 Let $k \in Nat$. Define the function $ident : Nat \longrightarrow Nat$ as follows:

$$ident(0) = 0$$

and

$$ident(s(k)) = s(ident(k)).$$

1. Show how this definition is an application of the Recursion Theorem (Theorem 3-4).

2. Compute the values of $ident(k)$ for $k = 0, 1, 2, 3, 4$.

3. *Theorem.* $ident$ is the identity function on Nat (see Definition 2-44).

E 3-14 Prove Theorem 3-15 using results that precede it.

E 3-15 (An important fact) *Theorem.* For any $k \in Nat$, $prv(s(k)) = k$. (Remark. This can be rewritten in the form, $(k + 1) - 1 = k$. It should be proved on the basis of the material in Section 3.2.)

E 3-16 For $i, k \in Nat$, define

$$r(k, i) = [\text{ if } k = 0 \text{ then } 0 \text{ else } r(k - 1, i) + i].$$

Prove that, for any $i, k \in Nat$, $r(i, k) = ki$. The proof should be based on the material in Section 3.2. (Caution: Be sure to keep track of the uses of i and k in the definition and in the assertion to be proved.)

E 3-17 Let a and r be positive natural numbers, and define the function g on Nat by

$$g(k) = [\text{ if } k = 0 \text{ then } a \text{ else } rg(k - 1)].$$

Prove that $g(k) = ar^k$. Note that this is the familiar geometric progression.

E 3-18 Let a be a positive natural number and define f on Nat by

$$f(0) = a, \text{ and } f(k + 1) = (f(k))^2.$$

Prove that

$$f(k) = a^{(2^k)}.$$

Assume the standard laws of exponents.

E 3-19 *Theorem.* Let G and H be functions from $Nat \times Nat$ to Nat. Define $r : Nat \times Nat \times Nat \longrightarrow Nat$ by

$$r(i, j, k) = [\ \text{if } k = 0 \text{ then } G(i, j) \text{ else } r(i, j, k - 1) + H(i, j)\ ].$$

Then, for all i, j, k in Nat, $r(i, j, k) = G(i, j) + kH(i, j)$.

E 3-20 Let d be a particular element of Nat, and define

$$r : Nat \times Nat \longrightarrow Nat$$

as follows: For any natural numbers i, k,

$$r(i, k) = [\ \text{if } k = 0 \text{ then } i \text{ else } r(i, k - 1) + d\ ].$$

Note the similarity between this function and the function defined in the previous exercise. Both of these are types of arithmetical progressions. Now prove that, for any i, k in Nat, $r(i + d,\ k) = r(i,\ k) + d$. Also, prove that, for all i, k in Nat, $r(i,\ k) = i + kd$.

E 3-21 Assume there exists a function $r : Nat \longrightarrow Nat$. Here are three conditions that r might satisfy:

C_1 For every i, $j \in Nat$, $r(i + j) = r(i) + r(j)$.

C_2 $r(0) = 0$, and for every $k \in Nat$, $r(k + 1) = r(k) + r(1)$.

C_3 For every $k \in Nat$, $r(k) = kr(1)$.

Prove that these three conditions are equivalent. Also, show that such functions exist by exhibiting a specific example and showing that it satisfies C_1. One way to establish the equivalence is to prove that C_1 implies C_2, that C_2 implies C_3, and that C_3 implies C_1.

E 3-22 Assume that there exists a function $\mathcal{F} : Nat \longrightarrow Nat$ that satisfies, for any $i \in Nat$,

$$\mathcal{F}(i) = [\ \text{if } i = 0 \text{ then } 1 \text{ else } 2i + \mathcal{F}(i - 1)\ ].$$

Prove the following:

1. $Img(\mathcal{F}, Nat)$ is a set of odd numbers. (Remark. The general form of an even natural number is $2a$, for some $a \in Nat$, and the general form of an odd natural number is $2b + 1$, for some $b \in Nat$.)

2. If $i > 0$, then $\mathcal{F}(i) > 1$.

3. $\mathcal{F} : Nat \xrightarrow{1-1} Nat$.

E 3-23 For this problem, *do not assume* that i^k has already been defined for *Nat*. You may assume the results in Section 3.2. For $i, k \in Nat$, define

$$r(i, k) = [\text{ if } k = 0 \text{ then } 1 \text{ else } ir(i, k - 1)\ ],$$

where $ir(i, k - 1)$ is the product of i with $r(i, k - 1$). Notice that $r(0, 0) = 1$. Prove the following:

1. For $k \in Nat^+$, $r(0, k) = 0$.

2. For any i, $r(i, 1) = i$.

3. For any $i, j, k \in Nat$, $r(i, j + k) = r(i, j)r(i, k)$.

4. For any i, j, k, $r(r(i, j), k) = r(i, jk)$.

(Remark. This problem defines i^k and establishes some of its basic properties. Intuitively, i^k equals the result of multiplying i by itself "k times," although the case $k = 0$ does not exactly fit this description. The case 0^0 is curious.)

E 3-24 *Theorem.* For any natural number k, $k^2 < 2^k$ if $5 \leqslant k$. (Hints. For this problem, freely assume the standard facts of arithmetic, including exponentiation and inequalities. Also, notice that if $k^2 < 2^k$, then $2k^2 < 2^{k+1}$. Also, for $5 \leqslant k$, we have $k(k - 2) > 1$, from which one can eventually obtain $2k^2 > k^2 + 2k + 1$. Work out the details.)

3.3   Extensions of Recursive Definition and Induction

3.3.1   Some Additional Applications of the Recursion Theorem

It was shown in Section 3.2.1 how recursive definitions can use for the base case the number 1 or 2, etc. It is also possible to start an induction proof with a nonzero base case. Let a be some particular nonzero number, and suppose that we wish to prove that ψ holds for a, $s(a)$, $s(s(a))$, $\cdots$. We can always define a new predicate ϕ by

$$\phi(k) \Leftrightarrow \psi(k + a).$$

An induction proof for ψ starting at a is equivalent to an induction proof for ϕ starting at 0. Thus, we are justified in proving ψ, starting at the base case a. We will see examples later in this chapter.

In Section 3.2.1, I defined a function r on *Nat* that returns the values

$$r(0) = \{0\}, \quad r(1) = \{0, 1\}, \quad \cdots.$$

A function of this kind is said to define a *sequence of sets*. More generally, a sequence of sets, of rational numbers, of real numbers, etc., can be thought of as the ordered set of values returned by a function defined on *Nat*, or on Nat^+. This function may, but need not, be defined recursively. The function itself is also sometimes called a "sequence." This subsection begins with some examples of sequences, especially sequences of sets, and their applications.

Examples.

1. Let $F : Rat^+ \longrightarrow Rat^+$, where Rat^+ is the set of positive rational numbers. Suppose that, for every $x \in Rat^+$, $F(x) = 1/x$, so that F is the reciprocal function. Now define $g : Nat^+ \longrightarrow Rat^+$ by

$$g(k) = F(k).$$

Then g returns the values (sequence),

$$1, \frac{1}{2}, \frac{1}{3}, \frac{1}{4}, \cdots .$$

2. Let

$$A_0 = \{3, 5\}$$

and

$$A_{k+1} = A_k \cup \{z \mid \text{there exist } x, y \in A_k \text{ such that } z = xy\}.$$

where xy is the product of x and y. Then

$$A_1 = \{3, 5, 3^2, 5^2, 3(5)\},$$

and A_2 has all elements of A_1 as well as all products of the elements in A_1. ■

To see that the second example is a good recursive definition, let

$$M = \mathcal{P}(Nat) - \{\varnothing\},$$

so that the elements of M are the nonempty subsets of *Nat*. Let $e = A_0$, and, for any $S \in M$, let

$$F(S) = S \cup \{z \mid \text{there exist } x, y \in S \text{ such that } z = xy \}.$$

Then we can define $r : Nat \longrightarrow M$ by

$$r(k) = [\text{ if } k = 0 \text{ then } A_0 \text{ else } F(r(k-1))].$$

This form of the definition is similar to that used in Section 3.2.1 to show that any nonzero natural number can be generated from 0 by repeatedly applying the successor function. When it is clear what M, e, and F are, it is customary to define a sequence of sets in the form illustrated in the preceding example.

Notice that, because of the way the sequence of sets, A_k, is defined, each set in the sequence is a subset of all later sets in the sequence. We can say that it is a *nested* sequence of sets; the sets are nested inside each other. If one imagines the sets as spheres, then we have one sphere inside of another one, and that one inside yet another sphere, etc., similar to the layers of an onion. However, unlike the onion with its finite number of layers, the sequence has an infinite number of sets. In the current example, each successive set is larger than all of its predecessors. On the other hand, some infinite sequences may have members that become constant after some value of k.

When working with sequences of numbers, one is often interested in possible sums or products of the elements of a sequence. Of course, we do not "add" or "multiply" sets. However, it is sometimes useful to form the union of a sequence of sets. For instance, let A_k be any element of the sequence defined in the previous example, and define

$$T = \bigcup \{ A_k \mid k \in Nat \},$$

which is often written in this form

$$T = \bigcup_{k=0}^{\infty} A_k .$$

One can just consider $\bigcup_{k=0}^{\infty} A_k$ to mean, 'the union of all A_k, for all $k \in Nat$.' By the definition of the generalized union, $x \in T$ iff there exists $k \in Nat$ such that $x \in A_k$. This is an important fact that is useful in proving certain kinds of theorems.

Example.

Let us assume the usual laws of exponents for natural numbers, and let T be the set just defined. Then we can prove that the elements of T have the form $3^i 5^j$ for some $i, j \in Nat$, where at least one of i or j is nonzero. The proof goes as follows:

As already seen, any element of T is in at least one A_k, so it suffices to prove that all elements of all the A_k have the specified form. We do this by induction on k. Let $\phi(k)$ be

$u \in A_k \Rightarrow ((\text{there exist } i, j \in Nat \text{ such that } (i \neq 0 \text{ or } j \neq 0)) \ \& \ u = 3^i 5^j).$

Case 0. $A_0 = \{3, 5\} = \{3^1 5^0, 3^0 5^1\}$, which establishes the base case.

Case $k + 1$. Assume $\phi(k)$. Let $u \in A_{k+1}$. We have

$$A_{k+1} = A_k \cup \{\, z \mid \text{there exist } x,\, y \in A_k \text{ such that } z = xy \,\},$$

so there are two cases to consider.

If $u \in A_k$, then, by $\phi(k)$, u has the specified form.

If $u \in \{\, z \mid \text{there exist } x,\, y \in A_k \text{ such that } z = xy \,\}$, then let $u = xy$, where x, $y \in A_k$. By the induction hypothesis, there exist a, $b \in Nat$ such that $x = 3^a 5^b$ and ($a \neq 0$ or $b \neq 0$). Similarly, $y = 3^c 5^d$, where c, $d \in Nat$ and at least one of c, d is nonzero. By commutativity and the laws of exponents, we obtain

$$u = xy = 3^a 5^b 3^c 5^d = 3^{(a+c)} 5^{(b+d)}.$$

Thus, u has the correct form. Also, at least one of $a + c$ or $b + d$ must be nonzero. If they were both 0, then we would also have $a = b = c = d = 0$, which contradicts $\phi(k)$.

Hence, $\phi(k)$ implies $\phi(k + 1)$. ∎

The preceding examples are contrived, but designed to be simple illustrations. I shall now discuss a useful application of sequences of sets, but first we must review some aspects of binary relations. Consider the graph of the relation R that is displayed in Figure 3.2.

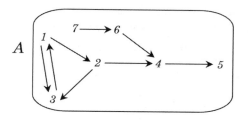

Figure 3.2.

The domain of R is $A = \{\, 1, 2, 3, 4, 5, 6, 7 \,\}$. The set of ordered pairs of R is indicated by the arrows: $< x, y > \in R$ iff there is an arrow starting at x and ending at y. Thus,

$$R = \{\, < 1, 2 >,\ < 2, 4 >,\ < 2, 3 >,$$

$$< 3, 1 >,\ < 1, 3 >,\ < 7, 6 >,\ < 6, 4 >,\ < 4, 5 > \,\}.$$

If $E \subseteq A \times A$ is any set of ordered pairs such that $R \subseteq E$, I shall call E an *extension* of R on A. Clearly, R has many such extensions. Some of these extensions are transitive relations, so they satisfy the condition

$$(Exy \ \& \ Eyz) \ \Rightarrow \ Exz,$$

for every x, y, and z in A (Definition 2-21). For example, suppose that T is a transitive extension of R; then the pair $< 1, 4 >$ must be an element of T, along with several other pairs. We may be interested in a transitive extension of R that has no pair that is not required for transitivity. For instance, the pair $< 5, 4 >$ is not required for transitivity, given the original relation R. At this point the reader should write down all pairs that are required to get a transitive extension of R, and no additional pairs. One way to do this is the following. Let $R_1 = R$. Let R_2 be the union of R_1 with this set,

$$Next = \{ < x, z > \mid \text{there exists } y \in A \text{ such that } R_1 xy \ \& \ R_1 yz \}.$$

One can check that the above set is

$$\{ < 1, 3 >, < 3, 2 >, < 3, 3 >, < 1, 4 >,$$

$$< 2, 1 >, < 1, 1 >, < 2, 5 >, < 6, 5 >, \ < 7, 4 > \}.$$

Yet this approach can yield a set *Next* that contains pairs that are already in R_1. In this example, $< 1, 3 >$ is already in R_1, as well as in *Next*. This is not an issue for the final result, but requires the extra work of checking for duplicate elements as we form the union of R_1 with *Next*. Therefore, we shall avoid this extra work by not putting duplicates in *Next* in the first place.[*] In order to do this, we just redefine

$$Next$$

$$= \{ < x, z > \notin R_1 \mid \text{there exists } y \in A \text{ such that } R_1 xy \ \& \ R_1 yz \}$$

$$= \{ < x, z > \mid \ < x, z > \notin R_1 \ \& \ \text{there exists } y \in A \text{ such that } R_1 xy \ \& \ R_1 yz \}.$$

---

[*]This issue about duplicate elements does not arise in set theory since sets have no duplicate elements. Yet, avoiding duplicates is a computational issue.

We then obtain

$$R_2 = \{\ < 1,2 >,\ < 2,4 >,\ < 2,3 >,\ < 3,1 >,$$

$$< 1,\ 3 >,\ < 7,6 >,\ < 6,4 >,\ < 4,5 >,\ < 3,2 >,\ < 3,3 >,$$

$$< 1,4 >,\ < 2,1 >,\ < 1,1 >,\ < 2,5 >,\ < 6,5 >,\ < 7,4 >\ \}.$$

But now some additional pairs are required, e.g., $< 2,2 >$, $< 1,5 >$, etc. Therefore, we repeat the procedure by forming the union of R_2 with the set,

$$\{\ < x, z >\notin R_2 \mid \text{there exists } y \in A \text{ such that } R_2 xy\ \&\ R_2 yz\ \},$$

and we call this new union R_3. If we continue repeating this procedure, eventually we obtain a set of pairs that no longer changes, that is, no additional pairs are required for transitivity. At this point we stop. However, suppose that our original $R_1 = R$ had been an infinite set. Then the procedure might continue indefinitely. This suggests using a recursive definition, followed by a generalized union. This is what is done in the next example.

Example.

Problem Statement. Let R be a nonempty binary relation on set A. Define a set that is a transitive extension of R on A and prove that it is indeed transitive.

Discussion. There are different approaches to this problem. We shall use the one that is motivated by the discussion immediately preceding this example. In doing so, we shall also use the technique of recursively defining a sequence of sets and then forming the union of all of these sets. The base set in the sequence is R, which we also set equal to R_1. The recursive definition follows the preceding discussion, and the sets in the sequence are sets of ordered pairs corresponding to extensions of the relation R. After forming the union of the sets in the sequence, we need to prove that this union is transitive. We do this by first using induction to prove that if $< x, y >$ and $< y, z >$ are elements of one of the sets in the sequence, then $< x, z >$ is an element in the next set in the sequence. This is easy to prove because of the way the sequence of sets is defined.

Written Solution.

Theorem. Let R be a nonempty binary relation on set A. Define $R_1 = R$ and, for $k \geqslant 1$,

$$R_{k+1} = R_k \cup \{\ < x, z >\notin R_k \mid \text{there exists } y \in A \text{ such that } R_k xy\ \&\ R_k yz\ \}.$$

Also define

$$R_{tc} = \bigcup \{\ R_k \mid k \in Nat^+\ \}.$$

Then R_{tc} is a transitive extension of R.

Proof. I shall first prove a couple of lemmas. The first one is obvious and can easily be proved by induction.

> *Lemma.* Let i, j be positive natural numbers. If $i \leqslant j$, then $R_i \subseteq R_j$.

> *Proof.* Since $i \leqslant j$, there is some n such that $i + n = j$. We can therefore rephrase the conclusion as $R_i \subseteq R_{i+n}$. Let $n \in Nat$ and $\phi(n)$ be the statement that, for all $i \in Nat^+$,

$$R_i \subseteq R_{i+n}.$$

We use induction on n.

Case $n = 0$. Then $R_i \subseteq R_{i+n}$ is $R_i \subseteq R_i$, which is true.

Case $n + 1$. Assume $\phi(n)$: $R_i \subseteq R_{i+n}$. Let $< a,\ b > \in R_i$, so by $\phi(n)$,

$$< a, b > \in R_{i+n}.$$

From the definition of the sequence, we have

$$R_{(i+n)+1} = R_{i+n} \cup \{ < x, z > \notin R_{i+n} \ |$$

there exists $y \in A$ such that $R_{i+n}xy$ & $R_{i+n}yz$ $\}$.

Thus, if a pair $< a, b > \in R_{i+n}$, then $< a, b > \in R_{(i+n)+1} = R_{i+(n+1)}$, so

$$R_i \subseteq R_{i+n} \text{ and } R_{i+n} \subseteq R_{i+(n+1)},$$

which completes the induction. Whence, $R_i \subseteq R_j$, as was to be proved.

> *Lemma.* For every k in Nat^+, and every x, y, z in A,

$$(< x, y > \in R_k \text{ and } < y, z > \in R_k) \Rightarrow < x, z > \in R_{k+1}.$$

Proof. Let x, y, z be in A, and assume that $< x, y > \in R_k$ and $< y, z > \in R_k$. From the stated definition, $R_k \subseteq R_{k+1}$. It is possible that $< x, z > \in R_k$. If so, then $< x, z > \in R_{k+1}$. On the other hand, if $< x, z > \notin R_k$, then we have the three conditions: $< x, z > \notin R_k$, $< x, y > \in R_k$, and

$< y, z >\in R_k$. Then, by the stated definition, $< x, z >\in R_{k+1}$. Therefore, in both cases,

$$(< x, y >\in R_k \text{ and } < y, z >\in R_k) \Rightarrow < x, z >\in R_{k+1},$$

which completes this lemma.

Proof of the main theorem. We now need to prove that R_{tc}, as defined above, is a transitive extension of R on A.

Suppose that we have $< x, y >\in R_{tc}$ and $< y, \ z >\in R_{tc}$. By the definition of R_{tc} and the meaning of the generalized union, there must exist positive numbers i, j such that $< x, y >\in R_i$ and $< y, z >\in R_j$.

Case 1. If $i = j$, then, from the second lemma, $< x, z >\in R_{i+1} \subseteq R_{tc}$.

Case 2. If $i \neq j$, we can assume, without loss of generality, that $i < j$. By the first lemma, if $i \leqslant j$, then $R_i \subseteq R_j$. Thus, if a pair is an element of R_i, then this pair is also an element of R_j. But then we also have $< x, y >\in R_j$. By the second lemma we have $< x, z >\in R_{j+1}$, so also $< x, z >\in R_{tc}$. So, in both Case 1 and Case 2, $< x, z >\in R_{tc}$, so R_{tc} is transitive. *Q.E.D.* ■

We have now defined a relation R_{tc} that is an extension of R and proved that this extension is transitive. Because of the way it is defined, the pairs in R_{tc} are either original pairs in R, or else are pairs that are required in order to achieve transitivity. R_{tc} is called a *transitive closure* of R. Transitive closures are important in certain types of application problems, and they are usually discussed in texts on discrete mathematics. Except for some exercises, we shall not pursue this topic further here.

There is another useful form of the Recursion Theorem that is sufficiently different from our original one that I will state and prove it as a theorem. However, like the other variations we have used, it is a consequence of the original one, Theorem 3-4. The proof begins with a lemma, the purpose of which will become clear shortly.

Lemma 3-17 Let S be a set, $e \in S$, and $F : Nat \times S \longrightarrow S$. Then there exists a unique function $r_1 : Nat \longrightarrow Nat \times S$ such that

$$r_1(0) =< 0, \ e >,$$

and satisfying, for all $k \geqslant 1$,

$$r_1(k) =< k, \ F(r_1(k - 1)) > .$$

Proof.

For any $k \in Nat$, $x \in S$, let $G : Nat \times S \longrightarrow Nat \times S$ be defined in terms of F by

$$G(k, x) = < k + 1, F(k, x) > .$$

G maps $Nat \times S$ to $Nat \times S$, so, by the Recursion Theorem, there exists a unique $r_1 : Nat \longrightarrow Nat \times S$, such that

$$r_1(k) = [\text{ if } k = 0 \text{ then } < 0, e > \text{ else } G(r_1(k - 1))].$$

From this recursive definition, $r_1(0) = < 0, e >$. We now need to prove that $r_1(k)$ satisfies the second equation in the statement of the lemma. We use induction on k starting with the base case of 1. The induction hypothesis $\phi(k)$ is

$$r_1(k) = < k, F(r_1(k - 1)) > .$$

Case 1. $r_1(1) = G(r_1(0)) = G(< 0, e >) = < 1, F(0, e) > = < 1, F(r_1(0)) >$, which is $\phi(1)$.

Case $k + 1$. $r_1(k + 1) = G(r_1(k))$. Using the induction hypothesis, we have

$$r_1(k + 1) = G(r_1(k)) = G(< k, F(r_1(k - 1)) >).$$

Applying the definition of G, we obtain

$$r_1(k + 1) = < k + 1, F(k, F(r_1(k - 1))) > .$$

Again using $\phi(k)$,

$$r_1(k + 1) = < k + 1, F(r_1(k)) >,$$

which is $\phi(k + 1)$. **Q.E.D.**

Theorem 3-18   (**Recursion Theorem in a New Format**) Let S be a set, $e \in S$, and $F : Nat \times S \longrightarrow S$. Then there exists a unique function

$$r : Nat \longrightarrow S$$

such that

$$r(k) = [\text{ if } k = 0 \text{ then } e \text{ else } F(k - 1, r(k - 1))].$$

Proof. By the previous lemma, there exists a unique function,

$$r_1 : Nat \longrightarrow Nat \times S,$$

satisfying

$$r_1(0) = < 0, e >,$$

and, for all $k \geqslant 1$,

$$r_1(k) = < k, F(r_1(k-1)) >.$$

Notice that $Img(r_1, Nat) \subseteq Nat \times S$, so the image of Nat under r_1 is a subset of the domain of F. Thus, we may define

$$r(k) = [\text{ if } k = 0 \text{ then } e \text{ else } F(r_1(k-1))].$$

Clearly, $r(0) = e$.

Applying the lemma to the definition of r, we have

$$r(1) = F(r_1(1-1)) = F(r_1(0)) = F(0, e) = F(0, r(0)).$$

So r satisfies the statement of the theorem when $k = 0$ and $k = 1$. For $k \geqslant 2$, again using the definition of r and the lemma, we have

$$r(k) = F(r_1(k-1)) = F(< k-1, F(r_1((k-1)-1)) >),$$

and hence,

$$r(k) = F(< k-1, r(k-1) >). \quad \textbf{\textit{Q.E.D.}}$$

Since Theorem 3-18 is a consequence of the original Recursion Theorem, we have not proved anything essentially new. Yet Theorem 3-18 is useful because it provides a *convenient* way to construct some forms of recursive definitions. As with the original Recursion Theorem, we can also apply the present theorem to define a function starting with a base case greater than zero. To illustrate the use of Theorem 3-18, I will now define the *sum* function, which opened this chapter, as well as the related factorial function.

Examples.

1. Let $F : Nat \times Nat \longrightarrow Nat$, where $F(i, x) = (i + 1) + x$, for $i, x \in Nat$. In this example, the S in Theorem 3-18 is Nat. Let $e = 0$. Then, by the theorem, we can define

$$sum(k) = [\text{ if } k = 0 \text{ then } 0 \text{ else } F(k-1, sum(k-1))],$$

or

$$sum(k) = [\text{ if } k = 0 \text{ then } 0 \text{ else } k + sum(k-1)].$$

This is equivalent to

$$sum(0) = 0$$

and

$$sum(k + 1) = sum(k) + (k + 1),$$

which is what we used in Section 3.1.1. In more traditional mathematical notation, one would write

$$sum(k) = \sum_{i=0}^{k} i.$$

2. Now modify the previous example by letting $F(i, x) = (i + 1)x$ and letting $e = 1$. Then we can define

$$factorial(k) = [\text{ if } k = 0 \text{ then } 1 \text{ else } F(k - 1, factorial(k - 1))],$$

or

$$factorial(k) = [\text{ if } k = 0 \text{ then } 1 \text{ else } k \, factorial(k - 1)].$$

When $k \geqslant 1$, this is traditionally expressed by

$$factorial(k) = \prod_{i=1}^{k} i.$$

The standard, short notation is $factorial(k) = k!$. ■

Using the format of Theorem 3-18 for recursion enables us to write simple definitions of the *sum* and *factorial* functions. Here is another illustration.

Example.

Problem Statement. Write a recursive definition of this function

$$f(k) = \prod_{i=2}^{k} \left(1 - \frac{1}{i^2} \right).$$

Show how your definition fits one of the forms of the Recursion Theorem, and then prove that

$$f(k) = \frac{k + 1}{2k}.$$

Discussion. If we expand the product formula, we have

$$f(k) = \left(1 - \frac{1}{2^2}\right)\left(1 - \frac{1}{3^2}\right) \cdots \left(1 - \frac{1}{(k-1)^2}\right)\left(1 - \frac{1}{k^2}\right).$$

It is natural to start our recursive definition with the number 2. Note that $f(2) = 1 - (1/4) = 3/4$. In addition, we can see that

$$f(k) = f(k-1)\left(1 - \frac{1}{k^2}\right),$$

which suggests that we try to apply Theorem 3-18, that is, look for a suitable function $F : (Nat - \{0, 1\}) \times Rat^+ \longrightarrow Rat^+$ in the definition form

$$f(k) = [\text{ if } k = 2 \text{ then } 3/4 \text{ else } F(k-1, f(k-1))].$$

A little thought suggests

$$F(i, r) = \left(1 - \frac{1}{(i+1)^2}\right) r.$$

Note that our definition does not exactly fit Theorem 3-18 since the base case is the number 2. But it is clear that Theorem 3-18 can be adjusted for base cases greater than 0 just as the previous forms of the Recursion Theorem were likewise adjusted. The rest of this problem uses induction and some simple algebra.

Written Solution. Let $F : (Nat - \{0, 1\}) \times Rat^+ \longrightarrow Rat^+$, where

$$F(i, r) = \left(1 - \frac{1}{(i+1)^2}\right) r.$$

Define $f : (Nat - \{0, 1\}) \longrightarrow Rat^+$ by

$$f(k) = [\text{ if } k = 2 \text{ then } 3/4 \text{ else } F(k-1, f(k-1))]$$
$$= [\text{ if } k = 2 \text{ then } 3/4 \text{ else } \left(1 - \frac{1}{k^2}\right) f(k-1)].$$

Theorem. For the previously defined function f, and $k \geqslant 2$,

$$f(k) = \frac{k+1}{2k}.$$

Proof. Let $\phi(k)$ be the statement: $f(k) = \dfrac{k+1}{2k}$. The proof is by induction.

Case $k = 2$. $f(2) = \frac{3}{4}$ by definition of f. Also, $\dfrac{k+1}{2k} = \dfrac{2+1}{2(2)} = \frac{3}{4}$. Thus, the base case is true.

Case $k + 1$. Assume $\phi(k)$. By the definition of f, we have

$$f(k+1) = \left(1 - \frac{1}{(k+1)^2}\right) f(k).$$

By $\phi(k)$, $f(k) = \dfrac{k+1}{2k}$. Substituting the latter equation into the former yields

$$f(k+1) = \left(1 - \frac{1}{(k+1)^2}\right)\left(\frac{k+1}{2k}\right) = \left(\frac{(k+1)^2 - 1}{(k+1)^2}\right)\left(\frac{k+1}{2k}\right)$$

$$= \frac{k^2 + 2k}{(2k)(k+1)} = \frac{k(k+2)}{(2k)(k+1)} = \frac{(k+1)+1}{2(k+1)},$$

which is $\phi(k+1)$. **Q.E.D.** ■

Exercises

General Instructions

In this exercise set, unless mentioned to the contrary, you may assume the basic principles of arithmetic and algebra as applied to *Nat*. Several of the problems have more specific instructions; please be sure to follow these instructions where they occur.

Throughout this book, if an exercise is simply the statement of a theorem or a metatheorem, then the task is to prove this theorem or metatheorem. In your proof, you may use the definitions given in the text. Unless stated to the contrary, you may also use theorems proved in the text.

Exercises for which answers are provided at the back of the book are marked with "Answer Provided."

Finally, please note that exercises that introduce important principles or useful applications are especially difficult, or are special in some way, are indicated by bold print. In some cases, parenthetical comments describe the special features of the problem. If such comments are missing, the exercise is nonetheless of special value in introducing a new concept, principle, or technique.

E 3-25 (Answer Provided) Write a recursive definition of a function, starting at 0, that returns values that match the first seven terms of this infinite sequence:

$$0,\ 1,\ 4,\ 25,\ 676,\ 458329,$$

$$210066388900,\ 44127887745906175987801,\ \cdots.$$

Let your function be $r(k)$ and let r_k abbreviate $r(k)$. Show that, for any $k \in Nat$,

$$r_{k+1} - r_k = r_k^2 + r_k + 1.$$

E 3-26 *Theorem.* Let $e \in M$, $F : Nat \times M \longrightarrow M$. Let $g : Nat \longrightarrow M$ and $h : Nat \longrightarrow M$, and suppose that $g(0) = h(0) = e$. Also, suppose that, for any $k \in Nat$, $g(k + 1) = F(k, g(k))$ and that $h(k + 1) = F(k, h(k))$. Then $g = h$. (Remark. Do not just use Theorem 3-18; give your own direct proof of uniqueness using induction.)

E 3-27 The positive even numbers can be represented as $2k$, for $k \in Nat^+$. Let

$$\Sigma(k) = 2 + 4 + \cdots + 2k$$

be the sum of the first k positive even numbers.

1. Write a recursive definition for Σ beginning with the base case of $k = 1$.

2. Using this recursive definition, prove that

$$\Sigma(n) = n(n + 1),$$

for every $n \in Nat^+$.

E 3-28 The odd natural numbers can be represented as $2k + 1$, for $k \in Nat$. Let

$$\Sigma(k) = 1 + 3 + 5 + \cdots + (2k + 1)$$

be the sum of all odd natural numbers from 1 up to and including $(2k + 1)$.

1. Write a recursive definition for Σ beginning with the base case $\Sigma(0) = 1$.

2. Using this recursive definition, prove that $\Sigma(n) = (n + 1)(n + 1)$, for every $n \in Nat$.

E 3-29 Let $SUMSQ(k)$ denote the sum of the squares of the natural numbers from 1 through k, inclusively. Write a recursive definition of $SUMSQ(k)$ and prove that

$$SUMSQ(k) = (k(k+1)(2k+1)\,)/6.$$

E 3-30 Let $r : Nat \longrightarrow Nat$, where r is defined by

$$r(0) = 0$$

and

$$r(k+1) = r(k) + 2k + 3.$$

Compute the first five values of this sequence, and then prove that, for any n,

$$r(n) = (n+1)^2 - 1.$$

E 3-31 (A familiar formula) Let $m \neq 1$ be a positive real number. The following is often called a *geometric series*:

$$g(k) = m^0 + m^1 + \cdots + m^k.$$

1. Write a recursive definition of $g(k)$.

2. Using your definition, prove that

$$g(k) = \frac{m^{k+1} - 1}{m - 1}.$$

(Remark. In the special case where $m = 2$, we get the nice result

$$2^0 + 2^1 + \cdots + 2^k = 2^{k+1} - 1.$$

This formula is sometimes handy when working with binary numbers.)

E 3-32 Let $r : (Nat - \{\,0,\ 1\,\}) \longrightarrow Rat$ be given by

$$r(k) = \prod_{i=2}^{k} \left(1 - \frac{1}{i}\right).$$

Write a recursive definition for r, and then prove that, for $k \geqslant 2$,

$$r(k) = 1/k.$$

E 3-33 *Theorem.* Define $\Sigma : Nat \longrightarrow Nat$ by

$$\Sigma(0) = 0$$

and

$$\Sigma(k+1) = \Sigma(k) + (k+1).$$

Then, for all n, $k \in Nat$,

$$\Sigma(n+k) = \Sigma(n) + \Sigma(k) + nk.$$

E 3-34 *Theorem.* Let $a \in Re$ be a constant real number, and define

$$H \ : \ Nat \longrightarrow Re$$

by

$$H(i) = \ [\ \text{if } i = 0 \text{ then } 1 \text{ else } a^i + H(i-1) \ ].$$

Then, for every $n \in Nat^+$, $H(n) = 1 + aH(n-1)$.

E 3-35 Define $f : Nat \longrightarrow Rat$: For any k in Nat,

$$f(k) = \left[\text{if } k = 0 \text{ then } 1 \text{ else } \frac{f(k-1)}{1 + (f(k-1))^2} \right].$$

Then, for all $n \in Nat$, $f(n) \leqslant 1$.

E 3-36 *Theorem.* Let $f : Nat^+ \longrightarrow Re$, and assume that, for every k in Nat^+, $f(k) < b$, where b is a particular real number. Define the function H on Nat^+ as follows: For every k in Nat^+,

$$H(k) = \ [\ \text{if } k = 1 \text{ then } f(1) \text{ else } f(k) + H(k-1) \ ].$$

Prove that, for all n in Nat^+, $H(n) < nb$.

E 3-37 (An interesting function) Let $f : Nat \longrightarrow Re$; f is a *sequence* (or *array*) which returns values in Re. Define another function,

$$max : Nat \longrightarrow Re$$

as follows: For any $k \in Nat$,

$$max(0) = f(0)$$

and

$$max(k + 1) = [\text{ if } f(k + 1) > max(k) \text{ then } f(k +1) \text{ else } max(k)],$$

where $>$ is the *greater than* relation on the real numbers. Prove that for any $j, n \in Nat$, if $j \leqslant n$, then $f(j) \leqslant max(n)$. (Remark. One can implement this *max* function as an inefficient computer program. Because the value returned by $max(n)$ is equal to the value returned by $f(j)$ for some $j \leqslant n$, this proof establishes that the program returns the maximum value of the array elements up to and including element n.)

E 3-38 (Uses a common form of induction proof) *Metatheorem.* Define sets of SC sentences as follows:

$$\Gamma_1 = \{(P \to Q_1)\},$$

and, for $k \geqslant 1$,

$$\Gamma_{k+1} = \Gamma_k \cup \{(Q_k \to Q_{k+1})\}.$$

Then, for any $n \in Nat^+$, $\Gamma_n \vdash_T (P \to Q_n)$.

E 3-39 (Uses a common form of induction proof) The domain of discourse is an imaginary world, the Universal Planetary Empire, U, which contains a large number of planets and other things. Some of the planets are inhabited (by at least some kinds of beings) and some are not. If x, y are planets, Fxy holds iff there is a nonstop spaceship flight from x to y. Let a denote the planet, Alpha-uglyon, and let $k \in Nat$. Define sets as follows:

$$A_0 = \{ a \},$$

$$A_{k+1}$$

$$= A_k \cup \{ y \mid y \text{ is a planet \& there exists a planet } x \in A_k \text{ such that } Fxy \},$$

and,

$$A = \bigcup \{ A_k \mid k \in Nat \}.$$

Each A_k, and hence also A, is a set of planets. Suppose that the following is true:

For any planets x, $y \in U$, if x is inhabited and Fxy, then y is inhabited.

Also assume that Alpha-uglyon is inhabited. Using all of these assumptions, prove the following:

1. For all $n \in Nat$, every planet in A_n is inhabited.

2. All planets in A are inhabited.

E 3-40 Figures 3.3 and 3.4 show three distinct (separate) graphs. Carefully and clearly sketch the transitive closure of each of these graphs.

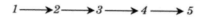

Figure 3.3.

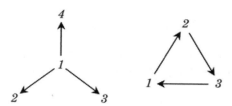

Figure 3.4.

E 3-41

1. Define the relation S on Nat by the following: For every i, $j \in Nat$, Sij iff $j = i + 1$. What is the transitive closure of S? (It is a familiar relation.) It is not necessary to prove your answer, but an informal justification should be given.

2. Chapter 2, Section 2, discusses an imaginary company called the "Omega Corporation." Figure 2.4 shows a graph of the organizational structure of this company in which the arrows represent the relation Dxy: y is the direct superior of x. Carefully and clearly sketch the graph of the transitive closure of the graph in Figure 2.4. In ordinary language, how would you describe or name this transitive closure relation?

E 3-42 Define $r : Nat \longrightarrow Rat$ by

$$r(k) = [\text{ if } k = 0 \text{ then } 1 \text{ else } r(k-1)/k \ ].$$

Make a table of the values r returns for $k = 0$, 1, 2, 3, 4, 5. Conjecture a simple, nonrecursive formula that equals $r(k)$ for all k in Nat. Prove that your formula is correct.

3.3.2   The Well-Ordering of the Natural Numbers

Theorem 3-15 states that $<$ is a strict simple order on Nat. This section will show that it also has another, very important property. We begin with some definitions.

Definition 3-19 Let R be a binary relation on $A \neq \varnothing$. Then x is an R-*first* element of A iff $x \in A$, and for every $y \in A$, if $y \neq x$, then Rxy.

Examples.

If we consider $<$ on Nat, then 0 is a $<$-first element of Nat, since if $y \in Nat$ and $y \neq 0$, then $0 < y$.

If we consider $A = \{ 23, 77, 101, 567, 13301 \}$, then 23 is a $<$-first element of A.

Let Int be the set of all integers. Then, if x is any integer, we can always find a smaller one, e.g., $x - 1$. Thus, Int has no $<$-first element.

Let Re be the real numbers and let

$$B = \{x \mid x \in Re \ \& \ 0 < x < 1\}.$$

Suppose (for RAA) that a is a $<$-first element of B. If a is a first element, then we must have $a < y$, for any $y \neq a$. But this is false, since it is not true that $a < y$ if we choose $y = a/2$. B has no first element with respect to $<$, i.e., no $<$-first element. ■

One might be tempted to consider $<$ on Nat, Int, and Re as one "relation." This is dangerous. These examples show that one cannot simply look at a relation like $<$ to determine whether or not it has a first element. The existence of a first element depends on both the relation and the domain in question. Strictly speaking, $<$ on Nat, Int, and Re are different relations because $<$ comprises different sets of ordered pairs on these different domains. The concept of R-first element is related, but not equivalent, to the following:

Definition 3-20 Let R be a binary relation on $A \neq \varnothing$. Then x is an R-*minimal* element of A iff $x \in A$, and for every $y \in A$, not Ryx.

Examples.

Let $y \in Nat$. Then not $y < 0$, so 0 is a $<$-minimal element of *Nat*. Thus, some minimal elements are also first elements, but this is not always the case.

Consider the relation $R = \{ < 0, 1 >, < 1, 2 >, < 3, 4 > \}$, on the domain $A = \{0, 1, 2, 3, 4\}$. Then, for any $y \in A$, not $Ry3$, so 3 is a minimal element. But also, not $Ry0$, so 0 is also a minimal element. Minimal elements need not be unique. In addition, notice that not $R03$ and not $R30$, so neither 0 nor 3 is an R-first element of A with respect to R.

Now consider $\leqslant$ on *Nat*. If $y \neq 0$, then $R0y$, so 0 is a $\leqslant$-first element of *Nat*. But it is not true that, for any $y \in Nat$, not $y \leqslant 0$, since $0 \leqslant 0$. Hence, 0 does not satisfy the definition of $\leqslant$-minimal element. Here we have a first element that is not minimal.

Like minimal elements, first elements need not be unique. Consider the relation $S = \{ < 0, 1 >, < 1, 0 > \}$. ■

Fortunately, *first* and *minimal* merge together for certain types of relations. The next lemma is especially useful.

Lemma 3-21 Let $\langle A, R \rangle$ be a relational system in which R is a binary relation that is asymmetric and connected on A. Let $m \in A$. Then m is an R-minimal element of A iff m is an R-first element of A. In addition, if a first element exists, then it is unique.

Proof. Suppose that m is minimal with respect to R. Let $y \in A$ and $m \neq y$. Since R is connected and $m \neq y$, we have Rmy or Rym. But because m is minimal, we have not Rym; therefore, Rmy. Hence, m is a first element with respect to R.

Now suppose that m is an R-first element. Let y be any element of A. I will consider two cases:

> Case $y = m$. Since R is asymmetric, it is also irreflexive. Thus, not Rmm, so not Rym.

> Case $y \neq m$. Since m is a first element and $y \neq m$, we have Rmy. But R is asymmetric, so we also have not Rym.

From the two cases, we have this result: If m is a first element and $y \in A$, then not Rym. Thus, m is also minimal.

Finally, suppose that $a \neq b$ are both first elements of A. Then we would have Rab and Rba, which contradicts the given fact that R is asymmetric. Thus, any first element must be unique. *Q.E.D.*

From this lemma, it is also obvious that any minimal element must be unique when R is connected and asymmetric. The next definition is widely used in set theory and in the theory of computation.

Definition 3-22 Let R be a binary relation on the nonempty set A. R is *well-founded* on A iff every nonempty subset B of A has an R-minimal element.

Of course, when we say that a subset B of A has an R-minimal element, we mean minimal with respect to those ordered pairs in R whose elements are from B.

Consider the relation $<$ on the set $A = \{\,2,\,3,\,5,\,7,\,20\,\}$. One can see that every nonempty subset of A has a minimal element, so $<$ is well-founded on this set. However, $<$ is not well-founded on *Int*, since there are nonempty subsets of *Int* that are infinite descending sequences with respect to $<$. One such sequence is

$$\cdots < -2 < -1 < 0 < 1,$$

so the subset $\{\cdots,\,-2,\,-1,\,0,\,1\}$ of *Int* has no minimal element. Intuitively, $<$ on *Nat* is well-founded. It will be proved shortly that this follows from Peano's Axioms. In fact, a stronger result will be proved.

Let R be well-founded on A, $B \neq \varnothing$, and $B \subseteq A$. Then there is some $m \in B$ such that, for every $y \in B$, not Rym. Notice that this definition is quite different from the previous definitions of relations that are transitive or symmetric or connected, etc. All of those earlier definitions characterized the property of the relation in terms of the individual elements of the domain of the relation in question. For example, in order for R to be symmetric on A, it must be the case that, for any x, y in A, if Rxy, then Ryx. The variables x, y range over elements of A; i.e., the possible values of x, y are elements of A. On the other hand, in the definition of well-foundedness, the variable B ranges over nonempty subsets of A. To determine that a relation is well-founded on A, it is not enough to consider only the elements of A; we must also somehow consider the subsets of A. Recall that the Induction Axiom also refers to subsets of *Nat*; it will be shown below that induction and well-foundedness can be related to one another. To facilitate this development, I will first introduce a restricted class of well-founded relations that is very important in spite of the restriction.

Definition 3-23 A relational system $\langle A,\,R \rangle$ is a (strict) *well-ordering* iff R is connected and well-founded on A. If $\langle A,\,R \rangle$ is a well-ordering, we also say that R *well-orders* A.

The reason for writing "strict" in parentheses will be explained after the next theorem.

Theorem 3-24 If $\langle A,\,R \rangle$ is a well-ordering, then $\langle A,\,R \rangle$ is a *strict simple order*.

Proof. By definition, R is connected, so we must prove that it is also asymmetric and transitive. Let $x, y \in A$ and suppose that Rxy. Consider the subset $\{x, y\}$ of A. It must have a minimal element. If y were this minimal element, then we would not have Rxy, which contradicts the supposition that Rxy. Thus, x must be the minimal element; hence not Ryx. So, R is asymmetric. Notice that this proof of asymmetry depends *only* on the fact that R is well-founded. We use connectedness in the proof that R is also transitive.

Let $x, y, z \in A$, and assume that Rxy & Ryz. I will show that Rxz. If we had $x = z$, then we would have Rxy & Ryx, which contradicts the asymmetry of R. Thus, $x \neq z$. But then, since R is connected, we must have Rxz or Rzx. Suppose that Rzx. Then, all together, we would have Rxy, Ryz, and Rzx. This puts an element before y (i.e., we have Rxy), and an element before z, and an element before x. Thus, the nonempty subset $\{x, y, z\}$ has no R-minimal element. Hence, Rzx cannot be true, so we have Rxz instead. So, R is transitive. **Q.E.D.**

In the definition of well-ordering, I wrote "strict" in parentheses. This is because our definition implies that a well-ordering is also a strict simple ordering. Some mathematicians use a definition of well-ordering that is similar to ours except that it requires that R be a (nonstrict) simple order. This is a mere terminological difference. I will follow the above definition and use the term *well-ordering* to refer to one that is a strict simple order. It is now easy to relate minimal and first elements for well-orderings.

Theorem 3-25 Let $\langle A, R \rangle$ be a binary relational system. Then R is a well-ordering of A iff R is connected, asymmetric, and every nonempty subset B of A has an R-first element.

Proof. Suppose that R well-orders A. By definition, R is connected and, by the previous theorem, it is asymmetric. Let $B \neq \varnothing$ and $B \subseteq A$. Since R is a well-ordering, B has an R-minimal element. Let R_1 be R restricted to B, i.e., R_1 consists of all the pairs in R whose elements are also members of B. Then R_1 is asymmetric and connected on B. Also, an element m of B is minimal with respect to R iff it is minimal with respect to R_1. By Lemma 3-21, any R_1-minimal element of B is also an R_1-first element, so it is also an R-first element of B.

Suppose that R is connected and asymmetric on A, and that every nonempty subset of A has an R-first element. Let B be a nonempty subset of A, and let m be an R-first element of B. Applying Lemma 3-21 again, it follows that m is also an R-minimal element of B. Thus, R is connected and well-founded on A, so it is a well-ordering of A. **Q.E.D.**

If R well-orders A, then any nonempty subset B of A has both a minimal and a first element. From Lemma 3-21, it also follows that these minimal and first elements of B are one and the same element of B. It is convenient, when working with well-orders, to make use of this fact that every nonempty subset has a unique first element. The previous theorem is also often useful when one is trying to prove that some given relation is a well-ordering. Several of the following proofs illustrate the convenience of working with first elements. The first task is to prove that the less-than relation (<) well-orders *Nat*.

Discussion of the next theorem. We want to prove the important result that the less-than relation (<) well-orders *Nat*. Notice the following elementary fact about < on *Nat*: If $i < k + 1$, then $i \leqslant k$. This is easy to prove from our previous development of the natural number system; it will be used in the proof of the next theorem.

By Theorem 3-15, < is connected and asymmetric on *Nat*. Therefore, by Theorem 3-25, we need only prove that every nonempty subset of *Nat* has a first element. The set of nonempty subsets of *Nat* is huge, and it contains many exceedingly complex sets of numbers. Thus, it is difficult to think of a *direct* way to prove that every nonempty subset must have a first element. Because of this, one is motivated to try an RAA proof: Assume the existence of a nonempty subset of *Nat* without a first element, and try to derive a contradiction.

How can one get a contradiction? Suppose that A is a *nonempty* subset of *Nat*, and A has no first element. There are no natural numbers before 0 with respect to <. Therefore, $0 \notin A$. Now suppose that $1 \in A$. Since $0 \notin A$, 1 would then be a first element of A. But, by hypothesis, A has no first element, so $1 \notin A$. Now both 0 and 1 are not elements of A. Suppose this goes on to some number $k \notin A$. Then $0, 1, \cdots, k \notin A$. But then, by the same kind of reasoning just used on $0, 1, \cdots, k$, it follows that $k + 1 \notin A$ (for if it were, then it would be a first element of A). This suggests an induction proof that all numbers are *not* elements of A, i.e., that $A = \varnothing$, which contradicts the hypothesis that $A \neq \varnothing$. This is the strategy that will be used in the next proof.

Theorem 3-26 (WO) $\langle Nat, < \rangle$ is a well-ordering.

Proof. By Theorem 3-15, < is connected and asymmetric. For RAA, suppose that A is a nonempty subset of *Nat* that has no first element with respect to <. Let $\phi(k)$ be

$$\text{for all } i \leqslant k, i \notin A.$$

We will induct on k. Notice that this induction hypothesis states that all numbers equal to or less than k are *not* elements of A. If we can prove this for all k, then A will be empty, a contradiction.

Case 0. The only i to consider is $i = 0$. If $0 \in A$, then 0 would be a first element of A. Hence, $0 \notin A$, so all $i \leqslant 0$ are not elements of A, which establishes $\phi(0)$.

Case $k + 1$. Assume $\phi(k)$. We must now prove that $\phi(k + 1)$, i.e., that

$$\text{for any } i \leqslant k + 1,\ i \notin A.$$

Suppose, for another RAA, that $k + 1 \in A$. Since A has no first element, $k + 1$ is not a first element. Thus, there exists $i \in A$ such that $i \neq k + 1$ and not $k + 1 < i$. Since $<$ is connected, we then have $i < k+1$, so $i \leqslant k$. But then, by $\phi(k)$ it follows that $i \notin A$. We now have both $i \in A$ and $i \notin A$, which is a contradiction. Thus, it is not the case that $k + 1 \in A$, so $k + 1 \notin A$.

By $\phi(k)$, all $i \leqslant k$ are not elements of A, and it was just proved that also $k + 1 \notin A$. Hence, for any $i \leqslant k + 1$, $i \notin A$. This is $\phi(k + 1)$.

We have now completed the induction proof that $\phi(n)$ is true for all $n \in Nat$, from which $A = \varnothing$. This contradicts the original RAA assumption that A is nonempty. By the previous theorem, $<$ well-orders Nat. **Q.E.D.**

This important theorem establishes that Peano's Axioms imply that the natural numbers are well-ordered by the less-than relation ($<$). In fact, one can axiomatize the natural numbers in terms of addition, multiplication, and $<$, and assume that $<$ is a well-ordering. From these assumptions, it is relatively easy to derive the Induction Axiom. I will not do that here. It is presented in a very readable form in the first few pages of Birkhoff and MacLane (1953). A somewhat different approach defines the numbers as special sets, shows that they are well-ordered, and then derives the Induction Principle; see, for instance, Chapter 5 of Suppes (1972). The fact that $<$ well-orders Nat is often called *The Least Number Principle* for obvious reasons. Using the well-ordering of Nat turns out to be very convenient for some proofs. This will be illustrated by an important theorem of number theory.

Definition 3-27 Let $i,\ j \in Nat$. We say that i *divides* j, or that i is a *divisor* of j, or that i is a *factor* of j, or that j is a *multiple* of i iff $i \neq 0$ and there exists $k \in Nat$ such that $j = ik$. The conventional notation for i *divides* j is $i|j$.

For example, $3|180$, since $180 = 3(60)$. The notation $i|j$ is a curious way to represent this important relationship, and $|$ is also used in other contexts with a different meaning. In particular, we must not confuse $|$ used in the definition of a set by abstraction, with $|$ used to denote divisibility. Nevertheless, the latter use of $|$ is traditional in number theory, so I will use it here. Context of use will distinguish the divisibility relation from set abstraction.

Divisibility has a fundamental role in the theory of numbers, and there are many surprising and arcane facts about divisibility. The next example illustrates one simple fact. It is presented primarily because it is another good example of the usefulness of mathematical induction, but it is intrinsically interesting.

Example.

Problem Statement. Theorem. For all natural numbers n, $3|2^{2n} - 1$.

Discussion. I shall define $\phi(k)$ to state: $3|2^{2k} - 1$. In the base case k is zero, $0 = 2^0 - 1$, and $3|0$. Assuming the induction hypothesis, we may write $3|2^{2k} - 1$, which means that there is some natural number, say m, such that $2^{2k} - 1 = 3m$. We want to prove $\phi(k+1)$, i.e., that $3|2^{2(k+1)} - 1$. How can we relate $\phi(k)$ to $\phi(k+1)$? A little algebraic scratch work helps. Since $2^{2k} - 1 = 3m$,

$$2^{2k} = 3m + 1.$$

Also,

$$2^{2(k+1)} - 1 = 2^{2k+2} - 1 = 2^2(2^{2k}) - 1.$$

If we multiply the former of these two equations by 2^2, we obtain

$$2^2(2^{2k}) = 2^2(3m+1) = 2^2(3m) + 2^2 = 2^2(3m) + 4.$$

Subtracting 1 from each side yields $2^2(2^{2k}) - 1 = 2^2(3m) + 3$. The right side of this equation is divisible by 3, and we can see how it is related to $\phi(k+1)$, so all we need to do is write up the proof.

Written Solution. Theorem. For all natural numbers n, $3|2^{2n} - 1$.

Proof. Let $\phi(k)$ state: $3|2^{2k} - 1$. We use induction on k.

Case $k = 0$. $0 = 2^0 - 1$, and $3|0$, so the base case holds.

Case $k+1$. Assume $\phi(k)$, so there exists some number m such that $2^{2k} - 1 = 3m$; whence $2^{2k} = 3m + 1$. Therefore,

$$2^{2(k+1)} - 1 = 2^{2k+2} - 1 = 2^2(2^{2k}) - 1 = 2^2(3m+1) - 1$$
$$= 12m + 3 = 3(4m + 1),$$

which is a multiple of 3. Hence, $3|2^{2(k+1)} - 1$, which is $\phi(k+1)$.

Therefore, the theorem is true for all natural numbers n. ***Q.E.D.*** ■

There are many other curious facts similar to the one in this example. Some others are in the later exercises. Although it is interesting to see how some numbers can be divided by others, the most interesting numbers are those that can be divided only by themselves and the number 1. These are the primes, which are defined precisely next.

Definition 3-28 A natural number p is *prime* iff $1 < p$ and p has no divisors other than 1 and p.

Thus, 2, 3, 5, 7, 11, 13, 17, $\cdots$ are primes. I will prove that every natural number equal to or greater than 2 has a prime divisor. At first glance, it is not obvious that this is true. It is also not obvious that it would be significant if it is true. In fact, it is both true and significant. We begin with some lemmas.

Lemma 3-29 The divisibility relation is transitive: Let i, j, $k \in Nat$, with $i \neq 0$, $j \neq 0$. If $i|j$ and $j|k$, then $i|k$.

Proof. We have $j = ix$ for some x, and $k = jy$ for some y, so $k = (ix)y$. Hence, $k = i(xy)$, so $i|k$. **Q.E.D.**

Lemma 3-30 Let i, j, $k \in Nat$, with $i < j$. Then

   (i) If $0 < k$, then $ik < jk$.

   and

   (ii) If $1 \leqslant k$, then $i < kj$.

Proof.

   (i) We have $j = i + x$, for some $x > 0$. Thus, $jk = ik + xk$. But $x > 0$ and $k > 0$, so it follows from statement (3.3) in Theorem 3-12 that $xk > 0$. Hence, $ik < jk$.

   (ii) We have $k = 1 + x$, for some x. Also, since $1 \leqslant k$, $0 < k$. By (i) we have, $ik < jk$, so $i(1 + x) < kj$. Hence, $i \leqslant i + ix < kj$, so $i < kj$. **Q.E.D.**

Lemma 3-31 If $i|j$ and $j > 0$, then $1 \leqslant i \leqslant j$.

Proof. From the definition of $|$, we have $1 \leqslant i$. For RAA, suppose that not $i \leqslant j$. By Theorem 3-14, $\leqslant$ is strongly connected. Therefore, we have $j \leqslant i$. But because not $i \leqslant j$, we also have $i \neq j$, hence $j < i$. Also, since $i|j$, we have $j = ik$, where $k \neq 0$, so $1 \leqslant k$. By the previous lemma (ii), $j < ki$. But $j = ki$, so we have a contradiction. Therefore, $i \leqslant j$. **Q.E.D.**

This is a simple but important lemma. In particular, if $i|j$ and $j < i$, then by this lemma, $j = 0$. This is a useful fact to remember. The lemma also implies that if $j > 0$ and $i|j$, then $i \leqslant j$. This shows that *the relationship of divisibility implies a relationship in size*. This fact plays an important role in the proof of the following theorem about the existence of a prime divisor. The overall strategy of the proof uses the well-ordering of *Nat*.

Theorem 3-32 (**Prime Divisor Theorem**) Every natural number $k > 1$ is divisible by at least one prime.

Proof. Let $k \in Nat$, and let

$$\phi(k) \text{ be as follows:}$$

If $k > 1$, then there exists $p \in Nat$ such that p is prime and $p|k$.

If $\phi(k)$ is false for a number k, let us say that k is a ϕ-*exception*. We can define the set of ϕ-exceptions by

$$V = \{\, i \mid \text{not } \phi(i) \,\}.$$

Thus, V is the set of all numbers greater than 1 which do not have a prime divisor. We want to prove that, for every natural number n, $\phi(n)$. This is true iff $V = \varnothing$. The proof is by RAA: We assume that V is nonempty, i.e., that $\phi(k)$ is false for some k. We now consider the elements of V.

By the well-ordering of Nat by $<$, V has a first element. Let K denote this first element. Then, for all i, if $1 < i < K$, then i has a prime divisor. We will use this fact to prove that K must also have a prime divisor, which will be a contradiction with $K \in V$. However, in the case $K = 2$, there is no i between 1 and 2. Therefore, we consider this case separately as a kind of "base case."

Case $K = 2$. Since 2 is prime and $2|2$, 2 has a prime divisor.

Case $K > 2$. K is either prime or nonprime.

> First Subcase. If K is prime, then $K|K$, so K has a prime divisor. (*Remark.* Notice that this includes the case $K = 2$. However, this is a special feature of this particular theorem; it is useful to see that a separate, explicit base case may be needed in some proofs.)
>
> Second Subcase. If K is nonprime, then by the definition of "prime," it has a divisor d other than 1 or K. By the previous lemma, $1 \leqslant d \leqslant K$. Therefore, $1 < d < K$. But, by hypothesis, all numbers less than K have a prime divisor. Thus, d has a prime divisor, say p (which might be the same as d). Therefore, $p|d$ and $d|K$, so by transitivity (Lemma 3-29), $p|K$. Thus, K has a prime divisor.

In both subcases, K has a prime divisor, which contradicts $K \in V$. Since we have a contradiction, the assumption that $\phi(k)$ is false is itself false, so $\phi(k)$ is true. *Q.E.D.*

This theorem is very important: Given any natural number $k > 1$, it is either prime or nonprime. If it is nonprime, we can factor out a prime, p, so that $k = pk_1$. Repeat this process on k_1. It is either prime or has a prime factor q (possibly the same as p). Intuitively, this process of repeatedly factoring out a prime cannot continue forever. In fact, a proof similar to the one just given establishes that any $k > 1$ is either prime or a product of primes. This is left as an exercise.

The Prime Divisor Theorem has been known since ancient times. It is Proposition 32 in Book VII of *Euclid's Elements*; see Heath (1956), Vol. 2, pp. 332–333. The proof given by Euclid is not the same as that used here, but both proofs have much in common. Among other things, Euclid takes for granted some basic properties of numbers, including the well-ordering of the positive integers. He also uses the fact that divisibility is transitive. From Peano's Axioms, all natural numbers greater than 0 are successors. From Theorem 3-32, we can see that any number greater than 1 is either prime or a product of primes. This fact also follows from Peano's Axioms; it is a striking and beautiful consequence of those simple axioms. We cannot prove within Peano's Axioms that they *really* characterize our intuitive system of arithmetic. The fact that these axioms imply this ancient Prime Divisor Theorem, and all of the other familiar properties of the natural numbers, is at least a partial justification for accepting Peano's Axioms as a "correct" set of axioms for the natural numbers.

In the next section, I will introduce another form of mathematical induction that is closely related to WO (the well-ordering of *Nat* by $<$). It will help to illustrate the relationship if we first outline the general form of proof used for the Prime Divisor Theorem. This form of proof is often used when well-ordering is invoked.

Schema for Proofs Using Well-Ordering (WO)

Let $\phi(k)$ be a predicate about natural numbers. The goal is to use well-ordering to prove that, for all $n \in Nat$, $\phi(n)$. (If the starting number is not 0, but a, then the goal is to prove that, for all $n \geqslant a$, $\phi(n)$.)

1. *Predicate and Set of Exceptions.* Precisely state $\phi(k)$ and define the set of ϕ-exceptions,

$$V_\phi = \{\, i \mid \text{not } \phi(i) \,\}.$$

2. *RAA Assumption.* Assume that $\phi(k)$ is false for some numbers, which is equivalent to $V_\phi \neq \varnothing$. By WO, V_ϕ has a first element K. Thus, $\phi(K)$ is false, but, for all $i < K$, $\phi(i)$ is true.

3. *Derive a Contradiction.* Use the previous information, together with special features of the particular problem, to try to prove that $\phi(K)$ must also be true.

    (a) When $K = 0$, we will often need a separate, explicit proof that $\phi(0)$ (similarly, for $\phi(a)$ if the starting number is a). (However, as the proof of the Prime Divisor Theorem shows, this "base case" may, in some proofs, be included under more general considerations.)

    (b) In the general case, one tries to prove $\phi(K)$ from the fact that ϕ holds for all numbers less than K. If this is accomplished, we have the contradiction: $\phi(K)$ & not $\phi(K)$.

This is only a heuristic outline for constructing such proofs; it is not guaranteed to be successful. However, it often works. One should check that the proof of the Prime Divisor Theorem fits this outline. In particular, the contradiction in that proof is that K both has and does not have a prime divisor. The latter came from the RAA assumption together with WO. The former was proved from the fact that all smaller numbers *do* have prime divisors. Now let us consider another important theorem that can be proved with the help of well-ordering.

In Chapter 2, Section 3, it was proved that $\sqrt{2}$ is irrational. The proof of this ancient theorem used the fact that any common fraction can be reduced to lowest terms. This will now be proved using the well-ordering (WO) of *Nat*. First recall the idea of a common factor or divisor. Let i and j be natural numbers. A natural number d is *a common divisor* of i and j iff both $d|i$ and $d|j$. For example, 1, 2, 3, 4, 6, 12 are all common divisors of 24 and 36. Notice that 12 is the greatest of all of these common divisors of 24 and 36.

In the following example, we shall be concerned only with positive-valued fractions. Let x, $y \in Nat^+$, and consider the fraction x/y. This fraction is said to be *in lowest terms* iff x and y have no common divisor greater than 1. For example, 2/3 is in lowest terms. Now suppose that we have a fraction, such as 30/126, that is not in lowest terms. We can start looking for common divisors (other than 1). Dividing the numerator and denominator by 2 yields the new fraction 15/63. Dividing again by 3 yields 5/21. Thus, we have

$$\frac{30}{126} = \frac{15}{63} = \frac{5}{21}.$$

By repeatedly dividing both numerator and denominator by common divisors, it appears that eventually this process must end, producing a fraction in lowest terms. This is what is to be proved.[*]

---

[*]Another, related procedure for reducing a fraction to lowest terms is to find the *greatest common divisor* (GCD) of the numerator and denominator and divide both by it. Theorem

Example.

Problem Statement. Theorem. Let i, $j \in Nat^+$. Then there exist a, $b \in Nat^+$ such that

$$\frac{i}{j} = \frac{a}{b},$$

and a and b have no common divisor greater than 1, i.e., a/b is in lowest terms.

Discussion. Let i, $j \in Nat^+$, and consider the fraction i/j. If i and j have no common divisor greater than 1, then i/j is already in lowest terms. If $d > 1$ is a common divisor of i/j, then we can divide numerator and denominator by d to obtain a new fraction equal to i/j. It is easy to show that the numerator and denominator of this new fraction are smaller than i and j, respectively. If the new fraction is not in lowest terms, this procedure is repeated.

This procedure of repeated divisions produces fractions with progressively smaller numerators and denominators. A process that produces successively smaller natural numbers must eventually stop because of WO. I shall apply WO to the denominators.

Written Solution.

Theorem. Let i, $j \in Nat^+$. Then there exist a, $b \in Nat^+$ such that

$$\frac{i}{j} = \frac{a}{b},$$

and a and b have no common divisor greater than 1, i.e., a/b is in lowest terms.

Proof. Let i, $j \in Nat^+$, and let $\phi(j)$ be

There exist a, $b \in Nat^+$ such that

$$\frac{i}{j} = \frac{a}{b},$$

and a and b have no common divisor greater than 1. Let

$$V_\phi = \{\, j \mid \text{not } \phi(j) \,\},$$

and assume that V_ϕ is nonempty. Then, by WO, V_ϕ has a smallest element, say, K. Thus, there is some fraction,

$$\frac{i}{K},$$

3-38, presented later in Section 3.3.4, presents the famous Euclidean Algorithm for finding the GCD of two numbers. At this time, however, we will not use the GCD method.

that *cannot* be reduced to lowest terms, and K is the smallest positive integer denominator in all such fractions with the numerator i.

Now i and K either have a common divisor greater than 1, or they do not. If they do not, then i/K is already in lowest terms, contrary to the assumption that $K \in V_\phi$. Therefore, i and K have a common divisor, say d, where $d > 1$. (Incidentally, notice that $K > 1$, since otherwise $1 = K$, and i/K would be in lowest terms.)

Since d is a common divisor, there are some x, $y \in Nat^+$, such that

$$\frac{i}{K} = \frac{xd}{yd} = \frac{x}{y}.$$

Now $K = yd$, so $y \mid K$. By Lemma 3-31, $1 \leqslant y \leqslant K$. Could $y = K$? If $y = K$, then

$$K = yd = Kd,$$

so that $d = 1$, which contradicts that $d > 1$. Therefore, $y < K$.

Now consider x/y. Either

    Case 1. There exists a fraction $a/b = x/y$, where a and b have no common divisor greater than 1, or

    Case 2. There does not exist a fraction $a/b = x/y$, where a and b have no common divisor greater than 1.

In Case 1, x/y can be reduced to lowest terms. But then i/K is also reducible, contrary to $K \in V_\phi$.

In Case 2, $y \in V_\phi$. But, since $y < K$, this contradicts that K is the minimal element of V_ϕ.

Therefore, the original assumption that V_ϕ is nonempty is contradictory, so $\phi(j)$ is true for all $j \in Nat^+$. ***Q.E.D.*** ▪

3.3.3 Course of Values Induction

We will now study another form of mathematical induction and see how it is related to the well-ordering of *Nat*. In fact, it will be proved from the well-ordering, which itself has already been proved from the original Induction Principle (IP). I will first state the new principle.

Course of Values Induction Principle (CVI). Let $\phi(k)$ be any predicate that applies to *Nat*.

IF: For any $k \in Nat$,

if (for every i, if $i < k$, then $\phi(i)$), then $\phi(k)$,

THEN: For all $n \in Nat$, $\phi(n)$.

Notice that, 'for every i, if $i < k$, then $\phi(i)$' can be paraphrased as 'for all $i < k$, $\phi(i)$'. Therefore, with the understanding that the variables range over *Nat*, CVI can be paraphrased as

IF IT IS TRUE THAT: For any k,

$$(\phi(i) \text{ is true for all } i < k) \;\Rightarrow\; \phi(k),$$

THEN: For all n, $\phi(n)$.

This principle is often called the "principle of strong induction" or the "principle of complete induction." However, it is no "stronger" than the original IP, as will soon be proved. Also, the term "complete" does not seem to have significant mnemonic suggestive power here. I prefer the name, "course of values induction." When using CVI, one makes the induction hypothesis that $\phi(i)$ is true for all $i < k$, and then one tries to prove $\phi(k)$. This is equivalent to assuming

$$\phi(0) \;\&\; \phi(1) \;\&\; \cdots \;\&\; \phi(k-1).$$

From this "course of values," one then tries to prove $\phi(k)$. If this is accomplished, then one infers that, for all $n \in Nat$, $\phi(n)$. Let us now prove that CVI follows from WO.

Theorem 3-33 (CVI) Let $\phi(k)$ be any predicate that applies to *Nat*.

IF: For any $k \in Nat$

$$(\text{for every } i,\; i < k \;\Rightarrow\; \phi(i)) \;\Rightarrow\; \phi(k),$$

THEN: For all $n \in Nat$, $\phi(n)$.

Proof. Assume the antecedent of the statement of CVI: For any $k \in Nat$,

if (for every i, if $i < k$, then $\phi(i)$), then $\phi(k)$.

Our goal is to prove that $\phi(n)$ is true for all n. I will follow the schema for WO proofs. For RAA, assume that $\phi(k)$ is false for some numbers, i.e., that

$$V_\phi = \{i \mid \text{not } \phi(i)\}$$

is nonempty. By WO, V_ϕ has a first element K. Thus, $\phi(K)$ is false, but ϕ holds for all numbers less than K, i.e., for every i, if $i < K$, then $\phi(i)$. The antecedent of CVI thus implies that $\phi(K)$, which is a contradiction. Hence, $V_\phi = \varnothing$, so for all n, $\phi(n)$. **Q.E.D.**

We now know that both WO and CVI are consequences of IP, the original Induction Principle. A careful study of the proofs of Theorems 3-26 and 3-33 should reveal how these three principles are related. In fact, a theorem that is proved with the help of WO can usually be proved in nearly the same way using CVI. Also, CVI proofs are very similar to IP proofs. Consider the antecedent of CVI again,

$$\text{if (for every } i, \text{ if } i < k, \text{ then } \phi(i)), \text{ then } \phi(k),$$

and remember that $i, k \in Nat$. When $k = 0$, $i < k$ is $i < 0$, which is false for any $i \in Nat$. In this case, the phrase (for every i, if $i < 0$, then $\phi(i)$) has a false antecedent for all possible values of i, so it is true. Thus, we obtain, 'if true, then $\phi(0)$', so $\phi(0)$ must be true, in order for the entire antecedent to be true when $k = 0$. Suppose that one proves *in general* that the antecedent is true. This amounts to proving that the assumption

$$\phi(0) \ \& \ \phi(1) \ \& \ \cdots \ \& \ \phi(k-1)$$

implies

$$\phi(k).$$

In the case $k = 0$, this reduces to

$$\text{If } \phi(0), \text{ then } \phi(0).$$

But this *does not prove* that $\phi(0)$ is true; yet, as just pointed out, we also need $\phi(0)$ true. Usually, we will need to give a separate, explicit proof that $\phi(0)$. This is similar to the base case in IP proofs. Hence, although the statement of CVI does not explicitly mention a base case, when giving a CVI proof, it is still necessary to show that the hypothesis holds for the starting, or "base," case. This will typically be for the case $k = 0$, although 1, 2, 3, etc., may be the starting number. It is interesting to notice that the preceding proof of CVI using WO did not require explicit mention of a starting case. This is because the antecedent of CVI implies that the base case must be true. Yet in particular applications, this fact must be separately proved. In fact, in some, more complex proofs, there may be more than one starting, or "base," case to consider. An example of this is presented at the end of this subsection.

Schema for Proofs Using Course of Values Induction (CVI).

The goal is to use CVI to prove that, for all $n \in Nat$, $\phi(n)$, where ϕ is a predicate applying to natural numbers. (If the starting ("base case") number is not 0, but a, then the goal is to prove that, for all $n \geqslant a$, $\phi(n)$.)

  1. *CVI Hypothesis.* Precisely state the course of values induction hypothesis: For every i, if $i < k$, then $\phi(i)$. More simply, $\phi(i)$ holds for all $i < k$.

  2. *Starting or Base Case.* Prove that $\phi(0)$ (or $\phi(a)$ if a is the first number to which ϕ is to apply).

  3. *Induction Step.* Assume the CVI induction hypothesis, together with special features of the particular problem, to prove that $\phi(k)$.

If the starting case and the induction step are accomplished, then from CVI it follows that $\phi(n)$ for all $n \in Nat$.

The similarity between this schema and the schema for WO proofs should be manifest. I will now prove the Prime Divisor Theorem again, this time using CVI.

Theorem 3-34 (Prime Divisor Theorem) Every natural number $k > 1$ is divisible by at least one prime.

Proof. Let $i, k \in Nat$. Let $\phi(k)$ be as follows: If $k > 1$, then there exists $p \in Nat$ such that p is prime and $p|k$. The proof is by CVI. Let the induction hypothesis be that $\phi(i)$ holds for all $i < k$. Since ϕ specifies that $k > 1$, we begin with 2.

Case $k = 2$. Since $2|2$ and 2 is prime, $\phi(2)$ is true.

Case $k > 2$. k is either prime or nonprime.

First Subcase. If k is prime, then $k|k$, so k has a prime divisor.

Second Subcase. If k is nonprime, it has a divisor d other than 1 or k. By Lemma 3-31, $1 \leqslant d \leqslant k$. Therefore, $1 < d < k$. Since $d < k$, by the induction hypothesis, d has a prime divisor p (which might be the same as d). Therefore, $p|d$ and $d|k$, so $p|k$. Thus, k has a prime divisor.

In both subcases, we find that k has a prime divisor. The theorem now follows by CVI. **Q.E.D.**

It can be seen that the two proofs of the Prime Divisor Theorem are very similar. They both use the fact that divisibility implies a size relationship. They both use the important implication that, if all numbers i such that $1 < i < k$ have prime divisors, then k must have a prime divisor. Also, they both use the *assumption* that it is true that, for $1 < i < k$, i has a prime divisor. In the WO proof, this assumption follows by applying WO to the RAA assumption. In the CVI proof, this assumption results from the induction hypothesis. Both proofs show that 2 has a prime divisor. Both proofs then use the assumption, for $k > 2$, to show that k must have a prime divisor.

We see that proving a theorem by CVI will usually be very similar to proving it by WO. There are situations in which the original Induction Principle (IP) is more convenient to use than CVI, and there are situations in which CVI (or WO) is more convenient. Here is a rule of thumb: If the proof involves induction on a function that has been defined for $k + 1$ in a simple, direct way in terms of the function applied to k, then IP will probably be most convenient. In effect, such a proof makes a "one-step jump" from k to $k + 1$, and for this reason is often called "stepwise induction." If the induction step of the proof cannot be expressed simply in terms of the predecessor value, and depends on one or more values that may be lower than the predecessor (as in the Prime Divisor Theorem), then usually WO or CVI is more convenient.

Just as induction proofs may have different forms, there are also many different forms of recursive definitions. In particular, a recursive definition may refer to several, rather than merely one, previous calls of the function being defined. Probably the best-known example of this is the famous Fibonacci Function, which may be defined by

$$fibo(k) = [\text{ if } k = 0 \text{ then } 0$$
$$\text{else if } k = 1 \text{ then } 1$$
$$\text{else } fibo(k - 1) + fibo(k - 2) \ ].$$

This, and some other definitions, are discussed in detail later, in Section 3.3.5. The point of that section is that some complex recursive definitions may be justified by special inductive arguments, including the use of CVI. Yet, one must be cautious in the use of suspicious-looking recursive definitions. The next example shows how CVI can be usefully applied to a function with repeated function calls in its definition.

Example.

Problem Statement. *Theorem.* For $k \in Nat$, define

$$g(k) = \left[\text{if } k = 0 \text{ then } 4 \text{ else if } k = 1 \text{ then } 8 \text{ else } \frac{g(k-1)g(k-2)}{2}\right].$$

Then, for all $k \in Nat$, $g(k)$ is a positive integer power of 2.

Discussion. Since the recursive step in this definition makes two calls of g, for the inputs $k - 1$ and $k - 2$, the definition requires two starting, or base, cases. We could try proving the result using IP with two earlier values of g, or we could try proving it with CVI. I shall write both proofs. They are quite straightforward and do not require further discussion.

Written Solution. *Theorem.* For $k \in Nat$, define

$$g(k) = \left[\text{if } k = 0 \text{ then } 4 \text{ else if } k = 1 \text{ then } 8 \text{ else } \frac{g(k-1)g(k-2)}{2}\right].$$

Then, for all $k \in Nat$, $g(k)$ is a positive integer power of 2.

First Proof. Let $\phi(k)$ be

$$g(k) \text{ is a positive integer power of 2.}$$

I shall use IP with a 2-step jump.

Cases $k = 0$ & $k = 1$. Since 4 and 8 are each positive integer powers of 2, these cases are true.

Case $k + 2$. Assume the induction hypothesis that both $\phi(k)$ and $\phi(k+1)$ are true. Thus, $g(k) = 2^a$ and $g(k+1) = 2^b$, where a and b are positive integers. By the definition of g, we have

$$g(k+2) = \frac{g(k+1)g(k)}{2} = \frac{2^b 2^a}{2},$$

and by the laws of exponents, we have

$$g(k+2) = \frac{2^b 2^a}{2} = \frac{2^{b+a}}{2^1} = 2^{b+a-1}.$$

Since $a, b \geqslant 1$, we have $b + a - 1 \geqslant 1$. Hence, $g(k+2)$ is a positive integer power of 2. ***Q.E.D.***

Second Proof. Let $\phi(k)$ be

$$g(k) \text{ is a positive integer power of 2.}$$

The proof is by CVI.

Cases $k = 0$ & $k = 1$. Since 4 and 8 are each positive integer powers of 2, these cases are true.

Induction Step. Assume the CVI hypothesis that $\phi(i)$ holds for all $0 \leqslant i < k$. Thus, $g(k-2) = 2^a$ and $g(k-1) = 2^b$, where a and b are positive integers. By the definition of g, we have

$$g(k) = \frac{g(k-1)g(k-2)}{2} = \frac{2^b 2^a}{2},$$

and by the laws of exponents, we have

$$g(k) = \frac{2^b 2^a}{2} = \frac{2^{b+a}}{2^1} = 2^{b+a-1}.$$

Since $a, b \geqslant 1$, we have $b + a - 1 \geqslant 1$. Hence, $g(k)$ is a positive integer power of 2. *Q.E.D.* ■

3.3.4 Two Arithmetical Algorithms

Before moving on to other topics, we should look at two more theorems of number theory. These are fundamental theorems that yield algorithms (procedures) for calculations. Every student of mathematics or computing should be familiar with them. The first brings back memories of elementary school. One of the more tedious school-day tasks (at least for this writer) was dividing one integer by another to obtain a quotient and a remainder. Although dividing numbers by hand is not very interesting, it *is very interesting* to understand how the division process works. Suppose that we do integer division with the dividend 20307 and the divisor 707. After some effort (which I turn over to a machine), one finds the quotient 28 and the remainder 511. Thus,

$$20307 = 707(28) + 511,$$

which fits the general form,

$$n = dq + r,$$

where n is the dividend, d the divisor, q the quotient, and r the remainder. Also, $r < d$. In elementary school, one learns how to divide numbers. Perhaps without conscious reflection, the student typically assumes that the process will always give unique results (q and $r < d$) for any n and any $d \neq 0$. There are two assumptions here: (1) The process will terminate and give at least one pair of results, and (2) the pair of numbers obtained from the process is actually unique. I will now prove that (1) and (2) are true. The theorem is often called the *Division Algorithm* because it suggests an algorithm (procedure) for division by repeated subtraction. The proof uses WO, but is simpler than many WO proofs and deviates from the general schema for WO proofs.

Theorem 3-35 (**Division Algorithm Theorem**) Let $n \in Nat$ and $d \in Nat^+$. Then there exist unique natural numbers q and r such that

$$n = dq + r$$

and

$$r < d.$$

Proof.

[Existence] As usual, the domain of discourse is *Nat*. Let

$$S = \{k \mid \text{there exists } i \in Nat \text{ such that } k = n - id\}.$$

For $i = 0$, we have $k = n$, so $n \in S$. Therefore, S is a nonempty subset of *Nat*. Hence, by WO, S has a first element. Let r denote this first element. Then

$$r = n - qd$$

for some $q \in Nat$. Thus,

$$n = dq + r.$$

Thus, we have a quotient q. Now we need to show that $r < d$. Suppose, for RAA, that it is not the case that $r < d$. Then $d \leqslant r$, so $r = d + x$, for some $x \in Nat$. Then we would have

$$n = dq + r = dq + (d + x) = (dq + d) + x = d(q + 1) + x.$$

Hence, we would also have

$$x = n - d(q + 1) \in S.$$

Moreover, since $r = d + x$ and $d > 0$, we have $x < r$. This contradicts the fact that r is the first element of S, so $r < d$. We have now proved both the existence of a quotient q and a remainder $r < d$.

[Uniqueness] Suppose there are two solution pairs,

$$n = dq + r = dq_1 + r_1.$$

The following argument uses some elementary facts about inequalities, divisibility, and the subtraction function on *Nat*. The reader should verify that the steps are justified. We consider two cases:

Case $r = r_1$. Then $dq = dq_1$. Since $0 < d$, $q = q_1$. Thus, the quotient and remainder are unique.

Case $r \neq r_1$. One of the remainders is larger than the other; we can assume $r_1 < r$. (Except for the symbols used, the proof would be the same if we assumed the opposite.) We have

$$dq + r = dq_1 + r_1.$$

Since $r_1 < r$, each side of this equation is a number that is not less than r_1, so we can subtract r_1 from both sides of the equation,

$$dq + r - r_1 = dq_1.$$

Also, since $r_1 < r$, we obtain $dq_1 > dq$, so by subtraction, we obtain

$$r - r_1 = dq_1 - dq = d(q_1 - q).$$

Since $r_1 < r < d$, we have $r - r_1 < d$. But the equation immediately above shows that $d \,|\, (r - r_1)$. It follows from Lemma 3-31 that

$$r - r_1 = 0,$$

so $r = r_1$. This contradicts the assumption that $r \neq r_1$, for this case of the proof. **Q.E.D.**

The algorithm suggested by this theorem is very simple (although not efficient) and is presented in Figure 3.5. Let n, d, q, and r be the same variables as in the theorem. The n and d in the first line of the procedure are the numerical inputs to the procedure. An expression of the form, "let $\alpha = \beta$" indicates that the current value of the variable β is assigned to the variable α. During the procedure, the values of n and d stay fixed; the values of the other variables may change.

Clearly, the key feature of this algorithm is repeated subtraction. From the theorem, $n = dq + r$, where $r < d$, and also q and r are unique. Because they

```
Procedure DIVIDE(n, d)
let q = 0
let r = n

while d ⩽ r do
        let q = q + 1
        let r = r − d
end while

return q
return r
end DIVIDE
```

Figure 3.5. *A Division Procedure*

are unique, there should be functions of n and d that return these values of q and r. Using the theorem as a guide, it is not difficult to write definitions for these functions. The *quot* function is defined here recursively.

$$quot(n, d) = [\text{ if } n < d \text{ then } 0 \text{ else } 1 + quot(n - d, d) \ ],$$

and

$$rem(n, d) = n - d \cdot quot(n, d),$$

where $d \cdot quot(n, d)$ is the product of d with $quot(n, d)$.

The second function, *rem*, is defined satisfactorily if *quot* is. It is defined this way here so that the exact connection with *quot* is explicit. Actually, if one wants only to compute *rem*, it is a bit silly to do so in terms of *quot*. It makes more sense to define *rem* in a simpler way that does not require computing *quot*. This can be done, but is left as an exercise.

The *quot* function is obviously recursive, but its recursion variable, n, steps down in decrements of d rather than in decrements of the number 1. Yet it is easy to see that *quot* is well-behaved. It returns 0 when $n < d$. If $n = d$, then

$$quot(n, d) = 1 + quot(0, d) = 1 + 0 = 1.$$

If $n > d$, then *quot* "calls itself" for the argument values,

$$< n, d >, < n - d, d >, < n - 2d, d >, \cdots.$$

The values of the first argument decrease just like the values of the elements of S in the proof of the theorem. The *quot* function increments its value by 1 for each subtraction that can be performed. By WO, the sequence of descending values of the first argument has a first element.* If this first element is 0, then $d|n$, and $rem(n,d) = 0$. Otherwise, rem returns a positive remainder. The functions $quot$ and rem (often with different names) are included in many programming languages, but are not necessarily defined by the formulas used here.

One often needs to divide two numbers by the same divisor. Sometimes we want to determine whether two numbers have any divisor in common, other than 1.

Definition 3-36 Let k, m, n, $d \in Nat$, with k, $d > 0$. If $d|m$ & $d|n$, then d is called a *common divisor* of m and n. The number d is a *greatest common divisor* (GCD) of m and n iff d is a common divisor of m and n and, in addition, for any k,

$$(k|m \ \& \ k|n) \ \Rightarrow \ k|d.$$

Suppose that d is a GCD of m and n. If there were another GCD, say d_1, then d and d_1 would divide each other, so we would have $d = d_1$. Hence, if a GCD exists, it is unique. The GCD is a familiar concept. For some pairs of numbers, e.g., 8 and 25, 1 is the greatest common divisor. Other cases are more interesting. For example, several numbers divide 30 and several numbers divide 186. In particular, the numbers 2, 3, and 6 divide both 30 and 186. By factoring 30 and 186 into their prime factors, one can check that 6 is the greatest common divisor. But factoring two numbers into primes can be a lot of work. There is a more direct procedure for finding the GCD of two numbers. First, let us note some simple facts. The pair $< 0, 0 >$ has no GCD, since any positive number divides 0. The GCD of $< m, 0 >$ is m. More generally, if $m|n$, then the GCD of m and n is m. In other words, if $rem(n, m) = 0$, then the GCD of $< n, m >$ is m. Because the GCD of two numbers is unique, from now on I will write it as a function value, $gcd(n, m)$. We know that, for any n, m, if $gcd(n, m)$ exists, then it is unique. We have just seen that it exists in some cases. We need to prove that it exists for any pair of positive natural numbers. The following is a key lemma.

Lemma 3-37 Let m, $n \in Nat^+$, with $n > m$. Suppose that

$$n = qm + r.$$

Then

$$gcd(n, m) = gcd(m, r).$$

---

*In computing theory, a sequence of values that indicates progress of a program toward a termination state is sometimes called a *metric*. The sequence n, $n - d$, $n - 2d$, $\cdots$ is a metric for the *quot* function. This sequence terminates because of WO. Other examples and discussion can be found in Chandy and Taylor (1992).

Proof. It is a simple proof and is left for practice.

This little lemma is the key to the algorithm for computing the GCD of two numbers. This is an ancient algorithm, called the *Euclidean Algorithm,* and is Proposition 2, of Book VII, of *Euclid's Elements* (Heath (1956), Vol. 2, pp. 298–299). Euclid describes the algorithm in terms of repeated subtractions. I will use a slightly different form that uses remainders. The two procedures are related by the Division Algorithm and give the same result. The Euclidean Algorithm is *the classical example* of a recursive procedure and is found in many books and articles on computer programming. Here is an illustration of how it works. Suppose we begin with 3490 and 1200. Using the Division Algorithm, we first divide 3490 by 1200 and get the remainder of 1090.

$$3490 = 2(1200) + 1090$$

We then divide this remainder into the divisor, to get a second remainder,

$$1200 = 1(1090) + 110.$$

We then divide this new remainder into the previous divisor, and so on.

$$1090 = 9(110) + 100,$$

$$110 = 1(100) + 10,$$

$$100 = 10(10) + 0.$$

Eventually, we get 0 as the remainder. The last equation (with the 0 remainder) tells us that $gcd(100, 10) = 10$ because it tells us that $10|100$. Moreover, 10 is the last nonzero remainder in the sequence of divisions. Because of the preceding lemma,

$$gcd(3490, 1200) = gcd(1200, 1090) = \cdots = gcd(100, 10) = 10.$$

Thus, $gcd(3490, 1200)$ equals the last nonzero remainder. The lemma and this example suggest how to formulate the desired theorem. The theorem applies to pairs of positive numbers; zero cases are handled after the theorem.

Theorem 3-38 (**Euclidean Algorithm Theorem**) Let $m, n \in Nat^+$, with $n \geqslant m$. Then there exists a unique $d \in Nat^+$ such that

$$d = gcd(n, m).$$

Proof. Let *rem* be the remainder function defined before, and let $r(0) = m$. By the Division Algorithm, we can write

$$r(1) = rem(n, m) = rem(n, r(0)),$$

where $r(1) < m = r(0)$. If $r(1) = 0$, then $m|n$ so (as noted in previous discussion) m is the GCD. If $r(1) \neq 0$, then let

$$r(2) = rem(m, r(1)) = rem(r(0), r(1)).$$

If $r(2) = 0$, then $r(1)|m$, so

$$gcd(m, r(1)) = r(1).$$

By the lemma,

$$gcd(n, m) = gcd(m, r(1)),$$

so again $gcd(n, m)$ exists, and it is $r(1)$. If $r(2) \neq 0$, we continue to compute remainders, obtaining a sequence as follows:

$$r(1) = rem(n, m) = rem(n, r(0)),$$

$$r(2) = rem(m, r(1)) = rem(r(0), r(1)).$$

For $k \geqslant 1$,

$$r(k + 2) = rem(r(k), r(k + 1)).$$

It follows from the Division Algorithm that

$$r(1) < m, \ r(2) < r(1), \text{ and } r(i + 1) < r(i),$$

for each $i \in Nat^+$.

But $rem(n, d)$ is not defined for $d = 0$, so if $r(i) = 0$ for some i, then this sequence stops. Let S be the set of all $r(i)$ that are defined. S is nonempty because at least $r(1) \in S$. By WO, S has a first element (with respect to $<$). For RAA, suppose this first element is not 0. In other words, it is $r(i) \neq 0$ for some i. Also, $r(i) < r(i - 1)$. Then we can continue to compute

$$rem(r(i - 1), r(i)) = r(i + 1),$$

contrary to the assumption that $r(i)$ is the first element of S. Thus, the first element of S is 0, and the sequence stops when we reach a 0 remainder.

Now, for $k \in Nat$, we state an induction hypothesis $\phi(k)$: Let S be any sequence of remainders as described above, with at least $k + 1$ nonzero terms. Then

$$gcd(n, m) = gcd(r(k), r(k + 1)).$$

Case $k = 0$. By the lemma and the previous calculations, we have

$$gcd(n, m) = gcd(n, r(0)) = gcd(r(0), r(1)).$$

This establishes the base case.

Case $k + 1$. Assume that $\phi(k)$ is true, so that

$$gcd(n, m) = gcd(r(k), r(k + 1)).$$

If we have a sequence with at least $k + 2$ nonzero terms, then we again use the lemma to obtain

$$gcd(n, m) = gcd(r(k + 1), r(k + 2)),$$

which establishes the induction step.

The sequence stops when it reaches a zero remainder. At that stage, the divisor is the last nonzero remainder, which is the GCD between itself and its dividend. Thus, $gcd(n, m)$ equals the last nonzero remainder. We have already seen that the GCD is unique. **Q.E.D.**

It is now easy to define the *gcd* function and also to include the cases where one of the arguments is 0.

$$gcd(n,\ m) = \quad [\quad \text{if } m = 0 \text{ then } n$$
$$\text{else if } n = 0 \text{ then } m$$
$$\text{else } gcd(m,\ rem(n,\ m))\ ].$$

There is no problem if $m > n$, since then

$$gcd(n, m) = gcd(m, rem(n, m)) = gcd(m, n).$$

This is a relatively complex recursive definition. The fact that the *gcd* function returns a unique value for all inputs with at least one of n or m not equal to zero is established by the proof of the previous theorem. The *gcd* function is included in some programming languages, and it can easily be defined if it is not built in. One word of caution: The definitions we have used for *quot*, *rem*, and *gcd* are motivated primarily by theoretical simplicity and the desire for these definitions to have forms similar to the constructions used in the proofs of the corresponding theorems. Our definitions are not necessarily the most computationally efficient ones possible.

3.3.5   Pitfalls of Recursion

The definitions of *quot* and *gcd* use recursive calls of the functions with larger decrements than the predecessor. It is easy to write even more complex forms of recursive definitions. For example, consider the familiar Fibonacci Function, *fibo*, defined by

$$fibo(k) = \;[\text{ if } k = 0 \text{ then } 0$$
$$\text{else if } k = 1 \text{ then } 1$$
$$\text{else } fibo(k-1) + fibo(k-2) \;].$$

Anyone who has ever examined this function certainly has no doubt that it is well-defined by this recursion equation. Yet, because of the two base cases in its definition, it is somewhat different from the previous forms of recursive definitions that we have used. This raises some interesting and challenging questions.

Let M be a given set. Let us say that a *set of recursion equations* for a possible function $r : Nat \longrightarrow M$ is any set of equations involving r such that, in some of these equations, r occurs on both the left and right sides. These equations will make use of previously given functions and relations. This is very general and somewhat vague, but it is good enough for present purposes. Here are some questions: How can we determine whether or not there exists a function r satisfying these equations? If a function does exist, is it unique? If it is not unique, can we determine all such functions? These are very difficult questions that are beyond the scope of this text. Because many computer programs use recursively defined functions, these questions relate to the problems of verifying that a program terminates and whether it satisfies certain specifications. These are important topics of current research. I will present a few examples to illustrate some aspects of these questions.

Why do we feel confident about *fibo*? Here is a sketch of an argument similar to the proof of the Recursion Theorem. Suppose that a relation $F \subseteq Nat \times Nat$, and define the predicate Δ as follows:

$$\Delta(F)$$

$$\text{iff}$$

$$F00 \;\&\; F11 \;\&\; ((F(k-2)i \;\&\; F(k-1)j) \Rightarrow Fk(i+j)).$$

The domain of any F satisfying Δ must be *Nat*. First, 0 and 1 are in the domain. Suppose that, for all x such that $x < k$, x is in the domain of F. If $k \geqslant 2$, then $k-1$ and $k-2$ will be in the domain of F, so by Δ, k will be in the domain of F. By CVI, all natural numbers will be in the domain.

If a relation, F, satisfying Δ is a function, then it must satisfy the definition of *fibo*. Also, if F_1 and F_2 are two such functions, then they return the same

values for 0 and 1. Assume that they return the same values for all $x < k$.
Then from Δ they will return the same value for k, so by CVI, $F_1 = F_2$. In
other words, if any function exists, there will be only one. By using CVI, one
can also show that if a function, F, satisfies Δ, then it must be a subset of
every relation that satisfies Δ, i.e., F must be the *smallest* such relation.

Let *fibo* be the intersection of all relations satisfying Δ, so it is the smallest
such relation. It would be tedious to prove that it is actually a function, but
this can be done as in the proof of the Recursion Theorem, except that CVI or
WO are more convenient to use than IP.

As a general guide, one can safely use a recursive definition of a function
f, where $f(k)$ is defined by calling f for smaller numbers, and in which the
values returned by these function calls are combined together by previously
defined functions. In addition, appropriate base-case values of the function
must be provided. *Appearances can be deceptive, so if it is not perfectly clear
that a function is really being defined, one must give a proof that the recursion
equations do define a function.* Let us consider a few more examples that
illustrate this point.

Define f on *Nat* as follows:

$$f(k) = [\text{ if } k = 0 \text{ then } 0 \text{ else } k + f(k-1) \text{ }].$$

It can be seen that f is just the *sum* function used at the beginning of this
chapter. It poses no problems.

Now define

$$g(k) = [\text{ if } k = 0 \text{ then } 0 \text{ else } g(k + g(k-1)) \text{ }].$$

This is similar to f except that g is applied to $k + g(k-1)$. This makes a big
difference! Observe that

$$g(1) = g(1 + g(0)) = g(1 + 0) = g(1),$$

so that $g(1)$ is defined in terms of itself. In a computer program this creates an
eternal loop, i.e., an endless cycle. The definition of g is faulty; this also blocks
attempted inductive proofs about g. One could say that this is all obvious,
because the definition of $g(k)$ applies g to k plus other terms. In programming
jargon, to compute $g(k)$, the function "calls" g on k (together with $g(k-1)$).
This is an appropriate observation, but things are not always so obvious.

One might get the impression that everything will be all right if we replace
the call to k plus other terms, with a call to $k-1$ plus the other terms. Those
other terms can become a problem.

$$h(k) = [\text{ if } k = 0 \text{ then } 0 \text{ else } h((k-1) + h(k-1)) \text{ }].$$

Now we have $h(0) = 0$ and $h(1) = h(0 + h(0)) = h(0) = 0$. It is easy to prove
by induction that $h(n) = 0$ for all n. But the fact that h always returns 0 is

fortuitous. Let us slightly change the definition of h to produce

$$p(k) = [\text{ if } k = 0 \text{ then } 1 \text{ else } p((k-1) + p(k-1)) \text{ }].$$

Notice that $p(1) = p(0 + p(0)) = p(0 + 1) = p(1)$, so the definition of p is circular, and attempting to compute it leads to an eternal loop. One might get the impression that we should never use iterated recursions such as p applied to a term that also contains p. But h is a good function, though it is trivial. There are less trivial examples of iterated recursions. One is John McCarthy's much discussed "91 function" defined by

$$q(k) = [\text{ if } k > 100 \text{ then } k - 10 \text{ else } q(q(k + 11)) \text{ }].$$

At first glance, it is not at all obvious that this is a satisfactory definition of a function, and it requires some effort to prove that it is. In fact, q returns $k - 10$ for $k > 100$ and 91 otherwise. This function has motivated much research on techniques for analyzing sets of recursion equations. Some advanced treatments are in Section 5-2 of Manna (1974), and in Feferman (1991) and Knuth (1991).

The moral of these examples is this: In general, one cannot simply look at a set of recursion equations and be sure whether or not they define any function or functions. It is often necessary to perform an extensive analysis of the equations. It is often helpful to look for some kind of well-ordering such that the recursions used in equations always apply the function to "smaller" values according to this ordering. This ordering need not always be the ordering of *Nat* by $<$.

Exercises

General Instructions

In this exercise set, unless mentioned to the contrary, you may assume the basic principles of arithmetic and algebra as applied to *Nat*, *Int*, *Rat*, and *Re*. Several of the problems have more specific instructions; please be sure to follow these instructions where they occur.

Throughout this book, if an exercise is simply the statement of a theorem or a metatheorem, then the task is to prove this theorem or metatheorem. In your proof, you may use the definitions given in the text. Unless stated to the contrary, you may also use theorems proved in the text.

Exercises for which answers are provided at the back of the book are marked with "Answer Provided."

Finally, please note that exercises that introduce important principles or useful applications, are especially difficult, or are special in some way, are indicated by bold print. In some cases, parenthetical comments describe the special features of the problem. If such comments are missing, the exercise is nonetheless of special value in introducing a new concept, principle, or technique.

E 3-43 *Theorem.* Define $f : Nat^+ \longrightarrow Rat^+$ by: For any $k \in Nat$,

$$f(k) = \Big[\text{ if } k = 1 \text{ then } 1 \text{ else } \frac{(k-1)f(k-1) + 2k - 1}{k} \Big].$$

Then, for any $k \in Nat^+$, $f(k) = k$.

E 3-44 *Theorem.* Define $f : Nat \times Re \longrightarrow Re$ as follows: For any $k \in Nat$, any $z \in Re$,

$$f(k, z) = [\text{ if } k = 0 \text{ then } 0 \text{ else } kz + f(k-1, z)].$$

Then, for any $k \in Nat$, any $x, y \in Re$,

$$f(k, x + y) = f(k, x) + f(k, y).$$

E 3-45 Let d be a particular element of *Nat*, and define

$$r : Nat \times Nat \longrightarrow Nat$$

as follows: For any natural numbers i, k,

$$r(i, k) = [\text{ if } k = 0 \text{ then } i \text{ else } r(i + d, k - 1)].$$

Prove that, for all i, k in *Nat*, $r(i, k) = i + kd$.

(Note: The variable i in $r(i, k)$ can be called an *accumulator* because it stores values that are computed by the function. Accumulators like this one are often used in programs to save memory and make more efficient programs.)

E 3-46 (Answer Provided) The factorial function is defined in Section 3.3.1. Prove the following:

1. *Lemma.* For any $k \in Nat$, $0 < k!$.

2. *Theorem.* Let $k \in Nat$ and $k \geqslant 4$. Then $2^k < k!$. (Hint. Lemma 3-30 may be useful.)

E 3-47 Define subsets of *Nat* as follows:

$$A_0 = \{0, 1\},$$

and, for $k \in Nat$,

$$A_{k+1} = \{x + 2 \mid x \in A_k\}.$$

Let

$$B = \bigcup\{A_k \mid k \in Nat\}.$$

Prove that $B = Nat$.

E 3-48 Let

$$\Gamma_0 = \{P, Q\},$$

$$\Gamma_{k+1} = \Gamma_k \cup \{\phi \wedge \psi \mid \phi, \psi \in \Gamma_k\},$$

and

$$\Delta = \bigcup\{\Gamma_k \mid k \in Nat\}.$$

Prove: If $\mathbf{I}$ is any $\mathcal{SC}$ interpretation, then $\mathbf{I}$ satisfies Γ_0 iff $\mathbf{I}$ satisfies Δ.

E 3-49 Define $f : Nat^+ \times Nat \times Nat \longrightarrow Nat$ by

$$f(i, k, n) = [\text{ if } k = 0 \text{ then } n \text{ else } f(i, k-1, ni) \,] \,.$$

Prove that $f(i, k, n) = ni^k$.

E 3-50 *Theorem.* The divisibility relation,

$$i \mid j,$$

on Nat^+ is a partial order on Nat^+.

E 3-51 *Theorem.* For all natural numbers n, $2 \mid n^2 + n$.

E 3-52 *Theorem.* For all natural numbers n, $3 \mid n^3 + 2n$.

E 3-53 *Theorem.* Let $n, i \in Nat^+$, and, for any i, let $r_i \in Re$. If at least one of $r_1, \cdots, r_n$ is nonzero, then

$$\sum_{i=1}^{n} r_i^2 > 0.$$

E 3-54 *Theorem.* Let $f : Nat^+ \longrightarrow Re$, and assume that, for every k in Nat^+, $f(k) < b$, where b is a particular real number. Define the function Σ on Nat^+ as follows: For every k in Nat^+,

$$\Sigma(k) = [\text{ if } k = 1 \text{ then } f(1) \text{ else } f(k) + \Sigma(k-1) \,] \,.$$

Prove that, for all n in Nat^+, $\Sigma(n) < nb$. (Hint. Use induction. You may assume elementary facts about inequalities, e.g., if $x < y$ and $u < v$, then $x + u < y + v$.)

E 3-55 Let $m \in M$ and $F : M \longrightarrow M$. By the Recursion Theorem, there is a unique function $r = r_m$ satisfying

$$r_m(k) = [\text{ if } k = 0 \text{ then } m \text{ else } F(r_m(k-1))\,]\,,$$

for each m in M. Define a new function $r : M \times Nat \longrightarrow M$ by

$$r(m,\, k) = r_m(k),$$

for each m in M.

1. Prove that, for each m in M, and each i, j in Nat,

$$r(m,\, i + j) = r(\,r(m,\, i),\, j).$$

2. Suppose that M is Nat, F is the successor function, s, and m is an element of Nat. Calculate the values of $r(m,\, 2 + 3)$ and $r(\,r(m,\, 2),\, 3)$.

E 3-56 *Theorem.* Let $\langle A, R \rangle$ be a well-ordering and B be a nonempty subset of A. Define R_B on B: For any x, y,

$$R_B xy \Leftrightarrow (Rxy \ \& \ x \in B \ \& \ y \in B).$$

Then $\langle B, R_B \rangle$ is a well-ordering.

E 3-57 *Nat* is well-ordered by $<$. Let $f : Nat \xrightarrow{1\text{-}1} Re$, and let $S = Img(f,\, Nat)$. Let R be a binary relation on S that satisfies $x < y \ \Leftrightarrow \ Rf(x)f(y)$, for every x, y in Nat. Prove that the relational system $\langle S,\, R \rangle$ is a well-ordering. (Hint: This is not an induction problem; it is just a problem about relations and functions.)

E 3-58 Prove that WO implies IP. (Hint: Review the proof of Theorem 3-33.)

E 3-59 Prove that CVI implies IP.

E 3-60 Prove directly, *without using* WO, that IP implies CVI. (Hint: Informally, the course of values is

$$(\phi(0) \ \& \ \phi(1) \ \& \ \phi(2) \ \& \ \cdots \& \ \phi(k-1)),$$

which can be paraphrased as

$$\text{for all } i < k,\ \phi(i).$$

322   Chapter 3   ■   Recursion and Mathematical Induction

You need to prove that

IF: (for all $i < k$, $\phi(i)$) implies $\phi(k)$,

THEN: For all n in *Nat*, $\phi(n)$,

using the assumption of IP. You might try defining a new predicate, something like

$$\psi(k) \iff (\text{for all } i < k, \phi(i)),$$

and applying IP to this predicate.)

E 3-61 Let *fibo* be the Fibonacci function previously defined. Prove that, for all natural numbers k, d,

$$(d \,|\, fibo(k) \,\&\, d \,|\, fibo(k+1)) \implies d = 1.$$

E 3-62 Let i, $k \in Nat$, let $A_0 = \{3\}$, and, for $k > 0$, let

$$A_k = \{3x \mid \text{there is } i < k \text{ such that } x \in A_i\}.$$

Prove that, for all natural numbers n, all elements of A_n are powers of 3 with exponents that are natural numbers.

E 3-63 (Answer Provided) *Metatheorem.* Let τ be a tautology, i, j, $m \in Nat$, and ϕ, ψ, χ be $\mathcal{SC}$ sentences. Define a sequence of sets of sentences as follows:

$$\Gamma_0 = \{\tau\},$$

and, for any $m > 0$,

$$\Gamma_m = \{(\phi \wedge \psi) \mid \text{there exist } i, j < m \text{ such that } \phi \in \Gamma_i \,\&\, \psi \in \Gamma_j\}.$$

Now let

$$\Delta = \bigcup \{\Gamma_k \mid k \in Nat\}.$$

Then for any $\chi \in \Delta$, χ is tautologous. (Hint: χ must be in Γ_k for some k. Why? Now use Course of Values Induction to prove that χ in Γ_k is tautologous.)

E 3-64 (Infinitude of primes) *Theorem.* Let $p_1 = 2$, $p_2 = 3$, $\cdots$, p_k be the first k prime numbers. Form the product of these primes and add 1 to yield the number

$$N = (p_1 p_2 p_3 \cdots p_k) + 1.$$

Then there exists a prime p, such that

$$p_k < p \leqslant N.$$

(Remarks. This is proved in Proposition 20, of Book IX, of *Euclid's Elements* (Heath (1956), Vol. 2, p. 412). Euclid's proof is clever and short. First note that none of the p_i, for $1 \leqslant i \leqslant k$, divides N. Justify this. Yet, by the Prime Divisor Theorem, there exists some prime, say p, that divides N. The rest is straightforward. This theorem obviously implies that there is no largest prime, so the set of primes is infinite.)

E 3-65 (An important theorem of number theory)
Theorem. Let $1 < k \in Nat$. Then k is either prime or is a product of primes. Prove this without using the Prime Divisor Theorem. (Remark. You may assume this: If M is a prime or a product of primes, and N is a prime or a product of primes, then NM is a product of primes. One might try CVI. The theorem is true for 2. Assume that it is true for all numbers i such that $2 \leqslant i < k$. Then prove it true for k. A similar proof uses WO.)

E 3-66 The text defined the remainder function, rem, in terms of the quotient function, $quot$. Use these definitions to show that

$$rem(n + d, d) = rem(n, d).$$

Now, without referring to $quot$, write a recursive definition for the remainder function.

E 3-67 (A useful fact of number theory) Let i, j, $d \in Nat$, with $i > j$ and $d > 0$. Prove that

$$d|(i - j) \Leftrightarrow (rem(i, d) = rem(j, d)).$$

(Hint. Use the Division Algorithm Theorem and a little algebra. Remember that a remainder is less than its corresponding divisor.)

E 3-68 Use induction on n to prove that, for any n, $d \in Nat$, with $d > 0$, there exist q, $r \in Nat$ such that

$$n = dq + r,$$

with $r < d$. This is an induction proof of the existence part of the Division Algorithm Theorem.

E 3-69 (Answer Provided) Prove that, for any $k \in Nat$, $3|k(k+1)(k+2)$. (Hint: By the Division Algorithm Theorem, $k = 3q + r$, where $r < 3$.)

E 3-70 Prove Lemma 3-37.

E 3-71 (Two useful arithmetical facts)

1. *Theorem.* Let n, d, $k \in Nat$, with $0 < d$, k. Then

$$rem(kn, kd) = k(rem(n, d)).$$

(Hint: This follows easily from the Division Algorithm Theorem; remember that quotients and remainders are unique.)

2. *Theorem.* Let n, m, $k \in Nat$, with $0 < k$. Then

$$gcd(kn, km) = k(gcd(n, m)).$$

(Hint: Make use of the equations in the proof of the Euclidean Algorithm Theorem.)

E 3-72 Let $fibo(k)$ be the Fibonacci function defined in Section 3.3.5. Write a recursive definition for the sum

$$\Sigma_{fibo}(k) = 0 + 1 + 1 + 2 + 3 + \cdots + fibo(k).$$

Using your definition and the definition of $fibo$, prove that

$$\Sigma_{fibo}(k) = fibo(k + 2) - 1.$$

(Hint: First make a table of values for k, $fibo(k)$, and $\Sigma_{fibo}(k)$; look for a pattern.)

E 3-73 (Uses a common form of induction proof) Let S_0 be a set containing five odd natural numbers and nothing else. For $k \in Nat$, define

$$S_{k+1} = S_k \cup \{xy \mid x, y \in S_k\},$$

where xy is product of x and y. Let

$$T = \bigcup \{S_k \mid k \in Nat\} = \bigcup_{k=0}^{\infty} S_k.$$

Prove that every number in T is odd. (Remark. The general form of an odd natural number is $2b + 1$, where $b \in Nat$.)

E 3-74 (Uses a common form of induction proof) *Metatheorem.* Let $\Gamma = \{\sigma_1, \sigma_2, \sigma_3\}$ be a set of $\mathcal{SC}$ sentences. For $i, j, k \in Nat$, define sets of $\mathcal{SC}$ sentences as follows:

$$\Gamma_0 = \Gamma,$$

and, for any $k > 0$,

$$\Gamma_k = \{(\phi \wedge \psi) \mid \text{there exist } i, j < k \text{ such that } \phi \in \Gamma_i \And \psi \in \Gamma_j\} \cup$$

$$\{(\phi \vee \psi) \mid \text{there exist } i, j < k \text{ such that } \phi \in \Gamma_i \And \psi \in \Gamma_j\},$$

and,

$$\Delta = \bigcup \{\Gamma_k \mid k \in Nat\}.$$

Then any interpretation **I** that satisfies Γ also satisfies Δ. (Hint: If $\chi \in \Delta$, then it must be in Γ_m for some m. Why? Now use CVI to prove that **I** satisfies Γ_k, for any $k \in Nat$.)

E 3-75 (Answer Provided) Let $\langle A, R \rangle$ be a relational system in which A has $n \neq 0$ elements. Let R be connected, symmetric, and irreflexive on A. Thus, R consists of all ordered pairs of distinct elements of A, i.e., all $< x, y >$ for $x \neq y$. It is clear that R contains $n(n-1)$ elements, since there are n possibilities for x and then $n-1$ possibilities for y in $< x, y >$. Obtain this result by a different method. Let $p(n)$ be the number of elements of R when A has n elements. If $n = 1$, then $p(n) = 0$. Suppose we have n elements in A and $p(n)$ in R. Then we add a new element to A, so A now has $n + 1$ elements. How many new elements are added to R? Let N be this number. Write a recursion formula for $p(n+1)$ in terms of $p(n)$ plus N, where N is expressed in terms of n. Use this formula to write a recursive definition of $p(n)$. Now use induction to prove that $p(n) = n(n-1)$. (Suggestion. After doing this exercise, try the next one.)

E 3-76 Let $\langle A, R \rangle$ be a relational system in which A has $n \neq 0$ elements. Let R be connected, symmetric, and irreflexive on A. Because R is symmetric, consider the undirected graph of $\langle A, R \rangle$, and assume that this graph has only the edges (arcs) required by the properties of R and no other edges. Prove that this undirected graph of $\langle A, R \rangle$ has $n(n-1)/2$ edges. (Hint: Because R is connected, any two distinct nodes of the graph have an edge between them. These are the only kinds of edges. Let $E(n)$ be the number of edges when A has n elements. If $n = 1$, the graph is just a single node with no edges, so $E(1) = 0$. If $n = 2$, the graph has two nodes and one edge, so $E(2) = 1$. For $n = 3$, the graph has three nodes and three edges, so $E(3) = 3$. Prove the theorem by induction. You may refer to graph diagrams in your argument. Compare this problem with the previous one.)

3.4   Nonnumerical Data

From the preceding parts of this chapter, one might get the impression that recursion and induction are only useful when one is working with numerical data and numerical functions. Actually, the same general techniques can be applied to nonnumerical objects and functions on such objects. There are different ways to extend recursion and induction into nonnumerical contexts. One simple approach begins by looking for numerical parameters associated with the nonnumerical data. If suitable parameters can be found, then one can often construct induction proofs using these numerical parameters. The induction still uses numbers, but the results apply to nonnumerical objects.

However, the construction of the nonnumerical objects is typically done in a more direct way. Recall from Chapter 1 that the set of all sentential calculus sentences was defined by starting with atomic sentences and recursively constructing compound sentences from previously given or constructed sentences. This type of construction is widely used in mathematics, linguistics, and computing theory. This section introduces these ideas and techniques with three kinds of constructed objects: *strings, lists, and sentential calculus sentences.* The aim of this section is not to develop any systematic theories; it is to show how recursive and inductive techniques can be used with nonnumerical data. These techniques are very useful in a variety of fields. In particular, they are applied in the formal verification of computer programs, and they have also been implemented in theorem-proving programs. In order to exhibit some computational applications of lists, I will also include a short discussion of stacks and queues.

3.4.1   Strings

Let us assume that we know what a *character* is. We can think of a character as a kind of mark that has a distinctive shape. Let *char* denote a nonempty set of characters. For example, *char* could be the set of lowercase letters of the English alphabet, $\{ a, b, c, \cdots, x, y, z \}$. Informally, a *string* is either an element of *char* or several elements of *char* concatenated together, such as '*abbaie*', '*dog*', '*xxxyya*'. In set theory, we can define *char* × *char*, *char* × *char* × *char*, etc., which are the sets of ordered pairs, triples, etc., of elements from *char*. We could identify strings with all of the possible ordered n-tuples (including 1-tuples, elements of *char*). Of course, a general definition of n-tuple can be based on the Recursion Theorem, which was proved using our informal set theory and Peano's Axioms. Thus, we are still working within the scope of the *Modus Operandi* of Chapter 2, and there should be no serious doubt about assuming the existence of strings. For our present purposes, I want to show how strings can be constructed in a recursive manner (analogous to the way one might recursively define n-tuples). The following development could be axiomatized within an appropriate formal language, but I will proceed in an informal style. A formalization can be found in Chapter 7 of Manna and Waldinger (1985).

Given a set of characters, *char*, we assume that it can be determined whether two instances of characters are the same (identical) or different. We also assume that we are given a function *prfx* that prefixes a character to the front of a preexisting string. For example, *prfx*('x', 'atyy') = 'xatyy'. The function *prfx* will not be defined; it is assumed to be given, along with the set *char*. The first argument to *prfx* must be a character. One might want to consider the analogous function for sets of ordered *n*-tuples of characters.

Definition 3-39 (**Informal Axioms for Strings**) Let *char* be a nonempty set of characters. We assume that there is a function *prfx* and a sequence of sets $string_k$ satisfying the following, for $k \geqslant 1$:

$$string_1 = char,$$

and

$$string_{k+1} = \{\sigma \mid \text{there exist } \alpha, \tau \text{ such that}$$

$$\alpha \in char \ \& \ \tau \in string_k \ \& \ \sigma = prfx(\alpha, \tau)\}.$$

The *set of all strings generated from char is*

$$string = \bigcup \{ \, string_k \mid k \in Nat^+ \, \} = \bigcup_{k=1}^{\infty} string_k.$$

If $\alpha_1, \alpha_2 \in char$ and $\tau_1, \tau_2 \in string$:

$$prfx(\alpha_1, \tau_1) \notin char,$$

and

$$\text{if } prfx(\alpha_1, \tau_1) = prfx(\alpha_2, \tau_2), \text{ then } \alpha_1 = \alpha_2 \text{ and } \tau_1 = \tau_2.$$

We will usually write $prfx(\alpha, \tau)$ in the form $\alpha\tau$. Thus, instead of

$$prfx('x', prfx('q', 'p')),$$

we will write 'xqp'.

As already mentioned, these informal axioms can be replaced by a formal, axiomatized theory of strings. Another approach is to give a very rigorous definition of strings within set theory. This would not be very difficult, but to keep this discussion as simple as possible, we will just assume that the above recursive definition indeed specifies a unique set, which is (intuitively) the set of all strings generated from *char*. The reason for using a set of axioms of this type is to have a precise definition of what counts as a string, a precise specification of the structure of strings, and a statement of the basic assumptions about the features of the *prfx* function. This information can be organized differently if one desires to do so.

It follows from the string axioms that any string is either a single character, or else it is in $string_k$ for some $k > 1$. In the latter case, it is constructed from two or more characters. This is a *decomposition property* of strings that will be used extensively in inductive proofs and recursive definitions. For some purposes, it is useful to include an "empty string," which has no characters. This is not done here, although our axioms could easily be modified to include such an object.

Suppose that we let *char* be the set of lowercase letters in the English alphabet. Then $string_1$ is just this set of letters. Also, $string_2$ is the set of all two-letter strings, such as '*aa*', '*ab*', $\cdots$, '*ba*', $\cdots$; $string_3$ is the set of three-letter strings, etc. It is often useful to think of decomposing a three-letter string into its first character plus the remaining two-letter string.

No character is constructed from another character and a string. Because *prfx* is a function, if $\alpha_1 = \alpha_2 \in char$ and $\tau_1 = \tau_2 \in string$, then $prfx(\alpha_1, \tau_1) = prfx(\alpha_2, \tau_2)$. The converse of this conditional is stated as one of the above axioms about *prfx*. These facts are used to prove the following theorem.

Theorem 3-40   Let $\sigma \in string$. Then there is exactly one $k \in Nat^+$ such that $\sigma \in string_k$.

Proof. Let $\sigma \in string$. From the definition of *string*, there exists $k \in Nat^+$ such that $\sigma \in string_k$. We need only prove that this k is unique.

Let $\sigma \in string$ and let $\phi(k)$ be

$$(\sigma \in string_k \ \& \ \sigma \in string_j) \Rightarrow k = j.$$

We use induction on k.

Case $k = 1$. If $\sigma \in string_1$, then $\sigma \in char$. Suppose also that $\sigma \in string_j$, for $j > 1$. Then there exist $\alpha \in char$, $\tau \in string_{j-1}$, such that $\sigma = prfx(\alpha, \tau)$. But this violates the axiom that $prfx(\alpha, \tau) \notin char$. Hence, it cannot be that $j > 1$, so $j = 1$.

Case $k + 1$. Assume $\phi(k)$ and suppose that $\sigma \in string_{k+1}$. Then $\sigma = \alpha\tau$, where $\alpha \in char$ and $\tau \in string_k$. Now suppose that $\sigma \in string_j$. Then $\sigma = \beta\eta$, where $\beta \in char$ and $\eta \in string_{j-1}$. Thus, $\alpha\tau = \beta\eta$, so it follows that $\alpha = \beta$ and $\tau = \eta$. Hence, $\tau \in string_k$ and also $\tau = \eta \in string_{j-1}$. Since $\tau \in string_k$, by the induction hypothesis $\phi(k)$, we have $k = j - 1$. So, $k + 1 = j$. **Q.E.D.**

This theorem is important because it enables us to define a new function from *string* to Nat^+. I will do this now, but only provisionally.

Definition 3-41 (Provisional Definition) For any $\sigma \in string$, let

$$length_s(\sigma) = k \iff \sigma \in string_k.$$

This definition is justified by the previous theorem; for any σ, there will be one and only one k such that $\sigma \in string_k$. The theorem, together with the $length_s$ function, facilitates induction proofs about strings. In other words, we can use the length of a string as a numerical parameter for induction proofs. Suppose that we want to prove that all strings have some property ψ. We let $\phi(k)$ be: All strings of length k have property ψ. We then use induction on k. It is possible to accomplish the same thing by using an ordering (similar to a well-ordering) that applies specifically to strings. An exposition of the general technique is in Burstall (1969). In order to minimize the addition of new concepts, I will continue to perform inductions on natural numbers k. However, I will use recursive definitions directly on strings. The preceding definition is called "provisional" because it will be replaced by an alternate definition that uses recursion on strings.

Definition 3-42 Let σ be a string.

$$head_s(\sigma) = \begin{cases} \sigma & \text{if } \sigma \in char, \\ \alpha & \text{if } \sigma = \alpha\tau = prfx(\alpha, \tau). \end{cases}$$

Let $\sigma = \alpha\tau$ be a string. Then $tail_s(\sigma) = \tau$.

The $tail_s$ function is defined only for strings that are not characters. The subscript 's' indicates that these are string functions. These two functions provide a way to decompose any string that is not a character into its first element and the remaining string after this first element. These functions facilitate recursive definitions on strings. I will not develop a general theory for this; we will just assume that it is justified for our simple examples. This section presents only a brief, informal introduction to strings and lists.

Definition 3-43 Let σ be a string. We define a function

$$length_s : string \longrightarrow Nat^+$$

as follows:

$$length_s(\sigma) = [\text{ if } \sigma \in char \text{ then } 1 \text{ else } 1 + length_s(tail_s(\sigma)) \,].$$

This definition of $length_s$ gives us a function that returns exactly the same values as those returned by the function defined before provisionally. If we define a relation, $L\sigma\chi$ iff $length_s(\sigma) = length_s(\chi)$, then L is an equivalence relation on the set of all strings. The equivalence classes of L are precisely the sets $string_k$. Henceforth, I will use this definition of the *length* function instead of the provisional definition. In definitions, I will continue to use recursion on strings, but inductive proofs will use the length of strings.

Definition 3-44 Let σ, χ be strings. *The append function for strings,*

$$append_s : string \times string \longrightarrow string,$$

is defined by

$$append_s(\sigma, \chi) = [\text{ if } \sigma \in char \text{ then } prfx(\sigma, \chi)$$
$$\text{else } prfx(head_s(\sigma), append_s(tail_s(\sigma), \chi))].$$

It is important to understand the difference between *prfx* and *append$_s$*. The arguments to the former are pairs whose first element must always be a character. The arguments to *append$_s$* are pairs of strings. Here is an example of how the *append$_s$* function operates.

Example.

$$append_s(\text{`}a\text{'}, \text{`}x\text{'}) = prfx(\text{`}a\text{'}, \text{`}x\text{'}) = \text{`}ax\text{'}.$$

$$append_s(\text{`}a\text{'}, \text{`}xy\text{'}) = prfx(\text{`}a\text{'}, \text{`}xy\text{'}) = \text{`}axy\text{'}.$$

$$append_s(\text{`}abc\text{'}, \text{`}xc\text{'}) = prfx(\text{`}a\text{'}, append_s(\text{`}bc\text{'}, \text{`}xc\text{'})),$$

so

$$append_s(\text{`}abc\text{'}, \text{`}xc\text{'}) = prfx(\text{`}a\text{'}, prfx(\text{`}b\text{'}, append_s(\text{`}c\text{'}, \text{`}xc\text{'}))).$$

Hence,

$$append_s(\text{`}abc\text{'}, \text{`}xc\text{'}) = prfx(\text{`}a\text{'}, prfx(\text{`}b\text{'}, prfx(\text{`}c\text{'}, \text{`}xc\text{'}))) = \text{`}abcxc\text{'}. \quad \blacksquare$$

In mathematics and in computing theory, it is often necessary to replace an intuitive concept or operation by a precisely defined counterpart. Definitions are not proved as theorems are, so one cannot *prove* that a definition is correct. But definitions can be rationally justified or criticized according as they do or do not serve certain intended purposes. Also, if the intuitive concept or operation exhibits some characteristic, then one normally expects the defined counterpart to exhibit the same, or some analogous, characteristic. If the defined counterpart fails to do this for some characteristic, then one *usually* either gives up the definition (at least for its original intended purpose), or else revises his view of the intuitive concept or operation. The former of these alternatives is the one typically selected. Whenever a definition is proposed, it should be carefully assessed in the ways just described. This is often done by proving theorems about salient features of the concept or operation that is defined. It must be noted, however, that there are situations in which the intuitive concept and the precise counterpart are not exactly analogous, although both are used in their respective contexts.

The intuitive operation of appending strings is associative: For strings σ, η, χ, if we append σ to the result of appending η to χ, we obtain the same result that is obtained by appending σ to η and then appending this combination to χ. If our defined function $append_s$ did not satisfy this condition, we would almost certainly conclude that we do not have an acceptable definition. Fortunately, it is not difficult to prove that the defined function is associative in this manner. One might consider this associativity to be obvious and not in need of proof. Yet the recursive definition is sufficiently complex that one should not rely on what *seems* obvious. To have confidence in the definition, we should prove associativity. Even this does not go very far towards justifying the definition, for there are other characteristics of the append operation to consider.

I will now prove that $append_s$ is associative. The proof illustrates how mathematical induction can be used on the lengths of strings, enabling us to apply induction proofs to obtain a result about nonnumerical objects. First recall that it was proved earlier that

$$i + (j + k) = (i + j) + k,$$

for i, j, $k \in Nat$. The recursive definition of $+$ used recursion on the second argument to $+$. For that reason, the earlier proof used induction on k because it is the only variable that occurs everywhere in the equation as the second argument to $+$. In the definition of $append_s$, the recursion is done on the first argument. For that reason, the next proof uses σ, which is the only variable (in the statement of the next theorem) that occurs everywhere as the first argument of $append_s$. However, σ is a string, not a number. I will not induct on the string σ, but rather on the length of σ.

Theorem 3-45 The $append_s$ function for strings is associative, i.e., for any strings, σ, η, χ.

$$append_s(\sigma,\ append_s(\eta,\ \chi)) = append_s(append_s(\sigma,\ \eta),\ \chi).$$

Proof. Let σ, η, χ be strings. We use induction on the length k of σ. Let the induction hypothesis $\phi(k)$ state:

$$\text{For any string } \sigma, \text{ if } length_s(\sigma) = k,$$

$$\text{then } append_s(\sigma, \, append_s(\eta, \, \chi)) = append_s(append_s(\sigma, \, \eta), \, \chi).$$

Case 1. In this case, $\sigma \in char$, and we have

$$append_s(\sigma, \, append_s(\eta, \, \chi)) = prfx(\sigma, \, append_s(\eta, \, \chi)).$$

Also,

$$append_s(append_s(\sigma, \, \eta), \, \chi) = append_s(\, prfx \, (\sigma, \, \eta), \, \chi).$$

From the definition of $append_s$,

$$append_s(\, prfx \, (\sigma, \, \eta), \, \chi) = prfx(\sigma, \, append_s(\eta, \, \chi)).$$

Hence,

$$append_s(\sigma, \, append_s(\eta, \, \chi)) = append_s(append_s(\sigma, \, \eta), \, \chi).$$

Case $k + 1$. The length of $\sigma > 1$, so there exist α, τ such that $\sigma = \alpha\tau$, where $\alpha \in char$ and τ is a string of length k. Assume $\phi(k)$. We have

$$append_s(\sigma, \, append_s(\eta, \, \chi)) = prfx(\alpha, \, append_s(\tau, \, append_s(\eta, \, \chi))).$$

Since $length_s(\tau) = k$, applying $\phi(k)$, we get

$$append_s(\sigma, append_s(\eta, \chi)) = prfx(\alpha, append_s(append_s(\tau, \eta), \chi)). \tag{3.5}$$

It is not obvious what to do next, so let us work with the right-hand side of the equation we are trying to establish.

$$append_s(\, append_s(\sigma, \, \eta), \, \chi) = append_s(\, prfx \, (\alpha, \, append_s(\tau, \, \eta)), \, \chi).$$

Applying the definition of $append_s$ to the right side of this equation yields

$$append_s(append_s(\sigma, \eta), \chi) = prfx(\alpha, append_s(append_s(\tau, \eta), \chi)). \tag{3.6}$$

From Equations (3.5) and (3.6), we obtain

$$append_s(\sigma, \, append_s(\eta, \, \chi)) = append_s(append_s(\sigma, \, \eta), \, \chi),$$

so $\phi(k)$ implies $\phi(k + 1)$. Hence, for all $n \in Nat^+$, if σ has length n, then

$$append_s(\sigma, \, append_s(\eta, \, \chi)) = append_s(append_s(\sigma, \, \eta), \, \chi),$$

so $append_s$ is associative for all σ (and all η, χ). **Q.E.D.**

This proof illustrates the technique of induction on the length of strings. Because strings have a simple structure, we will not consider them further. Let us move on to another type of object that has a more complex structure than strings.

3.4.2 A Simple Treatment of Lists

Suppose that we have a nonempty set, *atom*, of objects. This could be a set of characters, a set of numbers, a set of strings, or a set of some other kind of objects. A set of atoms could also be a combination of characters, numbers, etc. The only general assumption we make about the atoms is that they are integral units, i.e., they cannot be decomposed into smaller parts. Thus, if a string is an atom, it will be considered as a unit, without regard to the characters in it.

In addition to atoms, we will construct "lists." Intuitively, a *list* is a sequence of objects called its *members*. These members can be either atoms or other lists. A specific list is represented by writing its objects, separated by spaces, in order, and enclosing this sequence within parentheses. Here are some lists:

$$(a)$$

$$(b \; x \; dog \; bird \; zz)$$

$$(1 \quad 2 \quad -56 \quad 32.0 \quad 8)$$

$$(\; (a \; b \; c) \; x \; ((fat \; (cat) \; zyx)) \; y \; (1 \; 62 \; (a \; b)) \;)$$

The first has only one member, '*a*'. The second is a list of five strings and the third is a list of numbers. When I wrote specific characters and strings earlier, I put them in single quotes so they would not be confused with variables. In the present discussion, there is no chance of this kind of confusion when these elements are inside lists, so these single quotes are omitted in that context.

The last example has five members, three of which are sublists. The second of these sublists is a list with one element. This one element is itself a list, namely,

$$(fat \; (cat) \; zyx).$$

This list also has an element that is a list, (*cat*), which has one element, the atom, '*cat*'.

In addition to lists with members, we will also assume the existence of an *empty list*, denoted by ( ). There is an important difference between strings and lists: The elements of strings are all characters, whereas the elements of lists may be either atoms or other lists. The structure of lists is more complex than that of strings. Because of their surprising complexity, lists are very useful structures for representing a wide variety of data.

Lists are the main syntactic feature of the programming language, LISP, invented by John McCarthy, and widely used in artificial intelligence programming. Lists also play an important role in the language Prolog, which is used in artificial intelligence and general logic programming. If a programming language does not admit lists as built-in structures, it is not unusual to define lists in terms of more primitive components of the language. Lists can be used to represent sequences of numbers, such as (2.3 4 5.006 –23.4) and sequences of words, such as (The cow jumped over the moon). They can also be used to represent associations between data, e.g.,

$$(\ (\text{john } 5.9)\ (\text{mary } 5.6)\ (\text{jill } 6.0)\ (\text{bob } 5.8)\ ),$$

in which the numbers may represent the heights of the people in feet. Lists are especially useful for representing formulas. For example,

$$(+\ (*\ x\ x)\ (*\ y\ y)\ )$$

represents $x^2 + y^2$. The operators $+$ and $*$ (multiplication) are used in the list as prefix operators. Programs that do symbolic mathematics, i.e., solve problems in terms of symbols rather than specific numbers, use such representations. Programs that perform logical deduction usually use lists. For example, this conjunction of disjunctions,[*]

$$(A \vee (\neg B \vee C)) \wedge ((\neg A \vee D) \wedge (E \vee F)),$$

can be expressed more simply in the form

$$(\,A \vee \neg B \vee\ C\,) \wedge (\neg A \vee D) \wedge (E \vee F),$$

which can be represented by

$$(\,(\,A\ (\text{not } B)\ C\,)\ (\,(\text{not } A)\ D\,)\ (E\ F)\,).$$

In the list representation, it is understood that there are conjunctions between the main sublists and that there are disjunctions between the elements of these sublists. These examples illustrate the versatility of lists as a means of representing information. There are good practical reasons for studying lists; also, they are intrinsically interesting.

The theory of lists can be formally axiomatized in a manner similar to that used for strings; see Chapter 9 of Manna and Waldinger (1985). I will write informal axioms for lists, as was done for strings, and for the same general reasons. The axioms merely serve as a convenient way to codify the basic assumptions about the structure of lists. Lists are defined recursively in terms of atoms and other lists. I will not say exactly what may be atoms, although in particular applications the set of atoms will be specified. In order to construct

---

[*]A sentence of this form is said to be in *conjunctive normal form*. See Chapter 1, Section 7.

strings, we assumed that the *prfx* function was given, and we merely stated some of its important properties. A similar approach is used here for lists. The construction function is traditionally called the *cons* function (because it is used to *construct* lists). The *cons* function is like *prfx* in that both of them prefix a new element at the front end of a previously defined structure. But the first argument to *cons* may be either an atom or a list, whereas the first argument to *prfx* is always a character. Because of this added complexity, the recursive definition of lists is not as simple as that for strings. The following treatment introduces a theory of lists that is similar to the system used in LISP. There are some important differences, and LISP contains many additional features and complexities not mentioned here, including allowing atoms to have values assigned to them.

Definition 3-46 (**Informal Axioms for Lists**) Let *atom* be a nonempty set of atoms (e.g., numbers, characters, etc.). Let ( ) be the *empty list*. We assume that there is a function *cons* and a sequence of sets $list_k$ satisfying the following, for $k \in Nat$:

$$list_0 = \{(\ )\},$$

and

$$list_{k+1} = list_k \cup \{\lambda \mid \text{there exist } \alpha, \tau \text{ such that}$$

$$(\alpha \in atom \cup list_k) \ \& \ \tau \in list_k \ \& \ \lambda = cons(\alpha, \tau)\}.$$

The *set of all lists generated from atom is*

$$list = \bigcup \{\, list_k \mid k \in Nat \,\} = \bigcup_{k=0}^{\infty} list_k.$$

If $\alpha_1, \alpha_2 \in atom$ and $\tau_1, \tau_2 \in list$:

$$cons(\alpha_1, \tau_1) \neq (\ ),$$

and

$$cons(\alpha_1, \tau_1) \notin atom,$$

and

$$\text{if } cons(\alpha_1, \tau_1) = cons(\alpha_2, \tau_2), \text{ then } \alpha_1 = \alpha_2 \text{ and } \tau_1 = \tau_2.$$

When working with particular lists, it is customary to omit 'cons' and just write the elements within parentheses. Thus, instead of

$$cons(x, \ cons(q, \ cons(p, \ (\ )))),$$

we will usually write $(x \ q \ p)$.

From this characterization, we have $list_k \subseteq list_{k+1}$, from which $list_i \subseteq list_j$ for all $i \leqslant j$. Thus, the recursive definition works in a manner similar to Course of Values Induction, although it is stated here as a one-step form of recursion. Here are some examples of lists according to the definition of *list*.

Examples. Let *atom* be the set of lowercase letters of the English alphabet. The following are lists. For each one, I have indicated the smallest $list_k$ in which it is an element.

$$cons(y, \ (\ )) = (y) \in list_1.$$

$$cons(x, \ (y)) = (x \ y) \in list_2.$$

$$cons((\ ), \ (\ )) = ((\ )) \in list_1.$$

$$cons((x \ y), \ (\ )) = ((x \ y)) = list_3.$$

$$cons(a, \ (x \ y)) = (a \ x \ y) \in list_3.$$

$$cons((a \ x \ y), \ (x \ y)) = ((a \ x \ y) \ x \ y) = list_4.$$

$$cons(\ ((\ )\ ), (\ )) = (\ ((\ )\ )\ ) \in list_2. \quad ■$$

By custom, the "length" of a list is the number of elements on its "top level." This means: Count all of the atoms and lists that are its elements straight across the list, ignoring the internal parts of sublists. For example, the length of

$$(\ (a \ b) \quad x \quad (\ ) \quad (u \ (a \ b)\ ) \quad y \ )$$

is 5. An exact definition of this idea of length will be given shortly. For now, notice how complicated things have become. The length of a list may be different from the total number of elements of *atom* (e.g., a, y, etc.) in it. Also, either of these quantities may be different from the least k such that the list is an element of $list_k$. A simple example is $(\ (\ (a)\ )\ (\ )\ )$. The only element of *atom* in it is a, its length is 2, and the least k for which it is in $list_k$ is 3.

At this point, it may help to recall that $\mathcal{SC}$ sentences have a tree structure. They may contain subsentences just as lists may contain sublists. Figure 3.6 shows one way to visualize the tree structure of a list. In this figure, the

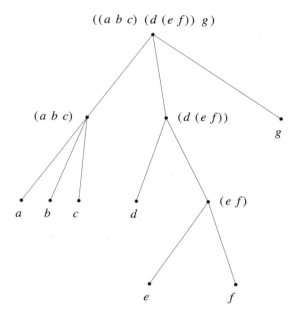

Figure 3.6. *Tree Structure of a List*

atoms are the terminal nodes of the tree. Using this type of representation, some trees may also have ( ) as a terminal node. Whereas strings have a simple linear structure, in general, lists have a nested tree structure. This raises an interesting question about how to construct recursive definitions of list functions and relations and how to construct inductive proofs about lists and list functions and relations.

It follows from the recursive definition of *list* that every list is either ( ), or else has the form $cons(\alpha, \tau)$ for some list τ and some atom or list α. This is a *decomposition property* of lists that is useful in induction. It is analogous to the decomposition property for strings, although considerably more complex. Some simple list functions depend only on the "top level" of the list. These functions have definitions similar to corresponding functions on strings. The most basic ones are the *head*, *tail*, and *length* functions. These were subscripted with an 's' in the case of strings; I will not use a subscript for the corresponding list functions. Using these, it is easy to define an *append* function for lists, which will also be used without a subscript.

Definition 3-47 Let λ, χ be lists.

$$head(\lambda) = \begin{cases} (\,) & \text{if } \lambda = (\,), \\ \alpha & \text{if } \lambda = cons(\alpha, \tau). \end{cases}$$

$$tail(\lambda) = \begin{cases} () & \text{if } \lambda = (\,), \\ \tau & \text{if } \lambda = cons(\alpha, \tau), \end{cases}$$

$$length(\lambda) = [\text{ if } \lambda = (\,) \text{ then } 0 \text{ else } 1 + length(tail(\lambda))].$$

$$append(\lambda, \chi)$$

$$= [\text{ if } \lambda = (\,) \text{ then } \chi \text{ else } cons(head(\lambda), append(tail(\lambda), \chi))] .$$

The *head* and *tail* functions are very similar to those for strings. Their definitions require that they both return the value ( ) when applied to ( ). This is a matter of convention, but it provides useful base cases for induction. Here are some examples.

Examples.

$$head((b)) = head(cons(b, (\,)) = b.$$

$$tail((b)) = (\,).$$

$$head((a\ b\ c)) = a.$$

$$tail((a\ b\ c)) = (b\ c).$$

$$head(\ ((a\ (xy))\ b)\ ) = (a\ (xy)).$$

$$tail(\ ((a\ (xy))\ b)\ ) = (b). \quad \blacksquare$$

The *length* and *append* functions also exhibit behavior similar to their counterparts for strings. Sometimes there is confusion between *cons* and *append*. A few examples should illuminate the differences.

Examples.

$$cons(a, (x\ y\ z)) = (a\ x\ y\ z).$$

But $append(a, (x\ y\ z))$ *is not defined*, since the definition of *append* requires that both of its arguments be lists, and a is a nonlist.

$$append((a), (x\ y\ z)) = cons(a, append((\,), (x\ y\ z))\ )$$

$$= cons(a, (x\ y\ z)) = (a\ x\ y\ z).$$

$$cons(a, (\ )) = (a).$$

$$cons((a), (\ )) = ((a)).$$

$$append((a), (\ )) = cons(a, append((\ ), (\ ))) = cons(a, (\ )) = (a). \blacksquare$$

One can use induction on the length of lists to prove that *append* is associative. The proof is similar to the one used to prove that $append_s$ is associative, but with the base case of length 0 corresponding to (\ ). As might be guessed, this proof is left for the Exercises.

The recursion in the definition of *length* calls $length(tail(\lambda))$. The recursion in the definition of *append* calls $append(tail(\lambda), \chi)$. In both recursions, the function being defined is recursively applied only to the tail of the list. This is why these functions behave very much like string functions. Yet the head of a list may also be a list, so there is the possibility of a recursive call of a function on the head. We have already seen that the length of a list may be different from the number of nonlists in it. It is natural to define a function that adds up all occurrences of these nonlists. In LISP implementations, it is often the case that (\ ) is also treated as an atom; I will not do that here. So an atom will just be any element in the original set *atom* of characters, numbers, etc. We want a function that counts up all occurrences of these, including repetitions. For example, this function should return 7 for the list

$$((a\ (b\ c))\ (\ )\ a\ ((x\ y))\ m).$$

The head of this list is $(a\ (b\ c))$. If we count up the atoms in this head and add this number to the count of the atoms in the tail, we will have the right answer. Thus, the function should be applied to the head and to the tail and the two results summed. The applications of the function to the head and to the tail will each behave recursively. Eventually, the function will reach a point when it calls a head that is an atom, so the function must be so defined that it applies both to atoms and to lists. Implementing this function is a standard LISP programming exercise. Here it is.

Definition 3-48 Let λ be an atom or a list.

$$count\text{-}atoms(\lambda) =$$

[if $\lambda = (\ )$ then 0

else if $\lambda \in atom$ then 1

else $count\text{-}atoms(head(\lambda)) + count\text{-}atoms(tail(\lambda))$].

In Section 3.3.5 we saw examples of recursive definitions that were flawed and fell into eternal loops for some values of their arguments. Given an input value λ, how do we know that *count-atoms* will return an output value? In other words, given an input value λ, will the recursion *terminate*, or will it fall into an eternal loop? In as much as we are treating strings and lists informally, this question will only be briefly, and informally, discussed here.

There is a simple, but vague plausibility argument for termination. If $\lambda = (\ )$, *count-atoms* returns 0. If $\lambda \in atom$, the function returns 1. So it certainly returns unique values if the input is the empty list or an atom. Now suppose that λ is a nonempty list. Then $\lambda = cons(\alpha, \tau)$ for some α that is either a list or an atom and for some list τ. Because λ is decomposable into α and τ, in some sense α and τ must be "simpler or smaller" than λ. Suppose that *count-atoms* returns unique values for α and τ; then it will return the sum of these two values for λ. But why should we believe that *count-atoms* returns unique values for α and τ? Because either they can be decomposed into further parts, or they are not decomposable (either an atom or $(\ )$). Finally, in that every list is only finitely long, it seems intuitive that repeated decomposition must eventually come to a halt.* Although it is vague, this argument has merit. If we could obtain a suitable well-ordering of lists (or an appropriately similar ordering), then we could justify a form of course of values induction on lists. Applications of this approach are described in Burstall (1969). In advanced set theory, there are some very powerful and abstract generalizations of the relationships between well-ordering and induction; see Dalen (1978), Chapter 2, or Suppes (1972), Chapter 7. Fortunately, we can continue to use the well-ordering of the natural numbers.

By the informal axioms for lists, each list is in $list_k$ for some $k \in Nat$. Also, if a list $\lambda \in list_k$, then $\lambda \in list_j$, for all $j > k$. Consider all numbers, k, such that $\lambda \in list_k$. This is a nonempty subset of *Nat*, so by WO it has a unique first element. Therefore, we can map any list to a unique natural number that is the least k such that the list is in $list_k$.

Definition 3-49 Let λ be a list and

$$N_\lambda = \{k \mid \lambda \in list_k\}.$$

We define

$$rank : list \longrightarrow Nat$$

by

$$rank(\lambda) = \text{the least } k \text{ such that } k \in N_\lambda.$$

---

*In LISP it is possible to create what are commonly called "circular lists," which repeat the same members indefinitely. The *count-atoms* function, and others, do not terminate and return values for these circular structures. Such circular structures are not lists according to our definition.

One may equivalently describe $rank(\lambda)$ as the least k such that $\lambda \in list_k$. From the definition, we have

$$rank((\ )) = 0, rank((a \ b)) = 2,$$

$$rank((((\ )))) = 2, rank(((a))) = 2,$$

and

$$rank(((a \ x \ y) \ x \ y)) = 4.$$

We already know that, in general, the rank of a list is not the same as its length, nor the same as the number of atoms in it. Yet for each $k \in Nat$, there exist lists of all ranks $i < k$. The empty list has rank 0, (a) has rank 1, etc. I have not given any explicit formula or algorithm for computing the value of the rank of a list. Fortunately, we do not need to compute these values (although it is an interesting exercise to develop a formula for computing $rank(\lambda)$). We do need the following theorem.

Theorem 3-50   Let λ be a list. If $\lambda = cons(\alpha, \tau)$, where α and τ are both lists, then

$$rank(\alpha) < rank(\lambda), \text{ and}$$

$$rank(\tau) < rank(\lambda).$$

If $\alpha \in atom$, then $rank(\tau) < rank(\lambda)$.

Proof. First suppose that both α and τ are lists. Since $\lambda = cons(\alpha, \tau)$, it follows from the list axioms that $\lambda \neq (\ )$. Therefore, $rank(\lambda) = k+1$ for some $k \in Nat$, so $\lambda \in list_{k+1}$. Hence, $\alpha, \tau \in list_k$, so $rank(\alpha) \leqslant k$ and $rank(\tau) \leqslant k$. Therefore, $rank(\alpha) < rank(\lambda)$ and $rank(\tau) < rank(\lambda)$.

The proof for the case where $\alpha \in atom$ is the same, except that $rank$ does not apply to α. **Q.E.D.**

We can now return to the *count-atoms* function. It returns unique values when given atoms or the empty list as input. Let λ be a nonempty list. Then $\lambda = cons(\alpha, \tau)$, where τ is a list and α an atom or a list. Suppose that *count-atoms* returns a value for α and one for τ; then its value for λ will be the sum of the former two values. By the theorem just proved, if α is a list, its rank is less than that of λ. The rank of τ is also less than the rank of λ, if α is either an atom or a list. If $\alpha \in atom$, then $count\text{-}atoms(\alpha) = 1$. Suppose that *count-atoms* returns values for all lists with rank less than that of λ. Then, regardless of whether α is an atom or a list, *count-atoms* is defined for α and τ, so it is also defined for λ. This is not a complete, rigorous proof that *count-atoms* is defined on all lists, but it is an adequate, intuitive argument for our purposes.

The *rank* function is useful for induction proofs. If we consider a list like $((a\ b\ c\ d\ e\ f\ g)\ x)$, it is clear that the length of the head of this list is greater than the length of the list itself. If we are doing an induction proof about lists or a list function, and the induction step applies to the heads of lists, then direct induction on length will usually not work. In such cases, we can do the induction on the rank of the list. Here is a simple theorem that illustrates this technique. To facilitate this theorem, we will say that a list is expressed in *cons format* iff it is the empty list or else all of its applications of *cons* are explicitly displayed. Thus, the cons format of $((a\ b)\ c\ (\ ))$ is

$$cons(cons(a,\ cons(b,\ (\ ))),\ cons(c,\ cons((\ ),\ (\ )))).$$

Any such expression has a definite number of left parentheses and a definite number of right parentheses.

Theorem 3-51 Let λ be a list. If λ is expressed in cons format, then the number of left parentheses in the expression equals the number of right parentheses.

Proof. For any list χ, let $left(\chi)$ be the number of left parentheses in it and $right(\chi)$ be the number of right parentheses in it. For any atom β, let $left(\beta) = right(\beta) = 0$. The proof is by CVI on the rank of λ. Let the induction hypothesis, $\phi(k)$ be

For any list χ, if $rank(\chi) < k$, then $left(\chi) = right(\chi)$.

Case 0. When $rank(\lambda) = 0$, $\lambda = (\ )$, so clearly the starting case holds.

Case $k > 0$. In this case, $rank(\lambda) = k$, for some $k \in Nat^+$, and $\lambda = cons(\alpha, \tau)$ for list τ, and α an atom or a list. From the previous theorem

$$rank(\tau) < rank(\lambda)$$

and, if α is a list, then $rank(\alpha) < rank(\lambda)$. Also, from the form of the expressions, λ must have one more left and one more right parenthesis than the total numbers in α and τ.

Subcase 1. Suppose that $\alpha \in atom$. Then $left(\alpha) = right(\alpha) = 0$. Hence,

$$left(\lambda) = 1 + left(\alpha) + left(\tau) = 1 + 0 + left(\tau) = 1 + left(\tau),$$

$$right(\lambda) = 1 + right(\alpha) + right(\tau) = 1 + 0 + right(\tau) = 1 + right(\tau).$$

By the induction hypothesis, $left(\tau) = right(\tau)$, so

$$left(\lambda) = right(\lambda).$$

Subcase 2. Suppose that $\alpha \in list$. Then

$$left(\lambda) = 1 + left(\alpha) + left(\tau),$$

$$right(\lambda) = 1 + right(\alpha) + right(\tau).$$

Since both α and τ have smaller rank than λ, we can apply the induction hypothesis to each of them, obtaining

$$left(\alpha) = right(\alpha)$$

and

$$left(\tau) = right(\tau).$$

Hence, $left(\lambda) = right(\lambda)$. **Q.E.D.**

If we had tried to prove this by induction on the length of λ, we would have encountered trouble in the second subcase. The tail of λ always has length one less than that of λ, so we could apply the induction hypothesis to τ. But the head of λ does not necessarily have a shorter length than the length of λ, so *induction on length* does not permit us to apply a simple induction hypothesis to the head of λ. By doing the induction on the rank of λ, the proof goes through smoothly.

Before starting an induction proof on lists, one should consider whether the induction step applies to the head. If not, then it is worth trying to induct on the length of the list. If the head is involved in the induction step, then try inducting on the rank. It may not be immediately obvious whether the head of the list is really involved, so one can start with the length, and see whether the proof goes through or gets blocked. In general, IP can be used for length inductions; CVI is usually the more convenient approach for inductions on rank.

3.4.3   Sentential Calculus Expressions

The beginning of Chapter 1 specified the syntax of sentential calculus, starting with a description of the symbols and then defining the set of all $\mathcal{SC}$ sentences. The definition of the sentences began with the atomic sentences, and then proceeded to specify how more complex sentences are constructed from previously given sentences. This is a recursive definition, but it was stated in a casual manner. Recursion was also used in defining the $\mathbf{V_I}$ function, which returns the truth value of a sentence under the interpretation $\mathbf{I}$. In Chapter 1 we relied on an intuitive understanding, and many examples, to see how these recursive definitions are used in practice. We are now in a position to see that recursion is related to the natural number system and, more generally, to well-ordered sets. Let us see how the set of $\mathcal{SC}$ sentences can be defined.

We build up the set of all $\mathcal{SC}$ sentences in a manner analogous to that used for lists, as follows:

Γ_0 is the set of $\mathcal{SC}$ letters, as given in Definition 1-1,

and, for $k \in Nat$,

$$\begin{aligned}
\Gamma_{k+1} = \Gamma_k &\cup \{\, \neg\phi \mid \phi \in \Gamma_k \,\} \\
&\cup \{\, (\psi \vee \chi) \mid \psi,\ \chi \in \Gamma_k \,\} \\
&\cup \{\, (\psi \wedge \chi) \mid \psi,\ \chi \in \Gamma_k \,\} \\
&\cup \{\, (\psi \rightarrow \chi) \mid \psi,\ \chi \in \Gamma_k \,\} \\
&\cup \{\, (\psi \leftrightarrow \chi) \mid \psi,\ \chi \in \Gamma_k \,\}.
\end{aligned}$$

The set of all $\mathcal{SC}$ sentences is then

$$\Gamma_{\mathcal{SC}} = \bigcup \{\, \Gamma_k \mid k \in Nat \,\} = \bigcup_{k=0}^{\infty} \Gamma_k.$$

When proving metatheorems, it is often necessary to use induction on the "size" of sentences. We need a size function that is similar to the *rank* function on lists, i.e., it should be a measure that corresponds to the way that a sentence is recursively constructed from "smaller" subsentences. We could define such a function in terms of the least k such that a sentence ϕ is an element of Γ_k. In mathematical logic, other measures are often used. The following is a typical one. To avoid confusion with the concept of rank, I will denote this function by *siz*.

Definition 3-52 Let ϕ be an $\mathcal{SC}$ sentence. The *size* of ϕ is denoted by $siz(\phi)$ and defined by

- If ϕ is atomic, then $siz(\phi) = 0$.

- If ϕ has the form $\neg\psi$, then $siz(\phi) = siz(\psi) + 1$.

- If ϕ has one of the forms $(\psi \vee \chi)$, $(\psi \wedge \chi)$, $(\psi \rightarrow \chi)$, or $(\psi \leftrightarrow \chi)$, then $siz(\phi) = siz(\psi) + siz(\chi) + 1$.

The value of $siz(\phi)$ is just the total number of occurrences of sentential connectives in ϕ. For any k, there exist sentences of size k. The size of any proper subsentence of a compound sentence ϕ is less than the size of ϕ. The siz function provides a convenient ordering on sets of $\mathcal{SC}$ sentences and facilitates CVI and WO proofs.

This concludes our brief consideration of $\mathcal{SC}$ sentences in this chapter, but the Exercises and Chapter 4 include additional examples of recursive definitions of syntactical structures. The significance of these examples extends well beyond formal logic. For instance, the syntax rules of a programming language are often given recursively using a special notation called the BNF, or Backus–Naur form. The recursive formation rules of the language are involved in the development of programs that interpret or compile the language. BNF is described in Appendix 1 of Gries (1981) and is discussed in Chapter 8 of Piff (1991). A detailed text on formal languages and their relations to automata theory is Lewis and Papadimitriou (1981).

3.4.4 Stacks and Queues

Algorithms and heuristic procedures, and therefore computer programs, often make use of various means for storing and retrieving information. For example, complex information may be stored using sentences of the predicate calculus, the subject of the next chapter. Simpler types of information are often stored in various formats called *data structures*. These structures are sometimes described in terms of the memory organization of the computer. But they can also be defined by general, abstract specifications and studied mathematically. When this is done, they are usually called *abstract data types*. In fact, strings and lists are data types that we have already considered in abstract formulations. These formulations specify sets of objects and functions that relate to these objects. The behavior of the objects and functions is specified by axioms. I used a somewhat informal, set-theoretical formulation of the axioms. Other styles of presentation are also used; see, for instance, Martin (1986), or Horowitz and Sahni (1982).

Many different data types are used besides strings and lists. Two commonly used structures are *stacks* and *queues*. They can be specified abstractly, but doing so would be a large digression in this book. Yet they can easily be defined in terms of lists, so it is worth considering them briefly at this point. They can also be defined in terms of sequences or in other ways, but lists provide especially elegant representations.

The concept of a stack is very simple. An example is a stack of books, e.g., *Moby Dick, Faust, Galois Theory*, and *The Nature of the Chemical Bond*. We can add another book, say *The Sound and the Fury*, to the top of the stack if we wish. Then another book may be added and so on. Adding an element to the top of a stack is called *pushing* it onto the stack. We can also remove one book at a time. So if *The Sound and the Fury* is currently the topmost book, it can be removed. Removing the topmost item is called *popping* the stack. Of course, if there are no books in the stack, it is empty and cannot be popped. In a computer program, trying to pop an empty stack usually calls an error message.

The easiest way to visualize a stack is this: Think of it as a possibly empty, changeable list of items. *Procedures* are used to change it from one *state* to another *state*. We may give a particular stack a name, say σ, but σ will not always denote the same object, because the stack is changeable. Suppose that initially $\sigma = (\ )$. We then push an object a onto σ. Then $\sigma = (a)$. We then push b onto σ, and σ now becomes $(b\ a)$. If we now pop σ, it returns to the state (a). The procedures, push and pop, are what cause σ to change state. Notice that these procedures are not functions. A list function accepts input arguments that are atoms or lists, and it returns some value (usually another atom or list). But the list function does not change the inputs in any way. A procedure, on the other hand, may change the object that it operates on. This, of course, assumes the existence of mutable, i.e., changeable, objects.

Stacks are also called *pushdown stores* and *Last In First Out* (LIFO) queues. They are used to store information in such a way that the last data item that is stored must be the first data item that is retrieved. There are many uses for this storage format; an example will be presented below. First, let us see how stacks and their associated procedures can be represented in terms of lists. There are two obvious approaches that one might use. (1) A stack can be considered *one* "mutable list" with different states and access procedures, in the manner just described. (2) A stack can be considered to comprise some special functions together with a sequence of lists to which these functions are applied. The former approach is perhaps closer to our intuitive idea of what a stack is, but the second is simpler to state in our mathematical language. For this reason, I use (2) in the following definition. It is assumed that a set of atoms, *atom*, is given.* As usual, *list* denotes the set of all lists that can be generated from these atoms.

Definition 3-53 Let $\lambda \in list$ and $\alpha \in atom$, and define functions as follows,

$$top(\lambda) = head(\lambda),$$

$$push(\alpha, \lambda) = cons(\alpha, \lambda),$$

$$delete(\lambda) = tail(\lambda).$$

Let $Seq = \lambda_0, \lambda_1, \cdots, \lambda_i, \cdots$ be a sequence of lists satisfying

$$\lambda_0 = (\ ),$$

---

*Although we use *atom* in this definition, the atoms can be any kind of object. We could interpret the set *atom* as any set of items, including lists, if desired.

and, for any $i \in Nat$, either

$$\lambda_{i+1} = delete(\lambda_i),$$

or there is some $\alpha \in atom$ such that

$$\lambda_{i+1} = push(\alpha, \lambda_i).$$

A *stack* is any such sequence, *Seq*, together with the functions *top*, *push*, and *delete*.

The three functions in this definition are just *head*, *tail*, and *cons* with new names. From the previous definitions of *head* and *tail*, *top* and *delete* return ( ) with the input ( ). In a computer program, they are often defined so as to return an error message with this input. To make things simple, we will depart from that practice here, although an error message would be required if ( ) is a possible element of a stack. According to the definition, a stack consists of the three functions *together with* a special kind of sequence of lists. The sequence must begin with ( ), and each element of the sequence is obtained from the previous element by applying either *push* or *delete*.

Our definition has only three basic stack functions, although stacks are often defined in such a way that additional functions or procedures are associated with them. For instance, some definitions include a procedure that clears (or initializes) a stack, i.e., assigns it the value ( ) regardless of what the list may have contained. However, our purposes will be served by the three functions in Definition 3-53, plus two procedures that will be defined in terms of those three functions. When accessing or changing a stack, one should use *only* the functions or procedures given in the definition. Some programming techniques implement stacks in such a way that this restriction is strictly enforced. In other words, the implementation simply does not allow a programmer any direct interaction with a stack that is not definable in terms of the basic stack functions.

As already discussed, a stack is usually visualized as a list that is subject to changes produced by various procedures. Suppose that we use σ to refer to elements of the *sequence* of a particular stack. Initially, σ has the value $\lambda_0 = $ ( ). Suppose that we push a onto σ. This corresponds to

$$push(a, (\,)) = (a),$$

which will be the next element, λ_1, in the sequence referred to by σ. Instead of writing λ_1, we adopt the convention of assigning the new value, (a), to σ. Thus, we may let the symbol σ refer to any list in the stack sequence, and we will *assign new values* to σ as required when the functions *push* or *delete* are applied. It is now convenient to introduce the *procedures* PUSH! and POP!. Let $\alpha \in atom$. Use the notation

$$\sigma \leftarrow newvalue$$

to indicate that the name σ is being assigned the new value, *newvalue*. Also, use 'return x' to indicate that a procedure returns the value assigned to x. Then we can use the procedure definitions:

procedure PUSH!(α, σ)

    $\sigma \leftarrow push(\alpha, \sigma)$

end PUSH!

procedure POP!(σ)

    $x \leftarrow top(\sigma)$

    $\sigma \leftarrow delete(\sigma)$

    return x

end POP!

    I have used uppercase, typewriter style font to distinguish the procedures from the functions. The PUSH! and POP! procedures modify σ; thus, σ denotes a mutable object. It is this kind of mutable object that is commonly called a "stack." Henceforth, the word *stack* will also be used to refer to such a mutable object. Context of use should distinguish this meaning from that given in Definition 3-53. The exclamation point is used to emphasize the fact that PUSH! and POP! can change a mutable data structure. This use of the exclamation point, "bang," is common in programs written in the language Scheme, which is a dialect of LISP. Notice that PUSH! just changes σ, whereas POP! does more: It temporarily stores $top(\sigma)$ in the variable x, it deletes the top element of σ, then it returns the value stored in x, namely, the element that was on the top of σ before POP! was applied. Thus, POP! can be used as a function to return this top value, but it also changes σ in the process. Whereas *top* does not affect σ, and returns only an atom, we do not need a procedure corresponding to it.

    Here is a simple example of stack operations using the definitions that have been introduced. Let $\sigma = (\ )$. Then

$$\text{PUSH!}(a, \sigma)$$

produces $\sigma = (a)$. Then

$$\text{PUSH!}(b, \sigma)$$

and

$$\text{PUSH!}(c, \sigma)$$

yield $\sigma = (c\ b\ a)$. Then

$$top(\sigma) = c,$$

without changing σ. Finally,

$$\mathrm{POP!}\,(\sigma) = c,$$

and $\mathrm{POP!}\,(\sigma)$ changes σ to $\sigma = (b\ a)$.

Stacks have many uses. A simple application is in the sequencing of calculations. We are accustomed to familiar arithmetical notations, like $3 + 4$, 3×2, in which the function symbol is placed between two arguments. This is called *infix* notation. If we put the function symbol in front, as in $+(3, 4)$ and $\times(3, 2)$, we have *prefix* notation. In the 1920s, Polish logician Jan Łukasiewicz invented a prefix notation for sentential calculus and showed that parentheses are not necessary in that notation.* For this reason, prefix notation is sometimes called "Polish notation." It is easy to reverse directions and use *postfix*. Let us do this for $+$, $\times$, and log (to the base e, of natural logarithms). Thus, instead of $2 + 3$, we will write $2\ 3\ +$; instead of -3.4×5.05, we will write $-3.4\ 5.05\ \times$; and instead of $\log(125)$, we write $125\ \log$. This postfix notation is sometimes called "reverse Polish notation" and abbreviated "RPN." This kind of notation is used in some calculators and in some programming languages, including PostScript®, which is used primarily for representing printing procedures and data. A stack is convenient when performing calculations specified by postfix notation.

We will represent a sequence of operations as a list of arguments and function symbols. For example,

$$(2.5\ 3 \times 8\ 3 + + 10 \times \log),$$

which is computed as follows:

$$2.5\ 3 \times = 7.5;$$

$$8\ 3 + = 11;$$

$$7.5\ 11 + = 18.5;$$

$$18.5\ 10 \times = 185;$$

$$185\ \log = 5.220356.$$

*Historical references are in Prior (1962).

Thus, the preceding list represents the same quantity as

$$\log (((2.5 \times 3) + (8 + 3)) \times 10).$$

Suppose that σ is an empty stack. It will be used to keep track of the calculations performed above. We begin at the front of the list. First 2.5 is pushed onto σ and removed from the list. Then 3 is pushed onto the stack and removed from the list, so $\sigma = (3\ 2.5)$. The next element of the list is $\times$. Unlike numbers, function symbols are not pushed onto the stack. Instead, we apply POP! to obtain 3. Because $\times$ requires two arguments, POP! is applied again to obtain 2.5. Then $\times$ is removed from the list, and the expression 2.5 3 $\times$ is evaluated to obtain 7.5. After 7.5 is computed, it is pushed onto σ so that $\sigma = (7.5)$. The next two elements of the list are the numbers 8 and 3, so they are removed from the list and pushed, so σ becomes $(3\ 8\ 7.5)$. The next element of the list is $+$. It is removed from the list, the stack is popped twice, the expression 8 3 $+$ is evaluated, and the result 11 is pushed onto the stack, so that $\sigma = (11\ 7.5)$. The next object in the list is another $+$. Thus, we pop the stack two more times and evaluate the expression 7.5 11 $+$, which returns the value 18.5. The stack is now empty, but we then push 18.5 onto it, so $\sigma = (18.5)$. The next list element is 10, which is pushed, yielding $\sigma = (10\ 18.5)$. The head of the list is now $\times$. We pop the stack twice and evaluate the expression 18.5 10 $\times$. This returns 185, which is pushed onto the stack. We now have $\sigma = (185)$ and are left with the one operator log in the list. We pop 185 from σ and apply log to it, which yields 5.220356. This number is pushed, so that $\sigma = (5.220356)$, and log is removed from the list. The list is now empty, so the stack is popped one last time to yield the final answer 5.220356.

This is what is happening in this procedure: We begin with an empty stack and a list of numbers (operands) and function symbols (operators). It is assumed that we are given the functions that correspond to these operators. It is also assumed that the elements of this list are so arranged that they determine a meaningful RPN calculation. In an actual computer program, an error message would result if the list does not determine a meaningful calculation. The calculation begins at the front part of the list and moves toward the end. If the head of the list is an operand, it is removed from the list and pushed onto the stack. Whenever the head is an operator, it is removed from the list, the appropriate number of operands is popped from the stack, and the operator is applied to them. The resulting value is pushed onto the stack. These operations are repeated as we move down the list. When the list is empty, the stack is popped to obtain the value of the entire calculation.

The above procedures will now be specified precisely. Assume that we have a programming language that has the functions $+$, $\times$, and (natural) log built in. For definiteness, suppose that these functions are used with prefix functional notation. Thus, to add x to y, we write $+(x, y)$. If a program containing $+(2, 3)$ is executed, this part of the program returns the value 5.

We can describe the RPN calculation procedure in detail as follows. Let σ be an initially empty stack; let λ be a list of *operator symbols* and operands that determines a meaningful calculation. The symbols, $+$, $\times$, and log, when they occur in λ, do not stand for the built-in arithmetical functions. As will be seen, they act as pointers to the following three subprocedures:

procedure ADD!

    $x \leftarrow$ POP!(σ)

    $y \leftarrow$ POP!(σ)

    PUSH! $(+(y, x), \sigma)$

end ADD!

procedure MULTIPLY!

    $x \leftarrow$ POP!(σ)

    $y \leftarrow$ POP!(σ)

    PUSH!$(\times(y, x), \sigma)$

end MULTIPLY!

procedure LOGARITHM!

    $x \leftarrow$ POP!(σ)

    PUSH!$(\log(x), \sigma)$

end LOGARITHM!

None of these three procedures accepts any operand; they just operate on the stack. When any one of these procedures is called, it pops the appropriate number of operands from the stack, computes a result, and then pushes this result onto the stack. Let σ be an initially empty stack, and let λ be a suitable list of operator symbols, say, $+$, $\times$, log, and operands. The next procedure accepts the list λ as input and returns the result of the calculation. It is a recursive procedure.

procedure CALCULATE!(λ)

    if $\lambda = (\ )$ **then** POP!(σ)

        else if $head(\lambda) = +$ **then**

            begin

            ADD!
            CALCULATE! $(tail(\lambda))$
            end

        else if $head(\lambda) = \times$ **then**
            begin
            MULTIPLY!
            CALCULATE! $(tail(\lambda))$
            end

        else if $head(\lambda) = \log$ **then**
            begin
            LOGARITHM!
            CALCULATE! $(tail(\lambda))$
            end

      else

            begin
            PUSH! $(head(\lambda), \sigma)$
            CALCULATE! $(tail(\lambda))$
            end

end CALCULATE!

    When the list λ is empty, the final result of the calculations is on top of the stack and CALCULATE!(λ) pops this result. When the head of λ is an operator symbol, the corresponding arithmetical procedure is called. Thus, the presence of an operator symbol in the list points to an arithmetical procedure such as ADD!. When the arithmetical procedure is called, it pops operands from the stack, computes a value, and pushes this value onto the stack. Then CALCULATE! is recursively called on the tail of λ. If the head of λ is not one of the given operators, then this head should be an operand. This operand is pushed onto the stack, and CALCULATE! is recursively called. It is left as an exercise to extend CALCULATE! so as to include subtraction, division, exponentiation, and other operations.

    We will now take a quick look at queues. The standard type of queue has a structure like that of a line of people waiting to buy ice cream from a vendor in a park. New customers enter the queue (line) at the end, and people leave from the front of the queue after being served. There are other kinds of queues, for instance, priority queues, in which the items are sorted according to some priority ranking. The item with highest priority is moved to the front of the list and is the next item to leave the queue. I will consider only the standard type of queue, in which the first item to go into the queue is the first item out of it. For definiteness, such a queue is sometimes called a *First In First Out* (FIFO) queue.

Queues will be defined in a manner analogous to that used above for stacks. I will first define a queue to be a sequence of lists together with certain functions; then I will introduce some procedures that will enable us to treat a queue as a mutable object. I will use three basic functions: *front*, *delete*, and *insert*. The first of these returns the head of the queue. The second returns the tail. The last returns a list with a new element added to the end of the queue. There is a special consideration regarding the *insert* function: We have not defined any list functions that apply directly to the end of a list. For a simple treatment, I will define *insert* using *append*. Thus, if λ is a list, and α an atom, we can return $append(\lambda, (\alpha))$. However, it is important to note that *append* is *not* computationally very efficient. In order to insert α into λ in this manner, *append* must separately *cons* each of the elements of λ into other lists beginning with (α). There are much more efficient ways of implementing queues using appropriate functions and procedures. A standard method used in LISP is presented in Abelson and Sussman (1985), pages 208–212. Having pointed this out, I will proceed with the simpler and less efficient approach using *append*.

Definition 3-54 Let $\lambda \in list$ and $\alpha \in atom$, and define functions as follows:

$$front(\lambda) = head(\lambda),$$

$$insert(\alpha, \lambda) = append(\lambda, (\alpha)),$$

$$delete(\lambda) = tail(\lambda).$$

Let $Seq = \lambda_0, \lambda_1, \cdots, \lambda_i, \cdots$ be a sequence of lists satisfying

$$\lambda_0 = ()$$

and, for any $i \in Nat$, either

$$\lambda_{i+1} = delete(\lambda_i),$$

or there is some $\alpha \in atom$ such that

$$\lambda_{i+1} = insert(\alpha, \lambda_i).$$

A *queue* is any such sequence, *Seq*, together with the functions *front*, *insert*, and *delete*.

As was the case with stacks, it is convenient to consider a queue to be a mutable object. We therefore define the following two procedures, in which $\alpha \in atom$ and κ is a mutable queue list.

procedure INSERT!(α, κ)

$\qquad \kappa \leftarrow insert(\alpha, \kappa)$

end INSERT!

procedure DELETE!(κ)

$\qquad \kappa \leftarrow delete(\kappa)$

end DELETE!

Notice that DELETE! removes only the front of the queue, whereas POP! returns the top of a stack and also removes this top element. The difference is conventional. Here is an illustration of how these functions and procedures work. Let $\kappa = (\ )$. Then

$$INSERT!(a, \kappa)$$

yields $\kappa = (a)$, and

$$INSERT!(b, \kappa)$$

produces $\kappa = (a\ b)$, and

$$INSERT!(c, \kappa)$$

yields $\kappa = (a\ b\ c)$. Then

$$front(\kappa) = a,$$

and

$$DELETE!(\kappa)$$

produces $\kappa = (b\ c)$.

Queues, like stacks, have many uses. For example, several queues can be used in a program to simulate the progress of customers through the checkout counters of a supermarket. Queues are also used in simulations of the progress of jobs through a shop in which several different machines and operations are needed in order to produce the final products. They are also used in some graph-searching algorithms, and priority queues are used to control processes on a time-sharing computer.

Exercises

General Instructions

In this exercise set, unless mentioned to the contrary, you may assume the basic principles of arithmetic and algebra for *Nat*, *Int*, *Rat*, and *Re*. Unless specifically stated otherwise, in the problems about strings, assume that *char* is the set of lowercase letters of the English alphabet. Similarly, for the list exercises, assume that the set *atom* is this set of letters, unless stated otherwise.

Throughout this book, if an exercise is simply the statement of a theorem or a metatheorem, then the task is to prove this theorem or metatheorem. In your proof, you may use the definitions given in the text. Unless stated to the contrary, you may also use theorems proved in the text.

Exercises for which answers are provided at the back of the book are marked with "Answer Provided."

Finally, please note that exercises that introduce important principles or useful applications, are especially difficult, or are special in some way, are indicated by bold print. In some cases, parenthetical comments describe the special features of the problem. If such comments are missing, the exercise is nonetheless of special value in introducing a new concept, principle, or technique.

E 3-77 (Answer Provided) *Theorem.* For any strings, σ, γ,

$$length_s(append_s(\sigma, \gamma)) = length_s(\sigma) + length_s(\gamma).$$

E 3-78 Define the function $rev_s : string \longrightarrow string$ by

$$rev_s(\sigma) = \ [\ \text{if } length_s(\sigma) = 1 \text{ then } \sigma \\ \text{else } append_s(rev_s(tail_s(\sigma)), \ head_s(\sigma)) \ ].$$

1. Compute $rev_s(\text{`abcd'})$. Show your work.

2. Let $\alpha_1\alpha_2\cdots\alpha_k$ be an informal description of an arbitrary string, where the α_i are elements of *char*. Using this kind of informal description, say what $rev_s(\alpha_1\alpha_2\cdots\alpha_k)$ is.

3. What is $rev_s(append_s(\text{`abc'}, \text{`xy'}))$?

4. *Theorem.* For any strings σ, γ,

$$rev_s(append_s(\sigma, \gamma)) = append_s(rev_s(\gamma), rev_s(\sigma)).$$

(Remarks. The text has already proved that $append_s$ is associative. Although it is easy to program the reversing function according to the above definition, the resulting procedure is computationally inefficient, and for this reason is often called *naive reverse*. There are faster ways to reverse strings.)

E 3-79 *Theorem.* Let rev_s be the function defined in the previous exercise. For any string σ, $rev_s(rev_s(\sigma)) = \sigma$. (You may assume the theorem in the previous exercise.)

E 3-80 Let σ_1, σ_2 be any two strings of length one (thus, they are just characters). Assume that the expression $\sigma_1 = \sigma_2$ returns the value **T** if σ_1 and σ_2 are the same character, and **F** otherwise. Now let $\chi \in char$ and $\sigma \in string$. With the help of the identity relation just described, write a recursive definition for the function, $occ(\chi, \sigma)$, which returns **T** if χ is one of the characters occurring in σ and **F** otherwise. Let σ, $\gamma \in string$. Prove that

$$\text{if } occ(\chi, \, append_s(\sigma, \, \gamma)) = \mathbf{T},$$

$$\text{then } occ(\chi, \, \sigma) = \mathbf{T} \text{ or } occ(\chi, \, \gamma) = \mathbf{T}.$$

E 3-81 *Theorem.* The *append* function on lists is associative.

E 3-82 (Answer Provided) The *tail* function strips off the first element of a list and returns a list of the remaining elements, so that

$$length(tail(\lambda)) = length(\lambda) - 1$$

for any nonempty list λ. Let λ be any list with length $n \geqslant 1$. Let $1 \leqslant k \leqslant n$. Define the function f as follows:

$$f(k, \, \lambda) = [\text{ if } k = 1 \text{ then } tail(\lambda) \text{ else } tail(f(k - 1, \, \lambda)) \,].$$

Prove that $k + length(f(k, \, \lambda)) = length(\lambda)$.

E 3-83 Let α be an atom or a list, and let λ be a list. Define the function *member*, which maps into $\{\mathbf{T}, (\,)\}$ by

$$member(\alpha, \, \lambda) = [\text{ if } \lambda = (\,) \text{ then } (\,)$$
$$\text{else if } (\lambda \neq (\,) \text{ \& } head(\lambda) = \alpha) \text{ then } \mathbf{T}$$
$$\text{else } member(\alpha, \, tail(\lambda)) \,].$$

Theorem. Let α be an atom or a list, and let λ, η be lists. Then

$$\text{if } member(\alpha, \, append(\lambda, \, \eta)) = \mathbf{T},$$

$$\text{then } member(\alpha, \, \lambda) = \mathbf{T} \quad \text{or} \quad member(\alpha, \, \eta) = \mathbf{T}.$$

E 3-84 Let α be an atom or a list, and let λ be a list. Define the function *remove* by

$$remove\,(\alpha, \lambda) = [\text{ if } \lambda = (\,) \text{ then } (\,)$$
$$\text{else if } head(\lambda) = \alpha \text{ then } remove(\alpha, \ tail(\lambda))$$
$$\text{else } cons(\ head(\lambda), \ remove(\alpha, \ tail(\lambda)))\ ].$$

Let *member* be the function defined in the previous exercise. Prove that $member(\alpha, \ remove(\alpha, \ \lambda)) = (\,)$.

E 3-85 Let λ be a list. Define g as follows:

$$g(\lambda) = [\text{ if } \lambda = (\,) \text{ then } 1$$
$$\text{else if } head(\lambda) = \alpha \text{ then } g(tail(\lambda)) \text{ else } 0\ ].$$

Define

$$f(\lambda) = [\text{ if } \lambda = (\,) \text{ then } (\,)$$
$$\text{else if } head(\lambda) = \alpha \text{ then } cons(\alpha, f(tail(\lambda)))$$
$$\text{else } f(tail(\lambda))\ ].$$

Prove that $g(f(\lambda)) = 1$.

E 3-86 Let λ be a list of real numbers. Define

$$sumsqr(\lambda) = [\text{ if } \lambda = (\,) \text{ then } 0 \text{ else } head(\lambda)head(\lambda) + sumsqr(tail(\lambda))\ ],$$

where $head(\lambda)head(\lambda)$ is the numerical product of the head of λ with itself. The *sumsqr* function returns the sum of the squares of all members of λ. Now let $\lambda = (x_1 \cdots x_k)$ be any nonempty list of real numbers. Prove that if $sumsqr(\lambda) = 0$, then every member of λ is 0. In other words, for $i = 1, \cdots, k$, $x_i = 0$. Use induction on $k = length(\lambda)$, and assume basic properties of the real numbers.

E 3-87 (Illustrates use of CVI on the recursive structure of expressions) Define a sequence of sets as follows:

$$A_0 = \{a, \ b\},$$

and, for $k > 0$,

$$A_k = \{\ (\alpha + \beta)\ |\ \text{there are } i, j < k \text{ such that } \alpha \in A_i\ \&\ \beta \in A_j\ \}.$$

Thus,

$$A_1 = \{ \, (a+a), (b+b), (a+b), (b+a) \, \},$$

and

$$A_2 \text{ contains } (a+a), (a+b), (a+(a+a)), ((b+a)+b),$$

$$((a+a)+(b+b)), \text{ etc.}$$

These are just algebraic expressions; they have no values. Now let

$$A = \bigcup_{k=0}^{\infty} A_k.$$

By the above definition, if $\gamma \in A$, then either γ is a or γ is b, or else there are $\alpha, \beta \in A$ such that γ is $(\alpha + \beta)$. Define the rank of γ as follows:

if (γ is a or γ is b), then $rank(\gamma) = 0$,

if γ is $(\alpha + \beta)$, then $rank(\gamma) = rank(\alpha) + rank(\beta) + 1$.

For any $k \in Nat$, there is an element of A with rank k. Use CVI on $rank(\gamma)$ to prove that, in any element of A, the number of left parentheses equals the number of right parentheses. (Remark. Compare with the next problem.)

E 3-88 Do the previous exercise again, but this time prove the result using the well-ordering of Nat. Use RAA. Define

$$U = \{ \, k \mid \text{there exists } \gamma \text{ such that}$$

$$(\, \gamma \in A \,\&\, \gamma \text{ has unbalanced parentheses} \,\&\, k = rank(\gamma)) \, \}.$$

Assume that U is nonempty, and apply WO to it.

E 3-89 (Generalized Association) Let $\otimes$ be an associative binary operation on a set S. For example, $\otimes$ could be $+$ on Nat, $\times$ on Re, $\wedge$ or $\vee$ on SC sentences, functional composition or any group operation (see Chapter 2, Section 2.4). Since $\otimes$ is associative, for any $x, y, z \in S$,

$$x \otimes (y \otimes z) = (x \otimes y) \otimes z.$$

The identity sign means that the *value* of the left side of the equation is the same as the *value* of the right side for any input values x, y, z. If we have an operation on numbers, then the number denoted by the left side is the same as the number denoted by the right side. If we have an operation on a group, then the group element denoted by the left side is the same as the group element referred to by the right side. If we are considering $\wedge$ or $\vee$ on $\mathcal{SC}$ sentences, then the truth value of the left side (under an interpretation) is the same as the truth value of the right side (this amounts to tautological equivalence). These instances should clarify what is meant by the preceding equation. We now introduce *abbreviated notation*: For $n \in Nat$,

$$x_1 \otimes \cdots \otimes x_{0+1} = x_1$$

and

$$x_1 \otimes \cdots \otimes x_{n+1} = (x_1 \otimes \cdots \otimes x_n) \otimes x_{n+1}.$$

This is just a recursive definition for the three-dot notation (ellipsis) in this context. Let $n \geqslant 3$ and prove that, for $1 \leqslant k < k+1 \leqslant n$,

$$(x_1 \otimes \cdots \otimes x_k) \otimes (x_{k+1} \otimes \cdots \otimes x_n) = x_1 \otimes \cdots \otimes x_n.$$

(Remark. This result is a simple generalization of association. A stronger generalization can be proved for arbitrary associative combinations.)

Chapter 4

Predicate Calculus

The sentential calculus enables us only to represent truth-functional connective relationships *between complete sentences.* Yet we often need a detailed representation of the internal logical structure of sentences, as illustrated by the following Classical Argument:

> All men are mortal.
>
> Socrates is a man.
>
> *Therefore*, Socrates is mortal.

Each of the lines of this argument consists of an independent, complete sentence. If we represent this argument in sentential calculus, it would have the following form:

> *A*
>
> *B*
>
> *Therefore, C*

This *SC* argument is not *tautologically sound*; its conclusion is not a *tautological consequence* of its premises. Nevertheless, the original Classical Argument is an intuitively sound argument. The situation we face is this: If a natural language argument is suitably represented in sentential calculus and the corresponding *SC* argument is sound, then the natural language argument is tautologically sound. If a natural language argument is suitably represented in sentential calculus and the corresponding *SC* argument is not sound, all we know is that the natural language argument is not *tautologically* sound. It is still possible that the natural language argument is deductively sound on the basis of some more powerful logical principles. A careful investigation of such a possibility requires developing a more powerful formal representation language and defining *logical consequence* for this language. The language we will study is the *predicate calculus*.

In predicate calculus, we can use a predicate, M, to mean *is-mortal*; an individual constant, a, to refer to Socrates; and another predicate, H, to mean *is-human* (i.e., is-a-man). We then represent the Classical Argument as follows:

$(\forall x)(Hx \rightarrow Mx)$

Ha

Therefore, Ma

The lowercase x is an *individual variable*, and $\forall$ is a *universal quantifier*. $(\forall x)$ can be read as 'for any x'.

More generally, we often want to represent information about some non-empty set of objects, the *domain* of discourse, e.g., a set of people, the set of all animals, a set of various parts in a bin, a set of blocks in an imaginary "blocks world," a set of events and time intervals. We want to be able to refer to *particular individuals* in the domain, *kinds* of things, *properties* of things, and *relations* between things. For example, suppose that the domain is a set of animals. We want to be able to refer to particular animals (King Kong), species (*Ursus horribilis*), properties (having feathers), and relations (x dominates y).

In predicate calculus, kinds and properties are denoted by 1-ary *predicates*, like M and H in the Classical Argument. Binary relations are denoted by 2-ary predicates, such as Pxy for x preys-on y. Individuals such as Socrates or King Kong are denoted by lowercase *individual constants*, such as a or k.

Other lowercase letters, such as x and y, are *individual variables* and are used in sentences like

$$(\forall x)(Hx \rightarrow Mx),$$

which represents

For any x, if x is human, then x is mortal.

We will consider three main aspects of predicate calculus: the *syntax* of the language, its *semantics*, and the proof theory for *derivations*. Some applications will be illustrated in the main text; others are introduced in the Exercises. All of this is only introductory. Although it is a simple language, there is much to learn about, and much that can be done with, the predicate calculus. This includes general theory of the language, numerous kinds of applications in various problem domains, and computational implementations. This is far too much material to put into one book, especially an introductory one. Some sources for further reading are given in the References, especially Enderton (2001), Genesereth and Nilsson (1987), Kowalski (1979), Manna and Waldinger (1985) and (1990), Martin (1989), and Mendelson (1987).

4.1   Syntax of the Predicate Calculus

Definition 4-1 The syntax of the predicate calculus ($\mathcal{PC}$) consists of *symbols* and *formulas* as follows:

Symbols:

>   *parentheses*: (,)
>
>   *sentential connectives*: $\neg$, $\vee$, $\wedge$, $\rightarrow$, $\leftrightarrow$
>
>   *quantifiers*: $\forall$, $\exists$
>
> $\mathcal{SC}$ *letters* (*sentential letters*): A, B, $\cdots$, Z, and any of these letters with a positive Arabic numeral subscript.
>
>   *predicate symbols*: An *n-ary predicate* is an uppercase letter, A, $\cdots$, Z, with the numeral n as a superscript, where n denotes the *arity* of the predicate and $0 < n$. These uppercase letters may also have numerical subscripts. Note: We will usually omit the superscript when we know the arity of a predicate.
>
>   *individual constants*: lowercase letters a, $\cdots$, r, with or without numerical subscripts.
>
>   *individual variables*: lowercase letters s , $\cdots$, z, with or without numerical subscripts.

Formulas: The set of all predicate calculus ($\mathcal{PC}$) formulas is defined recursively, beginning with the atomic formulas.

>   *Atomic Formula*: Any single $\mathcal{SC}$ letter, or an *n*-ary predicate followed by exactly n symbols, each of which is either an individual constant or a variable.
>
>   *Formula*: Any atomic formula, or any expression (finitely long string of symbols) that is obtainable by use of the following predicate calculus construction rules (PCCR):
>
>>   Let ϕ and ψ stand for previously given formulas. Then expressions of the following forms are formulas:
>>
>> $$\neg\phi, (\phi \vee \psi), (\phi \wedge \psi), (\phi \rightarrow \psi), (\phi \leftrightarrow \psi).$$

Let ϕ be a previously given formula and let ν be an individual variable. Then $(\forall\nu)\phi$ and $(\exists\nu)\phi$ are formulas.

Anything that is a formula is obtainable by use of the preceding rules.

In this definition, ϕ and ψ are metalinguistic variables referring to arbitrary $\mathcal{PC}$ formulas. Also, ν (Greek "nu") is a metalinguistic variable that refers to predicate calculus individual variables. The individual constants and individual variables will usually just be called 'constants' and 'variables.' There is no general one-to-one correspondence between ordinary language (e.g., English) sentences or phrases and $\mathcal{PC}$ formulas. Here are some typical relationships, presented informally:

Consider a bag of objects, including marbles and other things. Let M^1x stand for 'x is a marble', R^1x stand for 'x is red', B^2xy stand for 'x is bigger than y', G^1x stand for 'x is green', and b denote a particular marble named 'Ben'.

G^1b represents: 'Ben is green.' Since we know that the arity of the predicate is 1, we can write this in the abbreviated form, Gb. In the following, the superscripts will usually be omitted. Outer parentheses will often be omitted also, as was done with the sentential calculus.

$(\exists x)Rx$ represents: 'There is something that is red' and 'At least one thing is red' and 'There exists something that is red', etc.

$(\forall y)(My \rightarrow Ry)$ represents: 'All marbles are red', 'Any marble is red', 'Every marble is red', 'Anything that is a marble is red', 'For anything, if it is a marble, then it is red', 'For anything, it is a marble only if it is red', etc. I used the variable y in this example, but x, or any other variable, could have been used.

$(\forall x)(Rx \rightarrow (\exists y)(My \wedge Byx))$ represents: 'For any red thing, there is a marble that is bigger than the red thing', 'For any red thing, there is some bigger marble', 'For every red thing, there is some bigger marble' 'For anything x, if it is red, then there exists a y such that y is a marble and y is bigger than x', etc. This example requires using two variables (any two could be used), because we need to distinguish between two things. I use x to refer to (any) red thing and y to refer to the (bigger) marble.

Colloquial English rarely, if ever, uses variables such as x and y, but mathematical English frequently does. We have used variables in this way during the past two chapters. Working Definition 2-1 introduced the idea of an English predicate. According to this definition, an expression like 'x is red' is an English predicate. A predicate of this kind is not a complete sentence. Expressions such as 'Ben is red', 'There exists x such that x is red', and 'For any

x, x is red' are complete sentences. Predicates have no truth values; complete sentences do. In Chapters 2 and 3, we used these distinctions in an informal way; now we can make them precise in predicate calculus.

Definition 4-2 An occurrence of a variable ν in a formula ϕ is *bound* iff ϕ contains a formula θ with one of the forms,

$$(\exists\nu)\psi \text{ or } (\forall\nu)\psi,$$

and this occurrence of ν is within θ. An occurrence of ν is *free* iff it is not bound. A *sentence* is a formula with no free occurrences of any variable.

Examples.

In these examples, I have often dropped outer parentheses and superscripts. The following two expressions are formulas, but are not sentences. The first has a free occurrence of z; the second, a free occurrence of x:

$$Fazc$$

$$Pb \rightarrow (\exists y)(Gyx \wedge Py)$$

The next five examples are *not even formulas*:

x is not a formula, and, in particular, not an atomic formula. It is not a sentential letter or a predicate followed by variables or constants.

$x \vee y$ is not a formula because neither x nor y is.

$(\exists b)G^2 bb$ is not a formula because b is not a variable, so the expression $(\exists b)$ is not defined by the construction rules, PCCR.

In $P^2 x \rightarrow (A \wedge B)$, the subexpression $P^2 x$ is not a formula because P^2 must be followed by exactly two symbols that are variables or constants. In other words, the single x does not agree with the arity of 2 of this predicate.

The expression $(\forall x)\psi$ is not a formula because ψ is not a $\mathcal{PC}$ symbol. When I used ψ before, I was using it in the *metalanguage* (mathematical English) to *refer* to a $\mathcal{PC}$ formula.

In the following, A is a sentential letter. The next four examples are all sentences; no variable occurs free in any of them:

$$Fabc$$

$$A \rightarrow (\exists x)Fx$$

$$(\forall x)(Px \rightarrow (\exists y)(Gyx \wedge Py))$$

$$(\forall x)(\exists y)Lxy$$

Here is an example described in detail: $Fxyz$ and Gzu are atomic formulas. Therefore, $(\forall z)Gzu$ is a formula, and

$$Fxyz \rightarrow (\forall z)Gzu$$

is a formula. Then

$$(\exists y)(Fxyz \rightarrow (\forall z)Gzu)$$

is a formula. In it, all occurrences of y are bound, since they occur within $(\exists y)\psi$, where ψ is $(Fxyz \rightarrow (\forall z)Gzu)$. Adding on another quantifier yields

$$(\forall x)(\exists y)(Fxyz \rightarrow (\forall z)Gzu).$$

In this formula, all occurrences of x are bound, since they occur within $(\forall x)\psi$, where ψ is the formula: $(\exists y)(Fxyz \rightarrow (\forall z)Gzu)$. The one occurrence of u is free. The first occurrence of z is free; the second two occurrences are bound, since they are within $(\forall z)Gzu$. ■

In predicate calculus, the terms *negation, conjunction, disjunction, conditional, antecedent, consequent,* and *biconditional* are used just as in the sentential calculus, except that the subformulas may now be any $\mathcal{PC}$ formula. Thus,

$$((\forall x)\neg Exy \rightarrow (\exists z)(Az \wedge Bzc))$$

is a conditional formula, with antecedent $(\forall x)\neg Exy$, and consequent

$$(\exists z)(Az \wedge Bzc).$$

It is not a sentence, since y occurs free.

Definition 4-3 If ν is a variable, then an expression of the form $(\forall\nu)$ is the *universal quantifier on ν*, and $(\exists\nu)$ is the *existential quantifier on ν*. Let ϕ be a formula. Then $(\forall\nu)\phi$ is the *universal generalization* of ϕ with respect to ν, and $(\exists\nu)\phi$ is the *existential generalization* of ϕ with respect to ν. If κ is an individual constant or variable, then $\phi\left[\nu/\kappa\right]$ denotes the result of replacing all free occurrences of ν in ϕ with occurrences of κ. If ϕ has no free occurrences of ν, then $\phi\left[\nu/\kappa\right] = \phi$.

In the remainder of this chapter, I will often use *constant* to mean *individual constant*. The metalinguistic expression $\phi\left[\nu/\kappa\right]$ is used to refer to the results of substitutions, where ν and κ are variables or constants. We will mainly be interested in substitutions of a constant for the free occurrences of a variable in a formula. For example, let ϕ be

$$(\forall x)(\exists z)(Pxyz \rightarrow (\neg(\exists y)\,Qyyb \vee Qyxy)),$$

ν be y, and κ be c. Then $\phi\left[\nu/\kappa\right]$ is

$$(\forall x)(\exists z)(Pxcz \rightarrow (\neg(\exists y)\,Qyyb \vee Qcxc)).$$

The two occurrences of y in $(\exists y)\,Qyyb$ are bound, so the substitution does not apply to them.

It takes a little practice to became familiar with the preceding syntactical concepts, but they are not difficult to master. Our definitions of terms such as 'free occurrence' and $\phi\left[\nu/\kappa\right]$ are stated in the metalanguage in a rather casual style. Actually, such terms can be defined more precisely using recursive definitions. For example, one can recursively define a function that accepts two input values, a formula ϕ and a pair ν/κ, and returns $\phi\left[\nu/\kappa\right]$. One text that discusses such functions in detail is Enderton (2001). These functions are not just of theoretical interest. They are required in computer programs (such as some artificial intelligence programs) that construct logical deductions. The reasons for needing substitution operations will become clear in the following sections.

The next set of examples should be studied carefully before attempting the following exercises.

Examples.

Problem Statement. Consider the following English language interpretation of some $\mathcal{PC}$ symbols:

D = the domain of discourse = the set of rational numbers, *Rat*

Rx: x is a rational number

Ix: x is an integer

Lxy: $x < y$ (x is less than y, or y is greater than x)

a: 1

b: 2

m: 1,000,000

$Pxyz$: $z = x^y$

Using this interpretation, symbolize the following English sentences in $\mathcal{PC}$. Give brief explanations of your symbolizations.

1. 1 is an integer.

2. Every integer is a rational number.

3. Not all rationals (rational numbers) are integers.

4. There is no greatest rational number.

5. There is no greatest integer.

6. If x and y are rational numbers, with x less than y, then there is a number z larger than x and less than y.

7. There is a power of 2 greater than 1,000,000.

Discussion. The discussions of each part are included in the explanations of the symbolizations that are given in the written solutions. I shall add some additional comments comparing the symbolizations with some alternative formulations.

Written Solutions.

1. The number 1 is denoted by the (individual) constant a and the set of integers by I, so we have Ia.

2. "Every integer is a rational number." This sentence has the general form: "All As are Bs," in which A and B denote subsets of the domain (or properties of domain elements). Thus, if anything x is an A, then it, x, is a B. Therefore, we write $(\forall x)(Ix \rightarrow Rx)$.

   (Notice that this sentence just says that the set of integers is a subset of the set of rationals. In set theoretical notation, we would write $I \subseteq R$. A sentence of the form, "All As are Bs," and thus of the form,

$$(\forall x)(Ax \rightarrow Bx),$$

is called a *universally quantified conditional*, or a *universal conditional*. Many scientific laws have this form or have more complex variations of this form. For example, "All samples of (pure) copper are electrical conductors." $\mathcal{PC}$ has been used extensively to analyze issues in the philosophy of science.)

3. The sentence "Not all rationals are integers" is just the negation of "All rationals are integers." The former sentence is the negation of a sentence with the general form, "All As are Bs." Thus, we obtain

$$\neg(\forall x)(Rx \rightarrow Ix).$$

(Notice that, if it is not the case that all rationals are integers, then there must exist some rational that is not an integer, for example, 1/3. We can symbolize, "There exists a rational that is not an integer," by

$$(\exists x)(Rx \wedge \neg Ix).)$$

4. Consider: "There is no greatest rational number." We are not given an interpretation for "greatest x," so we must improvise. If we were using a comparison relation like $\leqslant$, we might define

    x is a greatest number iff for every number y, $y \leqslant x$.

Since **I** does not include $\leqslant$, but does have L (corresponding to $<$), we can say that a greatest number x is one such that no number y is greater than x. In other words, we can say that there does not exist a number y greater than x; hence, we can use the definition,

    x is a greatest number iff for every y, it is not the case that $x < y$.

In order to symbolize "There is a greatest (rational) number," we can write

$$(\exists x)(\forall y)\neg Lxy.$$

We want the negation, so we end up with

$$\neg(\exists x)(\forall y)\neg Lxy.$$

This is one way to symbolize the original English sentence, but there is another way to represent it. Intuitively, there is no greatest number iff for any number x there is a greater number y. This can be directly symbolized as

$$(\forall x)(\exists y)Lxy.$$

(Later, in Section 4.3.2, I shall prove that these two symbolizations are logically equivalent.)

5. The previous symbolization pertains to the set of rational numbers, which is the domain of **I**. Therefore, the domain of the "possible values" of the variables x and y is the set of rationals. Now we want to symbolize "There is no greatest integer," so we must restrict the possible values of the variables to integers. In ordinary mathematical language, one would most likely write a statement analogous to the second symbolization in the preceding part 4. One would say, "For any integer x, there exists a greater integer y." This means "For any x, if x is an integer, then there exists a y such that y is an integer and y is greater than x." Thus,

$$(\forall x)(Ix \rightarrow (\exists y)(Iy \wedge Lxy)).$$

Of course, there is an infinite set of equivalent symbolizations, but additional knowledge of logic is required to prove most of these equivalences. Partly for this reason, we shall not consider them here. Furthermore, the symbolization presented here closely mirrors the original English sentence when it is paraphrased in ordinary mathematical language. In general, when symbolizing ordinary language sentences in $\mathcal{PC}$, the easiest and safest method is to mirror the structure of the ordinary sentence as closely as possible. This is a rule of thumb, and it has exceptions. Unfortunately, the exceptions often depend on idioms of the natural language one is using and are mastered only after considerable experience with symbolizations.

6. "If x and y are rational numbers, with x less than y, then there is a number z larger than x and less than y." This is straightforward:

$$(\forall x)(\forall y)(Lxy \rightarrow (\exists z)(Lxz \wedge Lzy)).$$

We could have conjoined $Rx \wedge Ry$ to Lxy, but since the domain is the set of rationals, that would be redundant.

7. "There is a power of 2 greater than 1,000,000." Using the given interpretation, how does one say "power of 2"? If we write

$$z = 2^x,$$

then z is a power of two. But we do not know what x is, so instead we can write

$$(\exists x)(z = 2^x).$$

This says that there is some number x such that z is 2^x. This makes z a power of 2, and it is the only way that z can be a power of 2. Using the interpretation that is given, we can rewrite this sentence as

$$(\exists x)Pbxz,$$

since b denotes 2. Thus, this sentence says that z is 2^x. We now have z is a power of 2. To say that z is a power of 2 and z is greater than 1,000,000, we write

$$(\exists x)(Pbxz \wedge Lmz).$$

We now say that z exists: $(\exists z)(\exists x)(Pbxz \wedge Lmz)$. ■

EXERCISES

General Instructions

Exercises for which answers are provided at the back of the book are marked with "Answer Provided."

E 4-1 Using $(((\forall x)P^1x \wedge M^1b) \wedge \neg(\exists y)(\forall x)(R^2xy \vee S^2yx))$ as the root node, sketch the tree structure of this formula. This kind of tree should have a double branch for each occurrence of $\wedge$, $\vee$, $\rightarrow$, $\leftrightarrow$, and a single branch for each occurrence of $\neg$ and each occurrence of a quantifier. The leaves of the tree must be atomic formulas.

E 4-2 (Answers Provided for 1–5) According to the precise and strict application of the syntax rules, determine which of the following are $\mathcal{PC}$ formulas. Rewrite the ones that are genuine formulas without using unnecessary outer parentheses and without superscripts.

1. $(P_3 \rightarrow R^2xy)$

2. $(P_3^1 \rightarrow R^2xy)$

3. $(P^1a \wedge b)$

4. $(R^3axr \vee (\forall x)Q^2xy)$

5. $(\forall x)(\exists y)R^2xy$

6. $\neg(\exists z)(A^1z \rightarrow (B^1z \wedge \neg R^2xy))$

7. $(\forall x)(\exists y)R^3yx$

8. $(P^1 x \neg \rightarrow P^2 ab)$

9. $\neg(\exists z)(\exists y)R^2 yx$

10. $((\exists x)P^1 x \rightarrow (\forall x)a)$

11. $(\exists y)\neg(\forall x)M^3 xby$

12. $(P^1 x \rightarrow P^2 ab)$

13. $(\forall x)(M^1 x)$

14. $(\neg(\forall x)Rxh \leftrightarrow (\forall z)\neg Fz)$

15. $(\forall c)M^1 c$

16. $(\forall x)(R_4^2 xd \rightarrow (\exists y)(P^1 y \wedge R_4^2 qy))$

17. $(R \leftrightarrow (R^2 xzb \vee B^1 a))$

18. $\neg\neg(A \vee (\exists x)(\exists y)(\exists z)(B^2 xu \rightarrow A^1 z))$

19. $(\forall xy)M_1^1 x$

20. $(\forall x)(\forall y)(M_1^1 x \vee (\neg M_1^1 u \wedge M^1 y))$

E 4-3 After you have completed the previous exercise, examine the formulas that you have rewritten. For each formula, specify all free occurrences of variables in it. If it has no free occurrences of any variables, indicate that it is a sentence.

E 4-4 Give examples of $\mathcal{PC}$ formulas with the following features:

1. A sentence that is a universal generalization of a conditional, in which the consequent of this conditional is an existential generalization.

2. A formula that is not a sentence but an existential generalization of a conjunction.

3. A formula with at least three free occurrences of exactly two different variables.

4. A sentence that is a disjunction of an atomic sentence with a 3-ary predicate and an existential generalization.

5. A formula that is not a sentence but has occurrences of two universal quantifiers with different variables and one existential quantifier and, in addition, becomes a sentence when it is universally generalized with one additional quantifier.

E 4-5 (Answers Provided for 1-5) Consider a set of books in a library. Let the variables s, t, u, $\cdots$ pertain to these books and to human beings as we used variables in the earlier example about the bag of objects. Use predicates and constants to represent information as follows:

r : *The Rubáiyát of Omar Khayyám*

f : *Les Fleurs du Mal*

b : *Begriffsschrift*

e : *An Essay Concerning Human Understanding*

g : G. Frege

c : Charles Baudelaire

Lxy : x is a longer book than y

Wxy : x is a book written by y

$F_1 x$: x is a book written in French

Px : x is a philosopher

Mx : x is a mathematician

$F_2 x$: x is fun to read

We can now symbolize some English sentences, e.g.,

Les Fleurs du Mal is a book written in French.

$$F_1 f$$

An Essay Concerning Human Understanding is (a book) written by a philosopher and is a longer book than *The Rubáiyát*.

$$((\exists y)(Wey \wedge Py) \wedge Ler)$$

There is a book written by a philosopher and it is longer (i.e., a longer book) than *The Rubáiyát*.

$$(\exists x)((\exists y)(Wxy \wedge Py) \wedge Lxr)$$

Symbolizations are not unique because any $\mathcal{PC}$ sentence is logically equivalent to other $\mathcal{PC}$ sentences in other forms. Also, a set of English predicates can often be symbolized into more than one set of equally satisfactory $\mathcal{PC}$ predicates.

Using commonsense interpretations of the meanings of the following English sentences, symbolize each of them in predicate calculus. It does not matter whether the English sentences are true or false; just try to capture their meanings. Try to find *PC sentences* (no free occurrences of variables) that have structures close to the structures of the corresponding English sentences. Symbolizations can be difficult; a good understanding of logic and much practice are required to do them in an effective and elegant way.

1. There is a book written by Baudelaire.

2. *The Rubáiyát* is a book written by a mathematician.

3. If *The Rubáiyát* is a book written by a mathematician, then it (*The Rubáiyát*) is a book written by a philosopher.

4. There is a book that is longer than *Les Fleurs du Mal*.

5. For any book written in French, there is a longer book.

6. It is not the case that there is a book that is longer than itself.

7. No book is written by a philosopher.

8. Every book is written by a mathematician.

9. It is not the case that every book is written by a philosopher.

10. For every book written in French and every philosopher, it is not the case that the book is written by the philosopher.

11. Every book written by a mathematician is fun to read.

12. Anything that is fun to read is not a book written by a philosopher.

13. For any book written by Baudelaire, *Begriffsschrift* is a longer book written by G. Frege.

14. For any book written by Baudelaire, there is a longer book written by G. Frege.

15. There is someone who is both a philosopher and a mathematician and who has written a book in French.

16. Any book that is longer than *Begriffsschrift* or *An Essay Concerning Human Understanding* is longer than *Les Fleurs du Mal*.

E 4-6 Let the domain of discourse be the set of all animals, both real and imaginary, and interpret predicates as follows:

Ax : x is an anole

Bx : x is a bird

Fx : x is feathered (i.e., has feathers)

Lx : x is a lizard

Sx : x is a snake

Wx : x is a wasp

Ixy : $x = y$, the identity relation on the domain

Px : x is poisonous

Mxy : y is more poisonous than x

c :   a bird named 'Cassandra'

q :   a feathered snake named 'Quetzalcoatl'

Symbolize the following English sentences in $\mathcal{PC}$. Try to write $\mathcal{PC}$ sentences that are similar in structure to the corresponding English sentences. (It may help to look over the previous exercise first.)

1. Quetzalcoatl is a feathered snake, but he is not poisonous.

2. All anoles are lizards.

3. Cassandra is a bird, not a wasp.

4. It is not the case that every lizard is not poisonous.

5. No bird is poisonous and no bird is a lizard.

6. Some snakes are poisonous, but not all snakes are.

7. For any wasp, there is a more poisonous snake.

8. If an animal is not feathered, then it is not a bird.

9. If any animal is poisonous and is not a snake, then it is either a lizard or a wasp.

10. Although Cassandra and Quetzalcoatl are both feathered, they are not identical.

11. There are at least two animals. (Hint: There are x and y, with $x \neq y$.)

12. There are at least two feathered animals.

13. There are at most two lizards. (Hint: If x, y, z are lizards, then at least two of them are the same.)

14. For any poisonous lizard and any poisonous snake, some wasp is such that it is more poisonous than the lizard and such that the snake is more poisonous than it.

15. For any poisonous wasp, there is a more poisonous feathered lizard.

16. There are two distinct lizards, one more poisonous than the other.

17. If one animal is more poisonous than another, then they are not both birds.

18. If there exists a bird that is not feathered, then there exists an anole that is poisonous and feathered.

4.2   Semantical Aspects of the Predicate Calculus

4.2.1   Interpretations and Truth

For any set, X, let X^n denote the set of ordered n-tuples of X. Thus, X^n is the nth–dimensional Cartesian product of X. The power set, $\mathcal{P}(X^n)$, contains all n-ary relations on X.

Definition 4-4 Let $SCLETTERS$ denote the set of all $\mathcal{SC}$ letters in $\mathcal{PC}$. For each $k \geqslant 1$, let $PREDS_k$ denote the set of k-ary predicates in $\mathcal{PC}$. Also, let $CONST$ denote the set of individual constants. A $\mathcal{PC}$ *interpretation* $\mathbf{I}$ consists of

a nonempty domain $\mathbf{D}$ of elements

a function $\mathbf{I}_{CONST} : CONST \longrightarrow \mathbf{D}$

a function $\mathbf{I}_{SC} : SCLETTERS \longrightarrow \{\mathbf{T}, \mathbf{F}\}$

for each $k \geqslant 1$, a function $\mathbf{I}_{PRED,k} : PREDS_k \longrightarrow \mathcal{P}(\mathbf{D}^k)$

In other words, a $\mathcal{PC}$ interpretation consists of a nonempty domain $\mathbf{D}$, together with many functions that map certain parts of the $\mathcal{PC}$ language into aspects of the domain. The preceding characterization of these functions can be paraphrased as follows:

1. Each individual constant (in $CONST$) is assigned (mapped) to a unique element of $\mathbf{D}$ by the function $\mathbf{I}_{CONST}$. The constants serve as names for elements of the domain. Many constants may be mapped to the same element of $\mathbf{D}$, so $\mathbf{I}_{CONST}$ is, in general, not a one-one function. It also need not map onto $\mathbf{D}$.

2. Each $\mathcal{SC}$ letter is assigned the value $\mathbf{T}$ or $\mathbf{F}$. This is done by $\mathbf{I}_{SC}$. Thus, $\mathbf{I}_{SC}$ amounts to a standard $\mathcal{SC}$ interpretation of the $\mathcal{SC}$ part of $\mathcal{PC}$.

3. Each 1-ary predicate is assigned to a 1-ary relation on $\mathbf{D}$. Since a 1-ary relation is a subset, each 1-ary predicate corresponds to a subset of $\mathbf{D}$ under an interpretation. These assignments are given by $\mathbf{I}_{PRED,1}$.

4. For $k \geqslant 2$, each k-ary predicate is assigned to a k-ary relation on **D**. These assignments are given by $\mathbf{I}_{PRED,\,k}$.

The definition of truth will be stated shortly. As in sentential calculus, except for some special cases, the truth value of a sentence is always relative to a particular interpretation. The truth value of an $\mathcal{SC}$ sentence, under a $\mathcal{PC}$ interpretation, will be determined by the assignments of **T** or **F** to the letters (by $\mathbf{I}_{\mathcal{SC}}$) just as in the case of the $\mathcal{SC}$ language. Other truth value assignments can be illustrated informally by the following:

Example.

Imagine a simple world that has six blocks, Block 1, Block 2, etc., some of which are yellow, some blue, and some green. At certain times, a given block may be on top of another block. Let **D** be the domain consisting of this set of blocks. At any given time, we represent the state of this world in terms of certain individual constants and predicates. $\mathbf{I}_{CONST}$ assigns the following values: a: Block 1, b: Block 2, c: Block 3, d: Block 4, e: Block 5, f: Block 6. The binary predicate O is assigned by $\mathbf{I}_{PRED,\,2}$ in such a way that Oxy denotes the domain relation of 'x being on y' (i.e., block x is on top of block y). The 1-ary predicates are assigned as indicated here: Yx: x is yellow (i.e., the subset of yellow blocks), Bx: x is blue, Gx: x is green, Cx: x is clear (where x is clear iff no block is on x).

Here are some predicate calculus sentences with their corresponding *truth conditions* in the blocks world.

Atomic Sentences:

Gf is true (under our blocks world interpretation) iff Block 6 is in the subset of green blocks.

Oba is true iff Block 2 is on Block 1.

Cd is true iff Block 4 is clear.

Quantified Sentences:

$(\forall x)\,Yx$ is true iff all blocks (in **D**) are yellow.

$(\exists x)\,Gx$ is true iff some block is green.

$(\exists z)(Yz \wedge Ozc)$ is true iff there is a yellow block on Block 3.

$(\forall x)(Yx \rightarrow (\exists y)(Gy \wedge Oyx))$ is true iff for every yellow block, there exists a green block, such that this green block is on the yellow one.

$(\forall x)(\forall y)((Gx \wedge By) \rightarrow Oyx)$ is true iff for any blocks x, y, if x is green and y is blue, then y is on x.

$\neg(\forall x)\,Gx$ is true iff not all blocks are green.

$(\exists x)\neg Gx$ is true iff there is a block that is not green.

$(\forall x)\neg Gx$ is true iff all blocks are not green (in other words, no block is green).

$(\forall x)(Yx \rightarrow \neg Cx)$ is true iff for any block, if it is yellow, then it is not clear. ■

Unless specified otherwise, in the following, when I say that **I** is an interpretation, this will mean that **I** is an interpretation of the predicate calculus consisting of a nonempty domain **D** together with the functions $\mathbf{I}_{CONST}$, $\mathbf{I}_{SC}$, and $\mathbf{I}_{PRED,k}$. Relative to such an interpretation, **I**, we can define a valuation function, $\mathbf{V_I}$, as was done for SC interpretations. For a PC sentence, ϕ, defining $\mathbf{V_I}(\phi)$ is more complex than it was for sentential calculus.

The definition of $\mathbf{V_I}$ is mainly due to Alfred Tarski, who published an important metatheorem about arithmetical truth in 1936; see Mendelson (1987, p. 169). Before proceeding, notice that $CONST$ can be well-ordered. The most obvious way is a, b, $\cdots$, r, a_1, b_1, $\cdots$. Any formula ϕ has a finite number of symbols in it, so the set of individual constants that do not occur in ϕ is nonempty. Because the set of all constants is well-ordered, there will be a *first* constant that does not occur in ϕ.

Definition 4-5 Let **I** be a PC interpretation and let $\kappa \in CONST$. Then $\mathbf{I}_\kappa$ is *a variant of* **I** *with respect to* κ iff $\mathbf{I}_\kappa$ differs from **I** at most in what is assigned to κ.

In other words, if $\mathbf{I}_\kappa$ is a variant of **I** with respect to κ, then either they are exactly the same interpretation or else they are the same for everything except for what they assign to κ.

Definition 4-6 Let **I** be a PC interpretation. We define a function $\mathbf{V_I}$ that maps the set of PC *sentences* into $\{\mathbf{T}, \mathbf{F}\}$. Thus, if ϕ is any sentence, $\mathbf{V_I}(\phi)$ will have the value **T** or the value **F**. The definition is recursive.

1. If ϕ is an SC letter, $\mathbf{V_I}(\phi) = \mathbf{I}_{SC}(\phi)$. *variable for predicates* ↗

2. Suppose that ϕ is an atomic sentence of the form $\mathbb{R}\kappa_1 \cdots \kappa_n$ for predicate $\mathbb{R}$ and constants κ_i. Let

$$\mathbf{R} = \mathbf{I}_{PRED,n}(\mathbb{R})$$

denote the n-ary relation on $\mathbf{D}$ corresponding to $\mathbb{R}$. For each i, let

$$\mathbf{k}_i = \mathbf{I}_{CONST}(\kappa_i)$$

be the element of $\mathbf{D}$ corresponding to κ_i. Then

$$\mathbf{V_I}(\mathbb{R}\kappa_1 \cdots \kappa_n) = \mathbf{T} \;\Leftrightarrow\; <\mathbf{k}_1, \cdots, \mathbf{k}_n> \,\in \mathbf{R}.$$

3. If ϕ is a negation, disjunction, conjunction, conditional, or biconditional, then the truth value of ϕ is recursively given in terms of subsentences in the same manner as for sentential calculus. (Hence, the truth table calculations apply to these forms of sentences.)

4. Suppose ν is a variable and ϕ has the form $(\exists \nu)\rho$. Let κ be any constant not in ρ. (One can always be found by choosing the first constant not in ρ given by the well-ordering of *CONST*.) Then

$$\mathbf{V_I}(\phi) = \mathbf{V_I}(\,(\exists\nu)\rho\,) = \mathbf{T}$$

iff there is a variant $\mathbf{I}_\kappa$ of $\mathbf{I}$ with respect to κ such that

$$\mathbf{V}_{\mathbf{I}_\kappa}(\,\rho\,[^\nu\!/_\kappa]\,) = \mathbf{T}.$$

[handwritten margin note: For example, if there is some S const & then true exists some λ such that Sₓ is satisfied]

5. Suppose ν is a variable and ϕ has the form $(\forall \nu)\rho$. Let κ be any constant not in ρ. Then

$$\mathbf{V_I}(\phi) = \mathbf{V_I}(\,(\forall\nu)\rho\,) = \mathbf{T}$$

iff for every variant $\mathbf{I}_\kappa$ of $\mathbf{I}$ with respect to κ

$$\mathbf{V}_{\mathbf{I}_\kappa}(\,\rho\,[^\nu\!/_\kappa]\,) = \mathbf{T}.$$

We say that ϕ is *true under* $\mathbf{I}$ iff $\mathbf{V_I}(\phi) = \mathbf{T}$; if ϕ is not true under $\mathbf{I}$, then it is *false under* $\mathbf{I}$.

For practice, one should consider how this definition applies to the previous blocks world example. Here is another example; in this case the domain is infinite.

Example.

Let $\mathbf{I}$ be an interpretation with the following features:

- $\mathbf{D}$ is the set of all real numbers.

- Lxy represents the relation $x < y$.

- $Pxyz$ represents the relation that holds when z equals the product xy.

- Certain numbers are assigned to constants as follows: a: 0, p: $\pi = 3.14159 \cdots$, b: -23.045.

Here are some sentences and their truth values under **I**:

1. Lba is true, since -23.045 is less than 0. Lpa is false, since π is not less than 0. To determine the truth value of $(\exists x)Lpx$, we find a constant not in that sentence, for instance, a. Now consider Lpa. Is there some variant $\mathbf{I}_a$ of **I** with respect to a such that Lpa is true under $\mathbf{I}_a$? The answer is "yes," since we can let

$$\mathbf{I}_{CONST}(a) = 100,$$

   and π is less than 100.

2. Consider $(\forall x)(Lxa \rightarrow Lxp)$. In this case, we use the constant b and consider

$$Lba \rightarrow Lbp$$

   for *all* variants of **I** with respect to b. In **D**, if any number is less than 0, then it is also less than π, so $Lba \rightarrow Lbp$ is true under every variant with respect to b, so

$$(\forall x)(Lxa \rightarrow Lxp)$$

   is true under **I**.

3. For $(\forall x)(\exists y)Lyx$, we look at Lba. If any variant $\mathbf{I}_a$ assigns a to an element **a** in **D**, then we can use the variant of $\mathbf{I}_a$ that assigns b to the number $\mathbf{a} - 1$. Then $\mathbf{b} < \mathbf{a}$, so

$$(\forall x)(\exists y)Lyx$$

   is true under the interpretation. If the domain had been the set of non-negative reals, this sentence would have been false, since there would then be no element of the domain less than 0. The truth value of a sentence depends critically on the details of an interpretation, except for sentences that are true (or are false) under every interpretation.

4. Finally, $(\forall x)(\forall z)(\exists y)Pxyz$ is false under **I**. Informally, this sentence says that, for any x, z, there exists y such that $z = xy$. This is true for most x, z, but not true when $x = 0$ and $z \neq 0$. ■

This is a good place to mention that we are examining the *first-order* predicate calculus: The predicates apply only to individual constants and variables, and there are no quantifiers for predicates. Thus, the language does not include expressions like $(\forall P)(\forall x)Px$. This is a sentence. In English, it might be rendered as 'For any property P and for anything x, x has P,' or as 'For any set P and any object x, x is an element of P.' Notice how $(\forall P)$ quantifies over properties or at least sets. This is called a *second order quantifier*. More powerful forms of predicate calculus use such higher-order quantifiers, but we limit our study to the first-order predicate calculus, which has quantifiers only for the individual variables, s, t, u, $\cdots$ in our particular formulation. Nevertheless, first-order predicate calculus is much more complex and powerful than sentential calculus.

Now that interpretations and truth have been characterized for the predicate calculus, it is easy to generalize many important concepts previously defined for sentential calculus.

Definition 4-7 Let ϕ be a $\mathcal{PC}$ sentence, Γ be a set of $\mathcal{PC}$ sentences, and $\mathbf{I}$ be a $\mathcal{PC}$ interpretation. Then $\mathbf{I}$ *satisfies* ϕ (ϕ is *true under* $\mathbf{I}$) iff $\mathbf{V_I}(\phi) = \mathbf{T}$. $\mathbf{I}$ satisfies Γ iff it satisfies every sentence in Γ.

ϕ is *valid* or *logically true* iff it is true under every ($\mathcal{PC}$) interpretation.

ϕ *is a consequence of* Γ (denoted by $\Gamma \vDash \phi$) iff there is no interpretation that satisfies Γ and does not satisfy ϕ.

If sentences ϕ and ψ are consequences of each other, they are said to be *logically equivalent* or, more briefly, *equivalent*.

Valid is a generalization of *tautologous* and *consequence* is a generalization of *tautological consequence*. If $\{\psi\} \vDash \phi$, we may also write $\psi \vDash \phi$. If ϕ is a consequence of Γ, we may say that Γ *implies* ϕ. Generalizing from sentential calculus terminology, we also say that ϕ is *contingent* iff it is satisfiable, but not valid. An interpretation $\mathbf{I}$ is called a *model* of Γ iff it satisfies Γ. Γ is *satisfiable* iff it has a model. A single sentence ϕ is satisfiable iff $\{\phi\}$ is. Most of these concepts are illustrated in the following examples.

Examples.

1. $(P \vee \neg P)$ is an $\mathcal{SC}$ tautology, so it is true under every interpretation, so it is valid. An $\mathcal{SC}$ sentence is valid iff it is tautologous.

2. A constant not in $(\forall x)(Qx \to Qx)$ is a, so we consider $Qa \to Qa$. Under any variant of any interpretation, Qa is either true or false; in either case, $Qa \to Qa$ is true. Hence, for any interpretation, $(\forall x)(Qx \to Qx)$ is true, so this sentence is valid.

3. To show that

$$\{ (\exists x)Ax,\ (\exists x)Bx,\ (\forall x)(Ax \rightarrow (\exists y)(By \wedge Rxy)) \}$$

is satisfiable, we need to find an interpretation that makes all of its sentences true. The predicates A and B must be assigned to nonempty subsets of the domain. Let $\mathbf{D} = Nat$ (the set of natural numbers), and use the assignments, Ax: x is even, Bx: x is odd, and Rxy: $x < y$. Since there are even numbers (e.g., 2) and odd numbers (e.g., 3), the first two sentences are true under **I**. Informally, the third sentence says that for any even number x, there is an odd number y, such that $x < y$. That is true; if x is even, then $y = x + 1$ is odd and greater than x.

4. The Classical Argument is an example of consequence,

$$\{ (\forall x)(Hx \rightarrow Mx),\ Ha \ \} \vDash Ma.$$

Informally, the first sentence says that anything that is H is M, i.e., the set assigned to H is a subset of the set assigned to M under any interpretation **I**. The second sentence requires that the domain object assigned to a be an element of H. Because of the subset relationship, this object must be an element of M. Thus, any interpretation that satisfies the premises must satisfy the conclusion.

5. $(\forall x)Ox$ is not a consequence of $(\exists z)Oz$. To show this, we construct a *counterexample* in a manner similar to that used for the sentential calculus. But there is a big difference: unless we are working only with $\mathcal{SC}$ sentences, the counterexample must specify a domain and values for the constants and predicates in the sentences under consideration. In this case, we can use $\mathbf{D} = $ the set of people on the earth, and Ox: x is honest. The premise says that 'Someone is honest'. The conclusion says that 'Everyone is honest'. This is clearly a fallacious argument. For a more precise counterexample, use $\mathbf{D} = Nat$, and Ox: x is prime. The number 7 is prime, so the premise is true; the number 10 is not prime, so the conclusion is false. ■

Example.

Problem Statement. Show that

$$\Gamma = \{ \neg(\forall x)Bx,\ \neg(\forall y)(By \rightarrow Ay),\ (\forall x)Rxx,\ (\exists x)(\exists y)((Bx \wedge \neg By) \wedge Rxy) \},$$

is satisfiable.

Discussion. We need to construct an interpretation **I** that satisfies Γ, i.e., each sentence in Γ must be true under **I**. It is convenient to denote the sentences with metalinguistic symbols

$$\psi_1 \;:\; \neg(\forall x)Bx,$$
$$\psi_2 \;:\; \neg(\forall y)(By \rightarrow Ay),$$
$$\psi_3 \;:\; (\forall x)Rxx,$$
$$\psi_4 \;:\; (\exists x)(\exists y)((Bx \wedge \neg By) \wedge Rxy).$$

A $\mathcal{PC}$ interpretation always requires a nonempty domain. Let **D** be the domain of **I**, and interpret the $\mathcal{PC}$ predicates as follows:

$$A\colon \mathbf{A}, \quad B\colon \mathbf{B}, \quad R\colon \mathbf{R},$$

where $\mathbf{A} \subseteq \mathbf{D}$, $\mathbf{B} \subseteq \mathbf{D}$, and $\mathbf{R} \subseteq \mathbf{D} \times \mathbf{D}$. Three of the four sentences have intuitively clear meanings. The first sentence says that not everything is in **B**, so there is at least one domain element not in **B**. The sentence ψ_2 states that it is not the case that $\mathbf{B} \subseteq \mathbf{A}$, so there is a domain element in **B** that is not in **A**. When constructing $\mathcal{PC}$ interpretations, it is almost always helpful to draw a sketch of the interpretation one is trying to build. At this point I begin with the sketch in Figure 4.1.

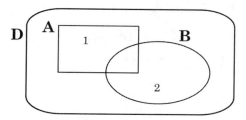

Figure 4.1.

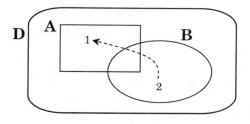

Figure 4.2.

I am using 1 and 2 as elements of **D**. Because 1 is not in **B**, ψ_1 is true. Since 2 is in **B** but not in **A**, ψ_2 is true. The fourth sentence states that there is an element in **B** that is related by **R** to some element not in **B**. Since we already have 1 and 2, we can use them, as indicated in Figure 4.2. The dashed arc represents **R**21. The third sentence states that **R** is reflexive, so we draw the reflexive loops in Figure 4.3.

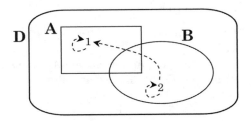

Figure 4.3.

Written Solution. Let

$$\Gamma = \{ \neg(\forall x)Bx, \ \neg(\forall y)(By \rightarrow Ay), \ (\forall x)Rxx,$$
$$(\exists x)(\exists y)((Bx \wedge \neg By) \wedge Rxy)\}.$$

Define the interpretation **I**, with domain **D**, as follows:

    $\mathbf{D} = \{1, 2\}, \quad A\colon \mathbf{A} = \{1\}, \quad B\colon \mathbf{B} = \{2\},$
    $R\colon \mathbf{R} = \{< 1, 1 >, < 2, 2 >, < 2, 1 >\}.$

Let

$$\psi_1 \ : \ \neg(\forall x)Bx,$$
$$\psi_2 \ : \ \neg(\forall y)(By \rightarrow Ay),$$
$$\psi_3 \ : \ (\forall x)Rxx,$$
$$\psi_4 \ : \ (\exists x)(\exists y)((Bx \wedge \neg By) \wedge Rxy).$$

The first sentence in Γ states that it is not the case that everything is in **B**. This is true since $1 \notin \mathbf{B}$. ψ_2 states that it is not the case that everything in **B** is in **A**. This is true since $2 \in \mathbf{B}$ & $2 \notin \mathbf{A}$. The only elements in **D** are 1 and 2, and we have **R**11 and **R**22, so ψ_3 is true. ψ_4 is true under **I** iff $(Ba \wedge \neg Bb) \wedge Rab$ is true for some possible values of a and b, i.e., some variant of **I** with respect to a and b. Use the assignments, a: 2 and b: 1. Then, $(Ba \wedge \neg Bb) \wedge Rab$ is true, so ψ_4 is true, and **I** satisfies Γ. ■

Example.

Problem Statement. Prove that $(\exists x)(\forall y)Rxy \to (\forall y)(\exists x)Rxy$ is valid.

Discussion. To show that a sentence is satisfiable, we must exhibit *one particular interpretation* under which the sentence is true. To show that a set of sentences is satisfiable (as in the previous example), we must exhibit *one particular interpretation* under which each sentence in the set is true. In the present problem, we must prove that a sentence is valid. In order to do this, we must write a *general proof* that this sentence is true under *every* $\mathcal{PC}$ interpretation. Any relevant interpretation **I** has a nonempty domain **D**, and R must denote some binary relation **R** on **D**. The antecedent of the sentence is $(\exists x)(\forall y)Rxy$. If this antecedent is false under **I**, then the given sentence must be true. Therefore, we are concerned only with interpretations that satisfy the antecedent. As usual, it helps to draw a sketch. The antecedent says that there is something in the domain that is related to everything in the domain. Whatever else may be in the domain, there must be some element like the one near the center of the next figure.

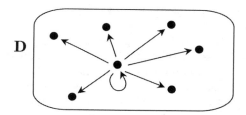

Figure 4.4.

In Figure 4.4, the black dots represent all of the elements in **D**. The dot in the middle is an element of **D** that is related to everything in **D** (including itself). There could also be other relations between various elements of **D**, but there must be at least one element like the one in the middle of Figure 4.4. By looking at this diagram, it is now obvious that the consequent, $(\forall y)(\exists x)Rxy$, must also be true under **I**. It is true because, for any particular element y in **D**, there is some element x in **D** (namely, the middle dot), that is related to y.

Written Solution. Prove that $(\exists x)(\forall y)Rxy \to (\forall y)(\exists x)Rxy$ is valid.

Proof. Suppose that interpretation **I**, with domain **D**, and **I**(R)=**R**, satisfies the antecedent, $(\exists x)(\forall y)Rxy$, of the given sentence. Consider the sentence Rab. Since **I** satisfies the antecedent, there is some variant $\mathbf{I}_a$ of **I** with respect to a such that Rab is true for every variant of $\mathbf{I}_a$ with respect to b. Let $\mathbf{I}_a(a) = \mathbf{a} \in \mathbf{D}$, and let **b** be any element of **D**. Then **Rab**. (The element **a** corresponds to the dot in the middle of Figure 4.4, and **b** corresponds to any of the dots in the diagram.) Now consider the consequent, $(\forall y)(\exists x)Rxy$. In order for this sentence to be true under **I**, $(\exists x)Rxb$ must be true for every variant $\mathbf{I}_b$ of **I** with respect to b. Let $\mathbf{I}_b$ be any such variant with respect to b, and let $\mathbf{I}_b(b) = \mathbf{b} \in \mathbf{D}$. Then, in order for $(\exists x)Rxb$ to be true, there must be some variant of $\mathbf{I}_b$ with respect to a such that Rab is true. If we choose the variant of $\mathbf{I}_b$ that assigns a to $\mathbf{a} \in \mathbf{D}$, then Rab is true because **Rab** holds as a result of the antecedent. Hence, any interpretation that satisfies the antecedent also satisfies the consequent, so the sentence is true under all interpretations and is therefore valid.   ***Q.E.D.*** ■

Example.

Problem Statement. Let

$$\Delta = \{\, (\exists x)(Ax \vee Bx), \neg(\exists x)(Ax \wedge Bx), (\forall x)(Ax \to (\exists y)(By \wedge Rxy)),$$
$$(\forall x)(Bx \to (\exists y)(Ay \wedge Rxy)), (\forall x)(\forall y)(Rxy \to \neg Ryx)\},$$

and ϕ be $(\forall x)(\exists y)(Rxy \vee Ryx)$. Show that ϕ is not a consequence of Δ.

Discussion. By definition, ϕ is a consequence of Δ ($\Delta \models \phi$) $\Leftrightarrow$ every interpretation that satisfies Δ also satisfies ϕ. To show not $\Delta \models \phi$, we must exhibit a particular (counterexample) interpretation that satisfies Δ and does not satisfy ϕ (ϕ is false under the interpretation). As usual, I shall start with a rough sketch of the domain **D** of an interpretation **I**. The first sentence in Δ states that the set $\mathbf{A} \cup \mathbf{B} \neq \varnothing$, where $\mathbf{A}$, $\mathbf{B} \subseteq \mathbf{D}$, and the second states that **A** and **B** are disjoint. Let $\mathbf{R} = \mathbf{I}(R)$.

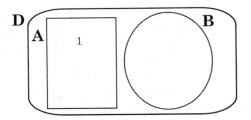

Figure 4.5.

The third sentence in Δ states that, for anything **a** in **A**, there is some **b** in **B** such that **Rab**. The fourth sentence makes the analogous assertion starting at **B** and going to **A**. In the diagram, I shall use numbers as elements of **D**. There is something in $\mathbf{A} \cup \mathbf{B}$, so I shall try starting with 1 in **A**, leading to Figure 4.5.

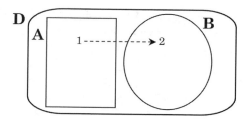

Figure 4.6.

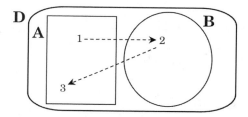

Figure 4.7.

Notice that **A** and **B** are disjoint in Figure 4.5. To satisfy the third sentence in Δ, we need something in **B** to which 1 is related. Let this element be 2, leading to Figure 4.6. To satisfy the fourth sentence in Δ, we need something in **A** to which 2 is related. The last sentence in Δ states that **R** is asymmetric. Hence, 2 cannot be related to 1. Therefore, we need another element in **A**. Let this element be 3, leading to Figure 4.7.

Now 3 must be related to some element in **B**. Because **R** is asymmetric, 3 cannot be related back to 2, so I shall put 4 in **B** and require **R**34. Now we need to relate 4 to an element of **A**, and we can set **R**41. We now have Figure 4.8.

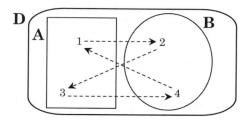

Figure 4.8.

The interpretation corresponding to Figure 4.8 satisfies Δ. Now add one more element 5 to **D** such that $5 \notin (A \cup B)$. Nothing in Δ requires that 5 be related to any element of **D**, and no element of **D** needs to relate to 5. There-fore, ϕ, $(\forall x)(\exists y)(Rxy \lor Ryx)$, is false under **I**, and so we have a counterexample interpretation.

Written Solution.

Let

$$\Delta = \{\,(\exists x)(Ax \lor Bx), \neg(\exists x)(Ax \land Bx), (\forall x)(Ax \to (\exists y)(By \land Rxy)),$$

$$(\forall x)(Bx \to (\exists y)(Ay \land Rxy)), (\forall x)(\forall y)(Rxy \to \neg Ryx)\},$$

and ϕ be $(\forall x)(\exists y)(Rxy \lor Ryx)$. Show that not $\Delta \models \phi$.

Define the interpretation **I**, with domain **D**, as follows:

   D = {1, 2, 3, 4, 5},

   A: **A** = {1, 3},

   B: **B** = {2, 4},

   R: **R** = {< 1, 2 >, < 2, 3 >, < 3, 4 >, < 4, 1 >}.

The sentence $(\exists x)(Ax \lor Bx)$ is true under **I** since 1 is in **A**. There is no element in both **A** and in **B**, so $\neg(\exists x)(Ax \land Bx)$ is true. The last sentence in Δ states that **R** is asymmetric, and this is true since no member of the converse of **R** is in **R**. The sentence $(\forall x)(Ax \to (\exists y)(By \land Rxy))$ states that every element in **A** is related to some element of **B**. The elements in **A** are 1 and 3, and we have **R**12 and **R**34. Since 2 and 4 are in **B**, the sentence in question is true under **I**. Similarly, we have **R**23, **R**41, **B** = {2, 4}, and **A** = {1, 3}, so $(\forall x)(Bx \to (\exists y)(Ay \land Rxy))$ is true. Now 5 is not in **A** nor in **B**, so no sentence in Δ requires that 5 be related to any element. Since 5 is not in any pair of **R**, there is no value of **a** that satisfies (**R**5a or **Ra**5). Hence, $(\exists y)(Ray \lor Rya)$ is false when **I**(a)=5, so ϕ is false under **I**. Thus, not $\Delta \models \phi$. ■

4.2.2   Tautologous Sentences in Predicate Calculus

A $\mathcal{PC}$ sentence, such as $(\forall x)(Fx \lor \neg Fx)$, may be valid (true under every $\mathcal{PC}$ interpretation), although it is neither an $\mathcal{SC}$ sentence nor a tautology. But

$$(\forall x)Fx \lor \neg(\forall x)Fx \tag{4.1}$$

is valid, and its validity depends only on its *truth functional structure*. It has the same *form* as the tautology $P \lor \neg P$ and is obtainable from $P \lor \neg P$ by substituting $(\forall x)Fx$ for P throughout this tautology. Under any interpretation, $(\forall x)Fx$ is either true or false, but in either case, Sentence (4.1) is true. Thus, some (but not all) valid $\mathcal{PC}$ sentences are valid because they have tautologous truth functional structures. We will make this observation precise.

Definition 4-8 Let σ be an $\mathcal{SC}$ sentence and ρ be a $\mathcal{PC}$ sentence. Suppose that $\lambda_1, \cdots, \lambda_n$ are $\mathcal{SC}$ letters in σ, and let $\phi_1, \cdots, \phi_n$ be $\mathcal{PC}$ sentences. Let

$$\sigma\left(\lambda_1/\phi_1, \cdots, \lambda_n/\phi_n\right)$$

be the result of replacing every occurrence of λ_i in σ with an occurrence of the corresponding ϕ_i. Then ρ is a $\mathcal{PC}$ *substitution instance* of σ if

$$\sigma\left(\lambda_1/\phi_1, \cdots, \lambda_n/\phi_n\right)$$

is the same sentence as ρ.

For example, suppose that σ is $(A \lor B) \to C$. Let

$$\phi_1 \text{ be } (\forall x)Fx,$$

$$\phi_2 \text{ be } \neg(\exists y)Gyy,$$

and

$$\phi_3 \text{ be } Pbd \to (\exists z)Gzd.$$

Then

$$\sigma(A/\phi_1, \ B/\phi_2, \ C/\phi_3)$$

is

$$((\forall x)Fx \lor \neg(\exists y)Gyy) \to (Pbd \to (\exists z)Gzd).$$

Definition 4-9 Let ρ be a $\mathcal{PC}$ sentence. Then ρ is *tautologous* iff there is an $\mathcal{SC}$ sentence σ such that ρ is a $\mathcal{PC}$ substitution instance of σ and σ is a tautology.

Examples. From the $\mathcal{SC}$ tautology $P \to P$, we obtain the tautologous $\mathcal{PC}$ sentence,

$$((\exists y)Ey \land M) \to ((\exists y)Ey \land M), \tag{4.2}$$

by substituting $((\exists y)Ey \wedge M)$ for P.

From $(P \wedge Q) \to Q$, we can obtain the next tautology by obvious substitutions,

$$(Wcd \wedge \neg(\forall x)(\exists z)Rxz) \to \neg(\forall x)(\exists z)Rxz. \qquad (4.3)$$

The sentence (4.3) is also a $\mathcal{PC}$ substitution instance of $(P \wedge \neg Q) \to \neg Q$, which is also a tautology. In addition, (4.3) can be obtained from $R \to S$ and from R, but these are not tautologies. ■

Some of the important facts can be summarized as follows:

- A valid $\mathcal{SC}$ sentence is an $\mathcal{SC}$ tautology.

- The $\mathcal{SC}$ tautologies are a proper subset of all tautologous $\mathcal{PC}$ sentences.

- Tautologous sentences are a subset of valid $\mathcal{PC}$ sentences.

- Some valid sentences are not tautologous.

   To show that a $\mathcal{PC}$ sentence ρ is tautologous, it is sufficient to exhibit *one* $\mathcal{SC}$ tautology σ of which ρ is a $\mathcal{PC}$ substitution instance. We will not need, and hence will not develop, a systematic procedure for proving that a $\mathcal{PC}$ sentence is *not* tautologous. Notice that no atomic sentence can be tautologous. Also, no universal or existential generalization is tautologous, although some are valid.

Examples.

Although valid, this sentence is not tautologous,

$$(\forall x)Px \to (\exists x)Px.$$

The only $\mathcal{SC}$ sentences of which this is a substitution instance are sentences of the same forms as A and $A \to B$, and none of these is tautologous. On the other hand, under any interpretation **I**, P is assigned to some subset **P** of the domain **D**, and either $\mathbf{P} = \mathbf{D}$ or else $\mathbf{P} \neq \mathbf{D}$. In the former case, the antecedent is true under **I**. Since $\mathbf{D} \neq \varnothing$ by the definition of *interpretation*, $\mathbf{P} \neq \varnothing$, so the consequent is true under **I**. In the case where $\mathbf{P} \neq \mathbf{D}$, the antecedent is false, so the sentence is true under **I**.

It is not difficult to verify that this sentence is contingent,

$$(\forall x)Px \vee (\exists x)Px,$$

and this one is unsatisfiable,

$$(\exists x)(Px \wedge \neg Px). \quad ■$$

4.2.3 Tautological Consequences in Predicate Calculus

I previously defined *tautological consequence* for $\mathcal{SC}$ sentences. This definition was stated directly in terms of $\mathcal{SC}$ interpretations, but the concept of $\mathcal{SC}$ interpretation does not apply to arbitrary sets of $\mathcal{PC}$ sentences. It would be convenient to have a new, natural extension of the idea of *tautological consequence* that applies to $\mathcal{PC}$ sentences. Because we already have a definition of what is a tautologous $\mathcal{PC}$ sentence, it is convenient to use the following:

Definition 4-10 Let θ be a $\mathcal{PC}$ sentence and let

$$\Gamma = \{\, \rho_1, \, \rho_2, \, \cdots, \, \rho_n \,\}$$

be a set of $\mathcal{PC}$ sentences. Then Γ *tautologically implies* θ (θ is a *tautological consequence* of Γ, $\Gamma \vDash_T \theta$) iff

$$(\cdots ((\rho_1 \wedge \rho_2) \wedge \rho_3) \cdots \wedge \rho_n) \to \theta$$

is tautologous. In addition, if θ is a tautologous $\mathcal{PC}$ sentence, we say that θ is a *tautological consequence* of $\varnothing$.

This definition is limited to finite sets of premises, but that is adequate for our purposes. Also, it can be proved that the exact grouping (by parentheses) of the sentences in a conjunction does not affect its truth value under an interpretation.* Hence, we can ignore the grouping of the ρ_i in the antecedent of the above sentence and write the following as an abbreviation of this sentence:

$$(\rho_1 \wedge \rho_2 \wedge \rho_3 \wedge \cdots \wedge \rho_n) \to \theta.$$

Example.

Let

$$\Gamma = \{\neg(\forall x)Px, \ (\forall x)Px \vee \neg(\exists y)\,Qy, \ Bac \to (\exists y)\,Qy\}.$$

Notice that

$$\{\neg A, \ A \vee \neg B, \ C \to B\} \vDash_T \neg C.$$

This is easy to see by performing the following inferences mentally: From $\neg A$ and $A \vee \neg B$, infer $\neg B$ by DS. From $\neg B$ and $C \to B$, one obtains $\neg C$ by MT.

---

*This fact is called the *general associativity of conjunction*. An analogous principle applies to disjunction and to other operations that are associative. See E 3-89 for a start on the proof.

Because of this tautological implication, it follows that

$$((\neg A \wedge (A \vee \neg B)) \wedge (C \to B)) \to \neg C$$

is a tautologous $\mathcal{SC}$ sentence. If we make the substitutions $(\forall x)Px$ for A, $(\exists y)Qy$ for B, and Bac for C, we obtain the $\mathcal{PC}$ tautology

$$((\neg(\forall x)Px \wedge ((\forall x)Px \vee \neg(\exists y)Qy)) \wedge (Bac \to (\exists y)Qy)) \to \neg Bac.$$

Hence, $\Gamma \vDash_T \neg Bac.$ ■

One can conveniently verify $\mathcal{PC}$ tautological inferences by applying the sentential calculus rules directly to premises and conclusion as was done in this example. Here is another example in a slightly different format.

Example.

Problem Statement. Show that $(\exists u)(\exists v)Euv$ is a tautological consequence of

$$\{ (\forall x)Px \to (\exists u)(\exists v)Euv, \ (\forall x)Px \vee \neg Eab, \ \neg(\exists u)(\exists v)Euv \to Eab \}.$$

Discussion. All we need to do is find a tautologically sound $\mathcal{SC}$ argument of which this argument is a substitution instance. Look at the sentences that do not contain $\mathcal{SC}$ connectives, namely, $(\exists u)(\exists v)Euv$, Eab, $(\forall x)Px$, and substitute them into corresponding $\mathcal{SC}$ sentences. The written solution is easy.

Written Solution.

Consider the $\mathcal{SC}$ argument:

1. $A \to B$.

2. $A \vee \neg C$.

3. $\neg B \to C$.

4. Therefore, B.

By Comm and CDis, 2 is equivalent to $C \to A$. The latter, together with 1, yields $C \to B$, using HS. Applying HS to 3 and $C \to B$ yields $\neg B \to B$. By the $\mathcal{SC}$ Clav Rule (or by CDis, DN, and Taut), the latter yields B. Therefore, sentences 1, 2, 3 tautologically imply B. (Remark. This result also could have been proved by using truth value assignments.)

Now make the following substitutions:

$$(\forall x)Px \text{ for } A, \quad (\exists u)(\exists v)Euv \text{ for } B, \quad Eab \text{ for } C.$$

The $\mathcal{SC}$ argument is transformed into

    i. $(\forall x)Px \rightarrow (\exists u)(\exists v)Euv.$

    ii. $(\forall x)Px \vee \neg Eab.$

    iii. $\neg(\exists u)(\exists v)Euv \rightarrow Eab.$

    iv. Therefore, $(\exists u)(\exists v)Euv.$

Hence, i, ii, and iii tautologically imply $(\exists u)(\exists v)Euv.$ ■

EXERCISES

General Instructions

Exercises for which answers are provided at the back of the book are marked with "Answer Provided."

When using $\mathcal{PC}$ interpretations, be sure that the domain is specified, as well as the values or meanings of all relevant sentential letters, individual constants, and predicates. Give informal reasons that justify the truth values that the interpretation imposes on the $\mathcal{PC}$ sentences that are of concern.

E 4-7 Take the blocks world described in the example in Section 4.2.1 and add to the domain one additional object, p, the floor. A block may now be on another block, or it may be on the floor, p. Thus, Oxy is now interpreted as block x is on y, where y may be either a block or the floor. Suppose that the following are true in this blocks world:

1. There is a blue block on the floor with a yellow block on this blue one and a green block on this yellow one.

2. There is a blue block on the floor with a yellow block on this blue one and a blue one on this yellow one.

3. No yellow block is clear.

4. There is a clear blue block and a clear green block.

5. For any yellow block, there is a block on this yellow one.

6. Not all blue blocks are clear.

7. Any green block is clear.

8. All blocks on the floor are blue.

9. Any block that is on a yellow one is either green or blue.

Symbolize these sentences. Draw a sketch of the arrangement of the floor and the six blocks in this world. Show the colors of the blocks.

E 4-8 Let B, G, Y, C, O, p have the same meanings as in the previous exercise. Symbolize the following sentences and construct a blocks world interpretation that satisfies them. Use as many blocks as you need, but try not to use more than are necessary. Name each of the blocks in your interpretation with an individual constant, say what colors they are, and describe their arrangements with respect to each other and the floor.

1. There is a green block on a green block.

2. For every blue block, there is a yellow one on this blue one.

3. There is a clear yellow block on the floor.

4. There is a blue block on the floor with a yellow block on it.

5. There is a green block on a yellow one.

6. There is a green block on the floor.

7. There is a yellow block such that it is on a blue one and a blue one is on it.

8. Every block is on something. (Remark. The only individual objects in the domain are the blocks [of types B, Y, G] and the floor.)

E 4-9 Let

$$\Gamma = \{ \, (\forall x)(\exists y)Rxy \, \},$$

$$\Delta = \{ \, (\forall x)\neg\, Rxx \, \},$$

$$\Lambda = \{ \, (\forall x)(\exists y)Ryx \, \},$$

$$\Sigma = \{ \, (\forall x)(\forall y)(\forall z)(\,(Rxy \wedge Ryz) \rightarrow Rxz) \, \}.$$

Also, let the following be two $\mathcal{PC}$ interpretations:

$\mathbf{I}_1 : \mathbf{D}_1 = Nat$ (the domain), Rxy is $x < y$.

$\mathbf{I}_2 : \mathbf{D}_2 = \{\Diamond,\ \heartsuit,\ \bowtie\}$, $R = \{\, <\Diamond,\ \heartsuit >,\ <\heartsuit,\ \bowtie >,\ <\bowtie,\ \Diamond >\,\}$.

Answer each of the following questions, and briefly justify your answers.

1. Does $\mathbf{I}_1$: satisfy Γ? satisfy Δ? satisfy Λ? satisfy Σ?

2. Does $\mathbf{I}_2$: satisfy Γ? satisfy Δ? satisfy Λ? satisfy Σ?

3. Is $\Gamma \cup \Delta \cup \Lambda \cup \Sigma$ satisfiable?

E 4-10 Give a counterexample to this assertion: Let ϕ, ψ be $\mathcal{PC}$ formulas, each with free occurrences of the variable ν. Then

$$((\exists \nu)\phi \wedge (\exists \nu)\psi) \rightarrow (\exists \nu)(\phi \wedge \psi)$$

is valid. You should exhibit specific $\mathcal{PC}$ formulas for ϕ and ψ, and also describe a specific interpretation, and use these together to show that the assertion is false.

E 4-11 (Answers Provided for 1 and 3) For each of the following sets of sentences, give a suitable interpretation to show that the set is satisfiable. Informally justify the truth values of the sentences under your interpretation.

1. $\{Ab,\ Bb,\ Cb,\ (\exists x)(Ax \wedge \neg(Bx \vee Cx)),\ (\forall x)(Cx \rightarrow Ax)\}$

2. $\{(\exists y)(By \wedge Cy),\ \neg(\forall z)(Cz \rightarrow Az),\ (\forall x)((Bx \wedge Cx) \rightarrow Ax)\}$

3. $\{((\exists x)Ax \wedge (\exists x)Bx),\ \neg(\exists x)(Ax \wedge Bx),\ (\forall x)\neg Rxx,\ (\forall x)(\forall y)(Rxy \rightarrow Ryx),$
   $(\forall x)(Ax \rightarrow (\exists y)(By \wedge Rxy)),\ (\forall x)(Bx \rightarrow (\exists y)(By \wedge Rxy))\}$

4. $\{(\exists x)(Ax \wedge \neg Bx),\ (\exists x)(Bx \wedge \neg Ax),\ (\forall x)(Ax \rightarrow (\exists y)(By \wedge Rxy)),$
   $(\forall x)(Bx \rightarrow (\exists y)(Ay \wedge Rxy)),\ (\exists x)(Ax \wedge Bx)\}$

5. $\{(\exists x)Ax,\ (\forall x)Rxx,\ (\forall x)(Ax \rightarrow \neg Bx),\ (\forall x)(Ax \rightarrow (\exists y)(By \wedge Ryx))\ \}$

6. $\{(\forall u)(\forall v)(\forall w)((Ruv \wedge Rvw) \rightarrow Ruw),\ \neg(\exists x)(\exists y)(Rxy \wedge Ryx),$
   $(\forall x)(\exists u)Rxu\}$

7. $\{(\forall u)(\forall v)(\forall w)((Ruv \wedge Rvw) \rightarrow Ruw),\ \neg(\exists x)(\exists y)(Rxy \wedge Ryx)\}$

8. $\{(\forall x)(Ax \rightarrow \neg Bx),\ (\forall x)(Ax \rightarrow (\exists y)(By \wedge Lxy)),$
   $(\exists x)(Ax \wedge \neg(\exists y)(Ay \wedge Lxy)),\ (\forall x)(\forall y)(Lxy \rightarrow \neg Lyx)\}$

9. $\{(\forall x)(Ax \rightarrow Bx),\ (\forall u)(Ax \rightarrow (\exists y)(Cy \wedge Rxy)),\ (\exists x)(\neg Ax \wedge Bx),$
   $(\exists x)(Bx \wedge (\forall y)(Cy \rightarrow \neg Rxy)),\ (\forall x)(\forall y)(Rxy \rightarrow Ryx)\}$

E 4-12 Show that the following set of sentences is satisfiable:

$$\{\,\neg(\forall x)Bx,\ (\forall y)(Ay \rightarrow By),$$
$$(\exists x)(Bx \wedge Ax),\ (\exists x)(\exists y)((Bx \wedge \neg By) \wedge Rxy)\,\}.$$

E 4-13 Show that the following set of sentences is satisfiable:

$$\{\,(Aa \wedge Rab),\ (\forall x)(\forall y)((Ax \wedge Rxy) \rightarrow (By \wedge Cy)),$$
$$(\forall x)(Ax \rightarrow \neg Bx),\ (\exists z)(Cz \wedge Az)\,\}.$$

E 4-14 Show that the following set of sentences is satisfiable:

$$\{\,(\exists x)Ax,\ (\forall x)(\,Ax \rightarrow (\exists y)(By \wedge Rxy)),$$
$$(\forall x)(Bx \rightarrow (\exists y)(Cy \wedge Sxy)),\ (\exists x)(Cx \wedge (\forall z)(Bz \rightarrow \neg Szx))\,\}.$$

E 4-15 Use the general definition of an interpretation to show that the following are valid:

1. $(\forall x)(Fx \vee \neg Fx)$.

2. $(\forall x)((Ax \wedge Bx) \rightarrow Bx)$.

3. $(\exists y)(Ay \vee \neg Ay)$.

E 4-16 Use the general definition of an interpretation to prove that

$$\neg(\forall x)Hx$$

is equivalent to

$$(\exists x)\neg Hx.$$

Notice that, if H means *honest*, then

>   *It is not the case that everyone is honest.*

is equivalent to

>   *Someone is not honest.*

E 4-17 (Answer Provided) Use a counterexample interpretation to show that

$$(\forall x)(Ax \rightarrow (Bx \vee \neg Cx))$$

is not valid.

E 4-18 Use a counterexample interpretation to show that

$$(\forall y)(\forall z)((Gy \wedge Syz) \rightarrow Gz)$$

is not valid.

E 4-19 Use a counterexample interpretation to show that $Lab \rightarrow (\exists x)Lxx$ is not valid.

E 4-20 Show that $(\exists x)Px \rightarrow Pa$ is not valid, but that $(\exists x)((\exists x)Px \rightarrow Px)$ is valid.

E 4-21 Show that the following are tautologous:

1. $Pa \rightarrow (Pa \vee (\forall x)Gx)$.

2. $(\neg(\forall x)(Ax \wedge Bx) \rightarrow (\forall x)(Ax \wedge Bx)) \rightarrow (\forall x)(Ax \wedge Bx)$.

3. $\neg((\forall x)Rxx \rightarrow (\forall x)(\forall y)(Rxy \rightarrow Ryx)) \rightarrow (\forall x)Rxx$.

E 4-22 Show that

$$(\neg(\exists x)(\forall u)Axu \rightarrow (\exists x)(\forall u)Axu) \rightarrow (\exists x)(\forall u)Axu$$

is tautologous.

E 4-23 Prove that $(\forall x)Px \rightarrow Pe$ is valid. Also, show that it is not tautologous.

E 4-24 (Answer Provided for 1) Use counterexample interpretations to show that

1. $(\forall x)Ax$ is not a consequence of $(\exists y)Ay$.

2. Gb is not a consequence of $(\exists x)Gx$.

3. $(\exists x)\neg Bx$ is not a consequence of

$$\{(\forall x)(Ax \rightarrow Bx), (\exists x)\neg Ax\}.$$

4. Rab is not a consequence of $(\exists x)Rax$.

5. $(\exists x)Bx \rightarrow P$ is not a consequence of $Bc \rightarrow P$.

6. $(\exists x)(\forall y)Lxy$ is not a consequence of

$$\{(\forall x)(\forall y)(\forall z)((Lxy \wedge Lyz) \rightarrow Lxz)\}.$$

7. $(\forall x)\neg Bx$ is not a consequence of $\{(\forall x)(Ax \rightarrow Bx), (\exists x)Ax\}$.

8. $(S \land (\exists x)Bx)$ is not a consequence of

$$\{(\forall x)(Ax \rightarrow Bx), \ (Bc \rightarrow S), \ (\exists x)Ax, \ R \lor Ac\}.$$

9. $(\exists x)(Ax \land Bx) \rightarrow (\forall x)Bx$ is not a consequence of

$$\{(\exists x)Ax, \ (\forall x)(Ax \rightarrow (\exists y)(By \land Rxy))\}.$$

10. Lcb is not a consequence of

$$\{\ (\forall x)(Rxa \rightarrow Lxb), \ (\forall x)\neg Rxx, \ (\exists x)Rxa\ \}.$$

11. $(\forall x)(\forall y)(Rxy \lor Ryx)$ is not a consequence of

$$\{\ \neg(\exists x)(Ax \land Bx), \ (\forall x)(\exists y)Rxy, \ (\exists x)(\exists y)(\ (Ax \land By) \land Rxy)\ \}.$$

E 4-25 Let

$$\Gamma = \{\ (\forall x)(\forall y)(Rxy \rightarrow \neg Ryx),$$

$$(\forall x)(\forall y)(\forall z)(\ (Rxy \land Ryz) \rightarrow \neg Rxz), \ (\forall x)(\exists y)Rxy\ \}$$

and let ϕ be $(\forall x)(\forall y)(Rxy \lor Ryx)$. Construct a $\mathcal{PC}$ interpretation that shows that Γ does not imply ϕ.

E 4-26 Let

$$\Gamma = \{\ Rab, \ (\forall x)(\forall y)(Rxy \rightarrow Ryx),$$
$$(\forall x)(\forall y)(\forall z)(\ (Rxy \land Ryz) \rightarrow Rxz)\ \},$$

and let ρ be $(\forall x)Rxx$.

1. Show that $\Gamma \cup \{\neg\rho\}$ is satisfiable.

2. Show that not $\Gamma \models \rho$.

E 4-27 Show that the following set of sentences is satisfiable:

$$\{\ Aa \land \neg Ab, \ (\forall x)Lax, \ (\exists x)Lxx \rightarrow (\forall y)Ryb,$$
$$(\exists z)(\exists u)Rzu \rightarrow \neg(\exists x)Sx\ \}.$$

E 4-28 Show that the following set of sentences is satisfiable:

$$\{\ (\forall x)(\forall y)(Rxy \rightarrow \neg Ryx), \ (\forall x)(\forall y)(\forall z)((Rxy \land Ryz) \rightarrow Rxz),$$
$$Rab, \ (\forall x)(\forall y)(Rxy \rightarrow (\exists z)(Rxz \land Rzy))\ \}.$$

E 4-29 Show that it is not the case that $\neg(\forall x)Mx \models (\forall x)\neg Mx$.

E 4-30 Is the set

$$\{\,(\forall z)\neg Bz,\ (\forall y)(Ay \rightarrow Cy),\ Ab,\ (\forall x)(Cx \rightarrow (\exists y)(By \wedge Rxy))\,\}$$

satisfiable? Either exhibit an interpretation that satisfies it, or else show that no such interpretation exists.

E 4-31 Show that $(\forall x)(Bx \rightarrow (\exists y)(Cy \wedge Rxy))$ is not a consequence of

$$\{\,(\forall x)(Ax \rightarrow Bx),\ (\forall x)(Ax \rightarrow (\exists y)(Cy \wedge Rxy))\,\}.$$

E 4-32 Let $\mathbf{R}$ be a binary relation on a domain $\mathbf{D}$. Express in $\mathcal{PC}$ the following conditions on $\mathbf{R}$:

1. $\mathbf{R}$ is transitive on $\mathbf{D}$.

2. $\mathbf{R}$ is strongly connected on $\mathbf{D}$.

3. $\mathbf{R}$ is symmetric on $\mathbf{D}$.

Now exhibit a $\mathcal{PC}$ interpretation that shows that 1 and 2 do not imply 3.

E 4-33 Show that $(\forall x)(\exists y)Rxy \rightarrow (\exists y)(\forall x)Rxy$ is not valid.

E 4-34 Sketch an $\mathcal{SC}$ argument (i.e., cite the rule[s] you use in order of their use) to show that

$$Pa \rightarrow (\exists x)Fx$$

is a tautological consequence of

$$\neg(Pa \wedge \neg(\exists x)Fx).$$

E 4-35 Sketch an $\mathcal{SC}$ argument (i.e., cite the rule[s] you use in order of their use) to show that

$$(\forall y)Ryy$$

is a tautological consequence of

$$\{\,(\forall z)Gz \rightarrow ((\forall y)Ryy \vee Tab),\ (\forall z)Gz,\ \neg Tab\,\}.$$

E 4-36 Consider the following: From

$$\Gamma = \{\, Ma \to \neg(\exists y)Fby, \ Ma \,\}$$

we may infer

$$\neg(\exists y)Fby$$

from which we may infer

$$\phi = (\forall y)\neg Fby.$$

Suppose that someone claims that this line of reasoning establishes that ϕ is a *tautological consequence* of Γ. Why is this claim **mistaken**?

4.3   Predicate Calculus Derivations

4.3.1   Derivation Rules

$\mathcal{PC}$ derivations (or proofs) are a generalization of $\mathcal{SC}$ derivations. The derivation rules we will use are standard; many sources use one or more of these rules, often in variant forms. The format of a $\mathcal{PC}$ derivation is exactly the same as that used for $\mathcal{SC}$ derivations, except that the sentences on the lines of the proof may now be $\mathcal{PC}$ sentences. The terms *derivation, premise, premise name, line number, derivation of ϕ from Γ*, etc., are used just as they were with the sentential calculus, except that they now apply to any predicate calculus sentences or derivations. If Γ is a set of $\mathcal{PC}$ sentences, and there is a derivation of sentence ϕ from Γ using the rules given below, we write $\Gamma \vdash \phi$. If $\varnothing \vdash \phi$, we say that ϕ is a *$\mathcal{PC}$ theorem* and also may write $\vdash \phi$. A set of derivation rules for predicate calculus is *sound* iff for any $\mathcal{PC}$ sentence ϕ and set of sentences Γ, if ϕ is derivable from Γ, then ϕ is a consequence of Γ. A set of rules is *complete* iff, if ϕ is a consequence of Γ, then ϕ is derivable from Γ by these rules. The concepts of *soundness* and *completeness* for predicate calculus are direct generalizations of the corresponding concepts for sentential calculus, except that we are now concerned with $\mathcal{PC}$ sentences, derivations, and consequences. Soundness will be proved in Section 4.3.3.

   In predicate calculus, we will use Rule P *and* Rule C *exactly as stated in Chapter 1, except that they may be applied to any $\mathcal{PC}$ sentences.* We add a Rule TC (next) that permits the inference of any tautological consequence of previous lines, including tautologies from $\varnothing$. Rules C, P, and TC do not permit any inferences that essentially depend on the roles of quantifiers in sentences, so we will need a few additional rules that pertain to the logical roles of the quantifiers.

RULE TC ($\mathcal{PC}$ Tautological Consequences). Let ϕ be a $\mathcal{PC}$ sentence and Λ (possibly empty) be a set of lines of a $\mathcal{PC}$ derivation. If the sentences on the lines Λ tautologically imply ϕ, then ϕ may be placed on a new line. The set of premise names of this new line is the union of the sets of premise names of Λ.

An application of Rule TC should be documented at least by

$$\text{TC } (i), (j), \cdots,$$

where (i), (j), $\cdots$, are the line numbers of Λ. In some cases, it is helpful to provide more specific information about the tautological inferences used. If $\Lambda = \varnothing$, then no existing lines are used, so ϕ must be a tautologous $\mathcal{PC}$ sentence (see Metatheorem 1-7). At this point, we have three $\mathcal{PC}$ derivation rules: P, C, and TC. We still need rules for inferences involving quantifiers.

Rule QE (Quantifier Exchange). If a sentence ρ, with a form specified in Table 4.1, is on a line of a proof, then the corresponding sentence θ, on the same row of the table, may be placed on a new line. Similarly, ρ may be placed on a new line if the corresponding θ is already on a line of the proof. The premise names of the new line are the same as those of the earlier line to which this rule is applied.

ρ	θ
$\neg(\exists\nu)\phi$	$(\forall\nu)\neg\phi$
$(\exists\nu)\neg\phi$	$\neg(\forall\nu)\phi$
$\neg(\exists\nu)\neg\phi$	$(\forall\nu)\phi$
$(\exists\nu)\phi$	$\neg(\forall\nu)\neg\phi$

Table 4.1. *Quantifier Exchange Rule*

Here is an easy way to remember Rule QE: Let $\mathbb{Q}$ denote either quantifier (universal or existential) and let $\overline{\mathbb{Q}}$ be the other (existential or universal) quantifier, respectively. Then $\neg(\mathbb{Q}\nu)\phi$ is equivalent to $(\overline{\mathbb{Q}}\nu)\neg\phi$, and $\neg(\mathbb{Q}\nu)\neg\phi$ is equivalent to $(\overline{\mathbb{Q}}\nu)\phi$. Always remember that Rule QE (as stated here) requires that the sentences in question (ρ and θ) be the *entire sentences* on lines of a derivation; this QE rule does *not* apply to proper subformulas or subsentences.

For example, by QE, we may infer

$$\neg(\exists x)(Ax \to Bx)$$

from

$$(\forall x)\neg(Ax \to Bx),$$

and vice versa. However, our QE rule does *not* permit going from

$$(\neg(\exists x)Ax \lor Ba)$$

to

$$((\forall x)\neg Ax \lor Ba) \quad \textbf{rule use error}$$

since, in this error example, the quantifiers do not apply to the entire sentences in question. In other words, QE can be used when the quantifiers are the last operations added in the recursive construction of the sentences.

It is true that these last two sentences are equivalent, but additional work is required to prove this. The inconvenience exhibited here is the result of the fact that we use a very simple set of $\mathcal{PC}$ rules. This simplicity makes it easier for beginners to learn the rules, and it has some theoretical advantages. The resulting inconveniences can be alleviated later by stating and proving some shortcut rules. How this can be done is partially, and briefly, indicated in the last section of this chapter. We should now move on to the next derivation rule.

RULE UI (Universal Instantiation). Let ν be a variable and κ be an individual constant. If a sentence of the form $(\forall\nu)\phi$ is on a line of a proof, then $\phi\,[^\nu/_\kappa]$ may be placed on a new line with the same set of premise names as $(\forall\nu)\phi$.

Here are some brief illustrations. If we have the line,

$\{\ Pr_1,\ Pr_2,\ Pr_4\ \}$ (7) $(\forall x)(\exists y)((Ax \land By) \to Rxy)$ [comment]

we may add the new line,

$\{\ Pr_1,\ Pr_2,\ Pr_4\ \}$ (19) $(\exists y)((Ab \land By) \to Rby)$ UI (7).

If we have the line,

$\{\ Pr_2,\ Pr_3\ \}$ (5) $(\forall x)(((\forall x)Bx \land Fx) \to (\forall y)Gyx)$ [comment]

we may add the new line,

$\{ Pr_2, Pr_3 \}$ (11) $(((\forall x)Bx \land Fa) \to (\forall y)Gya)$ UI (5).

We can now give a derivation for the Classical Argument:

$\{ Pr_1 \}$	(1) $(\forall x)(Hx \to Mx)$	P,
$\{ Pr_2 \}$	(2) Ha	P,
$\{ Pr_1 \}$	(3) $Ha \to Ma$	UI (1),
$\{ Pr_1, Pr_2 \}$	(4) Ma	TC (2), (3). ■

The basic idea of UI is simple: If a predicate ϕ is true for all x (i.e., all objects) in the domain of an interpretation, then ϕ must also be true for any *specific* object a in that domain. Applying the predicate ϕ to a specific object denoted by a is also called *instantiating* ϕ on a. This rule is sometimes called *universal specification* because the predicate ϕ is being applied to a specific object a.

It is important that the input sentence for the UI rule have the exact form, $(\forall \nu)\phi$. This means that prefixing $(\forall \nu)$ to ϕ was the last operation performed in the recursive construction of $(\forall \nu)\phi$. Thus, it would be *wrong* to try to apply UI to $((\forall x)Mx \to Rb)$, because it does not have the form, $(\forall \nu)\phi$. It is a conditional and has the form, $\phi \to \psi$. Similarly, it would be *wrong* to try to apply UI to the negation $\neg(\forall x)Rx$. It is a *serious mistake* to apply UI (or any other rule) to a sentence that does not exactly fit the form specified in the rule.

RULE EG (Existential Generalization). Let ν be a variable, κ be a constant, and ϕ be a $\mathcal{PC}$ formula. If $\phi[\nu/\kappa]$ is on a line of a proof, then $(\exists \nu)\phi$ may be placed on a new line with the same set of premise names as $\phi[\nu/\kappa]$.

Examples.

Using EG, we may make the following inferences:

From Raa, infer $(\exists x)Rxa$. Notice that Raa results from replacing all free occurrences of x in Rxa with occurrences of a.

From Raa, infer $(\exists y)Ryy$. Here, Raa results from substituting a for all free occurrences of y in Ryy.

From Rab, infer $(\exists y)Ray$. We may then go farther and infer $(\exists x)(\exists y)Rxy$ from

$(\exists y)Ray$. Notice that $(\exists y)Ray$ results from $(\exists y)Rxy$ by substituting a for all free occurrences of x. ■

The EG Rule is very natural: If a predicate ϕ is true of some particular object, then there exists something of which ϕ is true. In predicate calculus, this inference takes the general form, given $\phi[^{\nu}/_{\kappa}]$, infer $(\exists \nu)\phi$. The constant κ denotes some particular object in the domain of whatever interpretation is under consideration. Although EG is one of the simplest of the quantifier rules, it is occasionally misused.

Error Examples. The following inferences are erroneous:

$(\exists y)Fby$

Therefore, $(\exists y)(\exists y)Fyy$. **rule use error**

In this case, $(\exists y)Fby$ is *not* the result of substituting b for all free occurrences of y in $(\exists y)Fyy$ since the latter has no free occurrences. This argument not only violates the EG Rule; it is also fallacious. Let the domain of the interpretation **I** be the set of racing cars in Europe and b be Bob's racing car. Let Fxy hold iff y can go faster than x. The premise says that some car goes faster than b. From the definition of $\mathbf{V_I}$ for existential generalizations, it follows that the conclusion of the argument is true under **I** iff $(\exists y)(\exists y)Fyy$ is. But the latter is equivalent to $(\exists y)Fyy$, which says that some car goes faster than itself, which is false. We have a counterexample. Many other counterexamples exist; try $<$ on the real numbers.

$Dg \rightarrow A$

Therefore, $(\exists x)Dx \rightarrow A$. **rule use error**

This is a bad mistake: $(\exists x)Dx \rightarrow A$ is not even an existential generalization; it is a conditional. This argument is also fallacious. Let $\mathbf{D} = \{0, 1\}$, and use the assignments, D: $\{0\}$, g: 1 , and A: **F**. Then $Dg \rightarrow A$ is **F** $\rightarrow$ **F**, which is true; $(\exists x)Dx \rightarrow A$ is **T** $\rightarrow$ **F**, which is false.

Here is an ordinary language argument with the preceding form; it is intuitively unsound:

If Gerard dances (Dg, false), then everyone laughs (A, false).

Therefore,

if something dances ($(\exists x)Dx$, true), then everyone laughs (A). **error**

This example should be compared with the Existential Antecedent (EA) Rule, which will be introduced shortly. ∎

EG states that, if ϕ is true of some particular thing κ, then there exists something of which ϕ is true. From the premise that Keith is in Antarctica, it follows that somebody is in Antarctica. But it does not follow that everybody is in Antarctica. From the *mere fact* that ϕ is true of some particular thing κ, we may not infer that ϕ is true of everything. This would be a grossly unwarranted generalization. Yet under special conditions, a generalization to a universally quantified sentence is justified. The following rule states these special conditions.

RULE UG (Universal Generalization). Let ν be a variable, κ be a constant, and ϕ be a formula that has no occurrences of κ. Suppose that $\phi\left[{}^{\nu}\!/_{\kappa}\right]$ is on a line of a proof and that no premise of this line has an occurrence of κ. Then $(\forall\nu)\phi$ may be placed on a new line with the same set of premise names as $\phi\left[{}^{\nu}\!/_{\kappa}\right]$.

We have used UG many times in informal proofs; here is an example from set theory.

Theorem 4-11   For any set S, $S \subseteq S$.

Proof. Let S_0 be any set. (S_0 corresponds to κ.) Let e be any element in the domain of discourse. (e corresponds to another arbitrary constant, say, γ.) Then, if $e \in S_0$, then $e \in S_0$. Although not usually stated, from this we conclude that for any x, if $x \in S_0$, then $x \in S_0$. (We just applied UG with respect to the arbitrary element e, corresponding to γ.) Thus, from the definition of $\subseteq$, we obtain $S_0 \subseteq S_0$. But we started with the assumption that S_0 is a set, so by informal use of $\mathcal{SC}$ Rule C, we obtain

$$\text{If } S_0 \text{ is a set, then } S_0 \subseteq S_0.$$

Since S_0 is an arbitrary constant corresponding to κ, we apply UG again to get 'For any S, if S is a set, then $S \subseteq S$', which can be shortened to 'For any set S, $S \subseteq S$'. *Q.E.D.*

In this proof, since S_0 and e are *arbitrary*, we may *generalize* upon them. *Arbitrary* means here that the conclusion,

$$\text{if } e \in S_0, \text{ then } e \in S_0,$$

does not depend on what element e may be. Also,

$$\text{if } S_0 \text{ is a set, then } S_0 \subseteq S_0,$$

does not depend on what set S_0 may be. Another way of saying this is that any other constants could have been used in place of e and S_0.

The justification of Rule UG is based on the Generalization Theorem, which is given in Section 4.3.3. A little reflection can convince one that κ must not occur in any premise of $\phi\,[{}^\nu/_\kappa]$; otherwise, κ would not be arbitrary, and the conclusion $\phi\,[{}^\nu/_\kappa]$ might depend on which element of the domain κ denotes. A further illustration of this point can be given as follows:

Consider the Classical Argument: From

$$\Gamma = \{(\forall x)(Hx \rightarrow Mx),\ Ha\},$$

we may derive Ma. We have $\Gamma \vdash Ma$. But we may *not* use UG to generalize to the further inference of $(\forall x)Mx$. Just because all humans are mortal and Socrates is human, it does not follow that everything is mortal. Technically, the use of UG is blocked here because a occurs in a premise of Γ. The constant a is not arbitrary; for instance, we cannot derive Mg_{123} from Γ.

Contrast the Classical Argument with this: Let

$$\Delta = \{(\forall x)(Hx \rightarrow Mx), (\forall x)Hx\}.$$

By using UI twice, we obtain $Ha \rightarrow Ma$ and Ha. By MP we get Ma, so $\Delta \vdash Ma$. This time a occurs in no sentence of Δ nor in Mx, so we may apply UG to get the further result: $\Delta \vdash (\forall x)Mx$. In this example, a is arbitrary; we could have used any other constant, say, g_{123}, in place of a and derived Mg_{123} from Δ.

The constant κ should not occur in the final generalization, since this generalization should not depend on κ. For example, $(Pa \rightarrow Pa)$ is tautologous and does not depend on a. Yet $(\forall x)(Px \rightarrow Pa)$ does not follow, as one can show with a simple counterexample interpretation.* On the other hand, we may use UG to infer $(\forall x)(Px \rightarrow Px)$ from this tautology. To summarize: We can infer $(\forall \nu)\phi$ from $\phi\,[{}^\nu/_\kappa]$ *provided that*

(i) κ does not occur in ϕ, and

(ii) κ does not occur in any premise of $\phi\,[{}^\nu/_\kappa]$.

The conditions (i) and (ii) are called the *provisos* of the rule. When applying a rule with provisos, one must carefully check that they are satisfied.

---

*Hint: Try using the natural numbers for the domain and assigning P to the set of prime numbers.

The previous rules permit us to instantiate a universal and to generalize from an instance to an existential statement. Under special conditions, we can even generalize from an (arbitrary) instance to a universal statement. Can we instantiate an existential? The answer is clearly, 'No.' From a sentence like 'There exists a murderer', it does not deductively follow that 'John is a murderer', 'Ann is a murderer', 'Lizzie is a murderer', or that any specifically named person is a murderer. Nevertheless, it is often useful to introduce a *premise* that corresponds to an instance of an existential statement.

Suppose that Captain Nellie, who is an honest and informed space explorer, tells me that there are Saturnians, although she does not know any by name. In other words, *There exists x such that x is a Saturnian* is true, even though neither Nellie nor I know the name of any Saturnian.

If $(\exists x)Sx$ is true under an interpretation **I**, it does *not* follow that Sa, Sb, or $\cdots$, is true under **I**. We *may not* instantiate an existential generalization, but we can do something else. Consider this:

1. All Saturnians are logicians.

2. There is a Saturnian.

3. *Therefore*, there is a logician.

Intuitively, this argument is deductively sound, and it can be made more plausible as follows: Since there is a Saturnian, let us just make up a name for him (or her). Suppose that we decide to call him Sam. 'All Saturnians are logicians' means 'For any x, if x is a Saturnian, then x is a logician.' By UI, if Sam is a Saturnian, then Sam is a logician. Since we are assuming that Sam is a Saturnian, he is therefore a logician. Since Sam is a logician, there exists a logician. This pattern of reasoning is quite common in mathematics. To use an existential premise, one invents a *dummy name* that is assumed to stand for something that has the property asserted in the existential premise. The proof then proceeds with this new assumption. The end of the proof must not use this dummy name. The next rule may at first look complicated and awkward, but it will enable us to make use of dummy names and formalize this Saturnian argument.

RULE EA (Existential Antecedent). Let ϕ be a $\mathcal{PC}$ formula, σ a $\mathcal{PC}$ sentence, ν a variable, and κ a constant. Suppose that

$$\phi\left[{}^{\nu}\!/_{\kappa}\right] \to \sigma$$

is on a line of a derivation. Provided that κ does not occur in $\phi \to \sigma$ nor in any premise of the line with $\phi\left[\nu/\kappa\right] \to \sigma$, we may place

$$(\exists\nu)\phi \to \sigma$$

on a new line with the same set of premise names as those for $\phi\left[\nu/\kappa\right] \to \sigma$.

This rule will be justified in Section 4.3.3, along with others. For now, we will assume that it is sound. This is how it is typically used: We want to make use of a sentence of the form $(\exists\nu)\phi$, which is on a line of a derivation. We *assume* $\phi\left[\nu/\kappa\right]$ as a new premise on line (k) by Rule P, where κ is a constant that has not been previously used in the derivation. Line (k) has the premise name Pr_k. Suppose that, with the help of $\phi\left[\nu/\kappa\right]$, we derive σ, and suppose that the line with σ has premise names Γ and that κ does not occur in σ. We use Rule C to get $\phi\left[\nu/\kappa\right] \to \sigma$, and we delete Pr_k from Γ. Since κ did not occur in the derivation before line (k), if we have been careful not to introduce it again after line (k), then the provisos of Rule EA should be satisfied. We can therefore infer $(\exists\nu)\phi \to \sigma$. Using $(\exists\nu)\phi$ and Rule MP, we obtain σ. Proceeding along these lines will be called the *existential exploitation strategy* (ExEx Strategy). Choosing κ to be a 'new constant' that has not previously occurred in the proof is part of the strategy. This is not required by the EA Rule, but using a new constant can make it easier to satisfy the provisos of EA. Using the ExEx strategy, we can formulate the argument about the logical Saturnians as follows.*

PC Derivation Example.

$\{\ Pr_1\ \}$	(1) $(\forall x)(Sx \to Lx)$	P
$\{\ Pr_2\ \}$	(2) $(\exists x)Sx$	P
$\{\ Pr_3\ \}$	(3) Sa	P (for ExEx)
$\{\ Pr_1\ \}$	(4) $Sa \to La$	UI (1)
$\{\ Pr_1, Pr_3\ \}$	(5) La	TC (3), (4)
$\{\ Pr_1, Pr_3\ \}$	(6) $(\exists x)Lx$	EG (5)

---

*Warning: Some logic books introduce an "existential instantiation" or "existential exploitation" rule which *appears* to warrant instantiating an existential generalization. Such rules are not sound; from true premises (under an interpretation), they permit the inference of a false conclusion. The structure of proofs using them must be restricted in certain, rather awkward ways. These restricted proofs end up working in a manner similar to our ExEx strategy. All of our rules are sound, as is proved in Section 4.3.3. This has the great advantage, and added elegance, that any sentence on any line of a derivation is a consequence of the premises of that line.

$\{\,Pr_1\,\}$	(7) $Sa \to (\exists x)Lx$	C (3), (6)
$\{\,Pr_1\,\}$	(8) $(\exists x)Sx \to (\exists x)Lx$	EA (7)
$\{\,Pr_1,\,Pr_2\,\}$	(9) $(\exists x)Lx$	TC (2), (8) ∎

The EA Rule completes our set of rules for $\mathcal{PC}$ derivations. From now on, when we write $\Gamma \vdash \phi$, this asserts that there is a $\mathcal{PC}$ derivation of ϕ from Γ *using the rules summarized in* Table 4.2. Also, $\Gamma \vdash \phi$ is used to assert that ϕ is a $\mathcal{PC}$ *theorem*, i.e., it can be derived from the empty set of premises.

RULE NAME	ABBREVIATION
Introduction of Premises	P
Conditionalization	C
Tautological Consequence	TC
Quantifier Exchange	QE
Universal Instantiation	UI
Existential Generalization	EG
Universal Generalization	UG
Existential Antecedent	EA

Table 4.2. *$\mathcal{PC}$ Derivation Rules*

4.3.2 Proof Strategies and Examples

Section 1.6.2 lists several strategies for constructing $\mathcal{SC}$ derivations and for exploiting various forms of $\mathcal{SC}$ sentences. Because predicate calculus is a richer and more powerful language, one might expect even more strategies and techniques to apply to it. This is true in one sense, simply because sentential calculus is contained as a sublanguage of predicate calculus, so strategies for the latter include all of those of the former. Yet the predicate calculus does not require the addition of very many *new* strategies beyond those already available in sentential calculus. The following discussion assumes familiarity with all of the techniques and strategies described in Section 1.6.2. To construct proofs effectively, one must master tautological inferences (combined into Rule TC here). This holds for both formal and informal proofs. Of course, the very general strategy of *goal reduction* applies to predicate calculus problems as well as almost any other kind of problem.

Most of the $\mathcal{SC}$ rules of thumb were classified into exploitation techniques and goal-directed strategies. As will be seen, the derivation techniques pertaining to quantifiers can be used in different ways, and these techniques do not seem to fit nicely into the two main classifications used in sentential calculus. A little reflection on Table 4.2 reveals that the eight rules naturally fall into three sets: Rules P, C, and TC have their origin in sentential calculus. Rule P permits adding new premises; C permits removing a premise under certain conditions. TC encompasses all tautological inferences other than those by P and C. All of the sentential calculus strategies and techniques still apply to these three rules. The quantifier rules UI, EG, UG, and EA all pertain in some way to adding or removing quantifiers. EG and UG permit adding quantifiers under certain conditions. UI permits removing a universal quantifier. As already explained in Section 4.3.1, we cannot instantiate an existential generalization. But using EA in the ExEx strategy (explained earlier) enables us to work temporarily with some existential quantifiers removed. Finally, there is Rule QE, which helps to transform combinations of negations and quantifiers. The overall interaction of the $\mathcal{SC}$ rules and the quantifier rules leads to what I call the grand strategy.

The Grand Strategy. Typical $\mathcal{PC}$ derivations use three techniques:

- remove quantifiers,

- use tautological inferences by Rule TC,

- put quantifiers back on.

As we shall see, some proofs use all three of these techniques, and some use only one or two. These techniques may be combined with Conditional Proof (CP) and *reductio ad absurdum* (RAA) strategies, which are general $\mathcal{SC}$ strategies. The QE Rule is used to help implement the grand strategy in various ways that will be described.

Removing Universal Quantifiers. These are removed by Rule UI. For example, from $(\forall x)(\forall y)Pxy$, we can obtain $(\forall y)Pay$ by one application of UI, then obtain Pab by a second application. This is very easy, since any constants may be instantiated. For this reason, UI is one of the most flexible and powerful rules, but it does not tell us *what* constants to instantiate. Since there are no restrictions on these constants, it is often wise not to use UI until other steps of the proof have been made. At certain points in a proof, it often becomes clear that particular instantiations are needed. Then apply UI to get them. On the other hand, in many proofs it does not matter what constants are instantiated; in these cases, one may begin with UI. Here is a simple example:

{ Pr_1 }	(1) $(\forall x)Px$	P
{ Pr_1 }	(2) Pa	UI (1)
...	... ...	... ...
{ Pr_1 }	(7) $(\exists y)Py$	EG (2)

In this proof fragment, any constant would have worked as well as a.

> *A good rule of thumb is this: If all of the premises are universal generalizations, begin the proof with UI. If there are other types of premises, try to use them first, then use UI later when instantiations with particular constants are needed.*

When multiple applications of UI are used, sometimes different constants are instantiated with each use, and sometimes the same constant is used more than once. By two applications of UI, from $(\forall x)(\forall y)Rxy$ we may derive Rab. We may also derive Raa. The choice depends on exactly what one wants to obtain.

Removing Existential Quantifiers. We cannot instantiate an existential quantifier, so they cannot be directly removed. Instead, we obtain a kind of pseudo- or virtual removal by using the ExEx strategy. We add new premises that *temporarily play the role of* instantiated existentials. This strategy has already been described and exemplified (with the logical Saturnians derivation) in the previous section in connection with Rule EA. Needless to say, the extra premises that are added must not appear at the end of the derivation. If they are used at all, they can be removed only by using Rule C.

Using Tautological Inferences. Occasionally, an entire predicate calculus proof can be done with Rule TC. A trivial example is

{ Pr_1 }	(1) $(\forall x)Px \to (\forall y)Qy$	P
{ Pr_2 }	(2) $(\forall x)Px$	P
{ Pr_1, Pr_2 }	(3) $(\forall y)Qy$	TC (1), (2)

It is rare when Rule TC will do the entire job alone. Typically, one begins with some quantified premises and wants to end up with a conclusion containing quantifiers. It is usually necessary to remove most, or all, of the quantifiers on or in the premises, and then use TC to rearrange sentences with respect to the sentential connectives. One then often uses EG or UG to add new quantifiers back on, but this last step is not always required.

{ Pr_1 }	(1) $(\forall x)(Px \rightarrow (\forall x)(Qx \vee \neg Rx))$	P
{ Pr_2 }	(2) Pa	P
{ Pr_3 }	(3) Rc	P
{ Pr_1 }	(4) $Pa \rightarrow (\forall x)(Qx \vee \neg Rx)$	UI (1)
{ Pr_1, Pr_2 }	(5) $(\forall x)(Qx \vee \neg Rx)$	TC (2), (4)
{ Pr_1, Pr_2 }	(6) $Qc \vee \neg Rc$	UI (5)
{ Pr_1, Pr_2, Pr_3 }	(7) Qc	TC (3), (6)

Notice that the second quantifier in line (1) could not be removed until it appeared in a line of the form $(\forall \nu)\phi$, namely, line (5). Line (1) was instantiated with the constant a so that the antecedent of (4), Pa, would match line (2). Line (5) was instantiated with c so that (6) could be used together with (3).

This process of instantiating to constants that then enable pattern matches for tautological inferences is very important. Computer programs that perform logical deductions use a special procedure (the *Unification Algorithm*) for doing this. Among other applications, this algorithm is used in predicate calculus *resolution derivations*. Most texts on artificial intelligence programming in LISP or Prolog include discussions of substitution and unification procedures; see, for instance, Abelson and Sussman (1985), Tanimoto (1987), and Sterling and Shapiro (1986). A nice introduction to these topics is in the first few chapters of Genesereth and Nilsson (1987). Manna and Waldinger (1990) include an extended theoretical treatment of substitution and unification. An advanced survey of the problems and methods of computer deduction systems is in Robinson (1992). There is still much to learn about how to get computers to construct elegant proofs efficiently.

Adding Universal Quantifiers. Very often, the desired sentence (the goal sentence) is universally quantified, say, $(\forall \nu)\phi$. In rare cases, this may be derivable without using UG (for instance, by using TC, as in one of the previous examples). Usually, one first derives $\phi[^{\nu}/_{\kappa}]$ and then generalizes with UG. It is important that the UG provisos are satisfied: κ must not occur in ϕ nor in any premise of the line with $\phi[^{\nu}/_{\kappa}]$. There are some frequently encountered kinds of situations in which the second proviso is satisfied. One of these types of situations consists of arguments with universally quantified premises.

{ Pr_1 }	(1)	$(\forall x)(\forall y)(Cxy \leftrightarrow Ryx)$	P
{ Pr_2 }	(2)	$(\forall x)(\forall y)(Rxy \rightarrow \neg Ryx)$	P
{ Pr_1 }	(3)	$(\forall y)(Cay \leftrightarrow Rya)$	UI (1)
{ Pr_1 }	(4)	$(Cab \leftrightarrow Rba)$	UI (3)
{ Pr_2 }	(5)	$(\forall y)(Ray \rightarrow \neg Rya)$	UI (2)
{ Pr_2 }	(6)	$(Rab \rightarrow \neg Rba)$	UI (5)
{ Pr_1 }	(7)	$(\forall y)(Cby \leftrightarrow Ryb)$	UI (1)
{ Pr_1 }	(8)	$(Cba \leftrightarrow Rab)$	UI (7)
{ Pr_1 , Pr_2 }	(9)	$(Cba \rightarrow \neg Rba)$	TC (8), (6)
{ Pr_1, Pr_2 }	(10)	$(Cab \rightarrow \neg Cba)$	TC (4), (9)
{ Pr_1, Pr_2 }	(11)	$(\forall y)(Cay \rightarrow \neg Cya)$	UG (10)
{ Pr_1, Pr_2 }	(12)	$(\forall x)(\forall y)(Cxy \rightarrow \neg Cyx)$	UG (11)

In this derivation, the constants a, b are arbitrary; they do not occur in the premises. Any other two constants would have worked just as well. The UG provisos are satisfied in (10) and (11). This is a very typical proof. In fact, we have used such proofs informally for some time. According to the definitions in Section 2.2.2, premise (1) states that C is the converse of R, and premise (2) says that R is asymmetric. The conclusion, (12), says that C is asymmetric. This example uses predicate calculus to give a formal proof that the converse of an asymmetric relation is also asymmetric.

Informally, we might have argued like this: Let R be asymmetric on its domain A, and let $x, y \in A$. Suppose that $R^{-1}xy$. (The formal derivation uses C for R^{-1}.) Then Ryx, by the definition of converse. Since R is asymmetric, we obtain not Rxy. Therefore, not R^{-1} yx. Hence, for any x and y,

$$\text{if } R^{-1}xy, \text{ then not } R^{-1}yx.$$

In the informal proof, the variables x, y play the roles of the arbitrary constants in the formal proof. The fact that these variables are arbitrary is indicated by 'let $x, y \in A$'. The universal quantifiers are not even stated until the end, and some informal proofs omit them there. They are *implied*, however, by the fact that the variables are arbitrary. It should be clear that both proofs have essentially the same general structure: use UI to remove quantifiers, use TC to rearrange some things, then use UG to put the quantifiers back on. The TC part of the informal proof used CP; I avoided CP in the formal derivation. This is a minor difference; the formal proof could have used CP, and the informal proof could have used the tautological inferences used in the formal proof.

In the previous derivation, the constants a, b were arbitrary because they resulted from use of UI. Sometimes one can obtain arbitrary constants in a different way.

$\{ Pr_1 \}$	(1) $A \to Pd$	P
$\{ Pr_2 \}$	(2) A	P
$\{ Pr_1, Pr_2 \}$	(3) Pd	TC (1), (2)
$\{ Pr_1, Pr_2 \}$	(4) $Pd \vee Qd$	TC (3)
$\{ Pr_2 \}$	(5) $(A \to Pd) \to (Pd \vee Qd)$	C (1), (4)
$\{ Pr_2 \}$	(6) $(\forall z)((A \to Pz) \to (Pz \vee Qz))$	UG (5)

I say that the constant d in line (5) is arbitrary because it does not occur in any premise of that line. Of course, the reason it does not occur in any premise of line (5) is because the use of Rule C allowed us to drop Pr_1 from the premise names of line (4) when inferring line (5). In order to apply UG to line (5), d also must not occur in line (6), and it does not. Here is another situation in which there is an arbitrary constant.

| $\varnothing$ | (1) $Tb \vee \neg Tb$ | TC (a tautology) |
| $\varnothing$ | (2) $(\forall x)(Tx \vee \neg Tx)$ | UG (1) |

Adding Existential Quantifiers. Occasionally the goal sentence is existentially quantified, say, $(\exists \nu)\phi$. This may occur when the constant κ in

$$\phi \left[\nu/\kappa \right] \to \sigma$$

is not an arbitrary constant. This often happens when one is using an existentially quantified premise together with the ExEx Strategy, and κ results from the added premise. The constant a in the derivation about the logical Saturnians is an example of this situation. By using EG, we not only add a quantifier, but also get rid of this constant to form the sentence σ that is later used in an application of Rule EA to complete the ExEx Strategy. Here is another example.

In the next derivation, the problem is to derive (13) from (1) and (2). I use the ExEx Strategy. Since line (2) contains two existential quantifiers, two new constants, b, c, are used in the new premise (3). These constants later appear in line (7); they are removed by two applications of EG to obtain the sentence on (9), which plays the role of σ in the subsequent applications of EA. Note that Conditionalization produces (10), and then Rule EA is used twice.

$\{\,Pr_1\,\}$	(1)	$(\forall x)(\forall y)(\forall z)(Bxyz \leftrightarrow$	
		$\qquad (Oyx \wedge Ozy))$	P
$\{\,Pr_2\,\}$	(2)	$(\exists y)(\exists x)(Oxa \wedge Oyx)$	P
$\{\,Pr_3\,\}$	(3)	$(Oca \wedge Obc)$	P (for ExEx)
$\{\,Pr_1\,\}$	(4)	$(\forall y)(\forall z)(Bayz \leftrightarrow (Oya \wedge Ozy))$	UI (1)
$\{\,Pr_1\,\}$	(5)	$(\forall z)(Bacz \leftrightarrow (Oca \wedge Ozc))$	UI (4)
$\{\,Pr_1\,\}$	(6)	$(Bacb \leftrightarrow (Oca \wedge Obc))$	UI (5)
$\{\,Pr_1, Pr_3\,\}$	(7)	$Bacb$	TC (3), (6)
$\{\,Pr_1, Pr_3\,\}$	(8)	$(\exists u)Baub$	EG (7)
$\{\,Pr_1, Pr_3\,\}$	(9)	$(\exists v)(\exists u)Bauv$	EG (8)
$\{\,Pr_1\,\}$	(10)	$(Oca \wedge Obc) \rightarrow (\exists v)(\exists u)Bauv$	C (3), (9)
$\{\,Pr_1\,\}$	(11)	$(\exists x)(Oxa \wedge Obx) \rightarrow (\exists v)(\exists u)Bauv$	EA (10)
$\{\,Pr_1\,\}$	(12)	$(\exists y)(\exists x)(Oxa \wedge Oyx) \rightarrow$	
		$\qquad (\exists v)(\exists u)Bauv$	EA (11)
$\{\,Pr_1, Pr_2\,\}$	(13)	$(\exists v)(\exists u)Bauv$	MP (2), (12)

Using Quantifier Exchanges. Use Rule QE to help apply the techniques described above. For example, a negation in front of a quantified sentence can be moved inside the outermost quantifier to obtain a sentence starting with the opposite kind of quantifier. Sometimes the opposite action is performed if the resulting negated sentence can be used in a tautological inference. This is infrequent. If one wants to prove a quantified goal by RAA, this goal must be negated and used as a new premise. QE is often useful in such situations.

Example.

Problem Statement. Give a $\mathcal{PC}$ derivation of $(\forall x)Hx$ from the premises:

$$\{Pb,\ Rba \leftrightarrow (\forall x)Sx,\ (\forall z)(Sz \rightarrow Hz),\ (\exists x)Px \rightarrow (\forall y)Rya\}.$$

Discussion. The Grand Strategy for $\mathcal{PC}$ proofs is to remove quantifiers, use rule TC, then put quantifiers back on as needed. Of course, this strategy, like all strategies, can be used in combination with other strategies, including use within subproofs. We can remove the universal quantifier from the third premise by using UI. If we did that in the beginning of a proof, we would not know which constant, a, b, c, d, e, etc., to instantiate. In general, it is best not to use UI until it is clear that a certain constant (or constants) need to be instantiated, or it is clear that any constant can be used in the rest of the proof. Now let us consider the goal of the proof: to derive $(\forall x)Hx$. This sentence is not a subsentence of any of the premises. In fact, the only occurrence of H in the premises is in the third premise. This suggests that we try to obtain a sentence of the form $H\kappa$ for some constant κ that occurs in no premise of $H\kappa$. If we can do this, then we could use UG to obtain the goal sentence $(\forall x)Hx$. In order to obtain $H\kappa$, we clearly need to instantiate the third premise with the constant κ. If we could also obtain $S\kappa$, then we would be able to use MP to get $H\kappa$. The only obvious route to obtain $S\kappa$ is by using the second premise together with Rba. We can get the latter from the first and last premises. It emerges that the derivation will make some use of the Grand Strategy, along with other strategies. One of these other strategies is to look at the goal sentence and try to think of various ways it might be obtained.

Written Solution.

$\{Pr_1\}$	(1)	Pb	P
$\{Pr_2\}$	(2)	$Rba \leftrightarrow (\forall x)Sx$	P
$\{Pr_3\}$	(3)	$(\forall z)(Sz \rightarrow Hz)$	P
$\{Pr_4\}$	(4)	$(\exists x)Px \rightarrow (\forall y)Rya$	P
$\{Pr_1\}$	(5)	$(\exists x)Px$	EG (1)
$\{Pr_1, Pr_4\}$	(6)	$(\forall y)Rya$	TC (4), (5)
$\{Pr_1, Pr_4\}$	(7)	Rba	UI (6)
$\{Pr_1, Pr_2, Pr_4\}$	(8)	$(\forall x)Sx$	TC (2), (7)
$\{Pr_1, Pr_2, Pr_4\}$	(9)	Sc	UI (8)
$\{Pr_3\}$	(10)	$Sc \rightarrow Hc$	UI (3)
$\{Pr_1, Pr_2, Pr_3, Pr_4\}$	(11)	Hc	TC (9), (10)
$\{Pr_1, Pr_2, Pr_3, Pr_4\}$	(12)	$(\forall x)Hx$	UG (11)

(Remark. (11) can be generalized to (12) because the constant c does not occur in any premise of (11). Any constant could have been put into lines (9) and (10). However, I did not use either a or b, because they occur in premises of (11), and if (11) had contained a or b, then the use of UG in (12) would have been blocked. This illustrates that one should always be looking ahead toward later steps in a proof.) ■

Example.

Problem Statement. Write a $\mathcal{PC}$ derivation of

$$(\forall x)(\forall y)(((Ax \vee Cx) \wedge Rxy) \rightarrow ((Bx \vee Dx) \wedge Rxy))$$

from the set of premises $\{ (\forall x)(Ax \rightarrow Bx), \ (\forall x)(Cx \rightarrow Dx) \}$.

Discussion. We have two universally quantified premises and only those two premises. This suggests applying the Grand Strategy and instantiating them to $Aa \rightarrow Ba$ and $Ca \rightarrow Da$. The goal sentence, with the quantifiers removed, is a conditional with a disjunction conjoined with Rxy in the antecedent and a disjunction conjoined with Rxy in the consequent. The standard $\mathcal{PC}$ strategy for proving conditionals is the method of Conditional Proof, CP. Therefore, I shall try using CP within the Grand Strategy. Finally, the sentences $Aa \rightarrow Ba$ and $Ca \rightarrow Da$ seem to be related to the conditional form inside the goal sentence. This relationship should remind one of the Constructive Dilemma (CD) rule of sentential calculus. All of these observations lead to the next proof. Notice that the constants a and b do not occur in any premises of line (9), so it is permissible to generalize (apply UG to) each of them.

Written Solution.

$\{Pr_1\}$	(1)	$(\forall x)(Ax \rightarrow Bx)$	P
$\{Pr_2\}$	(2)	$(\forall x)(Cx \rightarrow Dx)$	P
$\{Pr_1\}$	(3)	$Aa \rightarrow Ba$	UI (1)
$\{Pr_2\}$	(4)	$Ca \rightarrow Da$	UI (2)
$\{Pr_5\}$	(5)	$(Aa \vee Ca) \wedge Rab$	P (for CP)
$\{Pr_5\}$	(6)	$(Aa \vee Ca)$	TC (5)
$\{Pr_1, Pr_2, Pr_5\}$	(7)	$(Ba \vee Da)$	TC (3), (4), (6)
$\{Pr_1, Pr_2, Pr_5\}$	(8)	$(Ba \vee Da) \wedge Rab$	TC (5), (7)
$\{Pr_1, Pr_2\}$	(9)	$((Aa \vee Ca) \wedge Rab) \rightarrow$ $((Ba \vee Da) \wedge Rab)$	C (5), (8)
$\{Pr_1, Pr_2\}$	(10)	$(\forall y)(((Aa \vee Ca) \wedge Ray) \rightarrow$ $((Ba \vee Da) \wedge Ray))$	UG(9)
$\{Pr_1, Pr_2\}$	(11)	$(\forall x)(\forall y)(((Ax \vee Cx) \wedge Rxy) \rightarrow$ $((Bx \vee Dx) \wedge Rxy))$	UG (10) ∎

Example.

Problem Statement. $\mathcal{PC}$ *Theorem.* $(\forall x)(\exists y)Lxy \leftrightarrow \neg(\exists x)(\forall y)\neg Lxy$. Write a $\mathcal{PC}$ derivation.

Discussion. Let ϕ denote the sentence to be proved. By definition, ϕ is a $\mathcal{PC}$ theorem iff it is derivable from the empty set of premises. Because our $\mathcal{PC}$ rules are sound (Metatheorems 4-19 and 4-20, proved in the next section), if ϕ is a theorem, then it is also valid. Whereas ϕ is a biconditional, if it is valid, then any interpretation that satisfies one side of this biconditional must also satisfy the other side. In that case, each side implies the other side, so the two sides are logically equivalent. In the discussion of Part 4 of an earlier example in this chapter, I promised to prove that $\neg(\exists x)(\forall y)\neg Lxy$ and $(\forall x)(\exists y)Lxy$ are logically equivalent. The present discussion shows that, if we prove that ϕ is a $\mathcal{PC}$ theorem, then it follows that $\neg(\exists x)(\forall y)\neg Lxy$ and $(\forall x)(\exists y)Lxy$ are logically equivalent. Of course, one could also prove the logical equivalence directly by considering interpretations. Before proceeding to the next paragraph, one should sketch graphs of binary relations satisfying each side of ϕ. These graphs should make it intuitively obvious that the two sides are logically equivalent. This is true for all binary relations satisfying the two sides of ϕ, not just ordering relations, as in the earlier example.

Because the task is to prove a $\mathcal{PC}$ theorem, one must derive the sentence ϕ from the empty set of premises. A standard strategy for proving biconditionals is to derive each side from the other using the CP strategy in each direction; I shall do that. First, assume $(\forall x)(\exists y)Lxy$ and try to derive $\neg(\exists x)(\forall y)\neg Lxy$. We can instantiate $(\forall x)(\exists y)Lxy$ to $(\exists y)Lay$. I then add the ExEx premise, Lab. The latter is equivalent to $\neg\neg Lab$. By EG, we obtain $(\exists y)\neg\neg Lay$, which, by QE, is equivalent to $\neg(\forall y)\neg Lay$. At this point it should be clear how to complete the ExEx and CP strategies. The proof that the right side of ϕ implies the left side is simpler.

Written Solution.

$\{Pr_1\}$	(1)	$(\forall x)(\exists y)Lxy$	P (for CP)
$\{Pr_1\}$	(2)	$(\exists y)Lay$	UI (1)
$\{Pr_3\}$	(3)	Lab	P (ExEx)
$\{Pr_3\}$	(4)	$\neg\neg Lab$	TC (3)
$\{Pr_3\}$	(5)	$(\exists y)\neg\neg Lay$	EG (4)
$\{Pr_3\}$	(6)	$\neg(\forall y)\neg Lay$	QE (5)
$\varnothing$	(7)	$Lab \rightarrow \neg(\forall y)\neg Lay$	C (3), (6)
$\varnothing$	(8)	$(\exists y)Lay \rightarrow \neg(\forall y)\neg Lay$	EA (7)
$\{Pr_1\}$	(9)	$\neg(\forall y)\neg Lay$	TC (2), (8)
$\{Pr_1\}$	(10)	$(\forall x)\neg(\forall y)\neg Lxy$	UG (9)
$\{Pr_1\}$	(11)	$\neg(\exists x)(\forall y)\neg Lxy$	QE (10)
$\varnothing$	(12)	$(\forall x)(\exists y)Lxy \rightarrow \neg(\exists x)(\forall y)\neg Lxy$	C (1), (11)
$\{Pr_{13}\}$	(13)	$\neg(\exists x)(\forall y)\neg Lxy$	P (for CP)
$\{Pr_{13}\}$	(14)	$(\forall x)\neg(\forall y)\neg Lxy$	QE (13)
$\{Pr_{13}\}$	(15)	$\neg(\forall y)\neg Lay$	UI (14)
$\{Pr_{13}\}$	(16)	$(\exists y)Lay$	QE (15)
$\{Pr_{13}\}$	(17)	$(\forall x)(\exists y)Lxy$	UG (16)
$\varnothing$	(18)	$\neg(\exists x)(\forall y)\neg Lxy \rightarrow (\forall x)(\exists y)Lxy$	C (13), (17)
$\varnothing$	(19)	$(\forall x)(\exists y)Lxy \leftrightarrow \neg(\exists x)(\forall y)\neg Lxy$	TC (12), (18) ■

With a little practice, these strategies and techniques will become familiar and $\mathcal{PC}$ derivations fairly easy. Unfortunately, the derivations can be tedious. The important facts are that these proofs can be formalized and that informal proofs use the same rules and strategies. When constructing proofs, it sometimes helps to be familiar with some standard $\mathcal{PC}$ theorems which, by soundness, are valid sentences. Here are a few that are especially interesting and useful. Their proofs are left for practice. Soundness is proved in the next section. Remember, however, that our rules do *not* allow direct application of these theorems. If one is trying to derive, within our system of rules, a sentence similar to one of these theorems, it is still necessary to construct a derivation using only those rules.

Selected Predicate Calculus Theorems

1. $(\exists x)(Px \rightarrow A) \leftrightarrow ((\forall x)Px \rightarrow A)$

2. $(\forall x)(Px \rightarrow A) \leftrightarrow ((\exists x)Px \rightarrow A)$

3. $(\forall x)(Px \wedge A) \leftrightarrow ((\forall x)Px \wedge A)$

4. $(\forall x)(Px \vee A) \leftrightarrow ((\forall x)Px \vee A)$

5. $(\exists x)(Px \wedge A) \leftrightarrow ((\exists x)Px \wedge A)$

6. $(\exists x)(Px \vee A) \leftrightarrow ((\exists x)Px \vee A)$

7. $(\exists x)(Px \vee Qx) \leftrightarrow ((\exists x)Px \vee (\exists x)Qx)$

8. $(\forall x)(Px \wedge Qx) \leftrightarrow ((\forall x)Px \wedge (\forall x)Qx)$

9. $(\exists x)(\forall y)Rxy \rightarrow (\forall u)(\exists v)Rvu$

4.3.3 Adequacy of the Predicate Calculus Rules

A large number of interesting metatheorems can be proved for $\mathcal{PC}$. As with sentential calculus, some of these metatheorems are used to help justify rules for derivations. The metatheorems in this section are especially important. A few of the proofs use mathematical induction and presuppose familiarity with Chapter 3. Some readers may wish to skip these inductive proofs. The other proofs should be easier to understand.

In the rest of this section, let ν be a variable, κ a constant, ϕ a $\mathcal{PC}$ formula, $(\forall \nu)\phi$ a sentence, and $\phi[^\nu/_\kappa]$ the result of replacing all free occurrences of ν in ϕ by occurrences of κ. At the end of Chapter 3, I defined a size function, *siz*, for $\mathcal{SC}$ sentences. This function is now redefined so that it applies to any $\mathcal{PC}$ formula.

Definition 4-12 Let ϕ be a $\mathcal{PC}$ formula. The *size* of ϕ is denoted by $siz(\phi)$ and defined by

If ϕ is atomic, then $siz(\phi) = 0$.

If ϕ has the form $\neg\psi$, $(\forall\nu)\psi$, or $(\exists\nu)\psi$, then $siz(\phi) = siz(\psi) + 1$.

If ϕ has one of the forms, $(\psi \vee \chi)$, $(\psi \wedge \chi)$, $(\psi \rightarrow \chi)$, or $(\psi \leftrightarrow \chi)$, then $siz(\phi) = siz(\psi) + siz(\chi) + 1$.

For a formula ϕ, this siz function returns the total number of occurrences of connectives and quantifiers in ϕ. The first metatheorem uses induction on the size of a sentence.

Metatheorem 4-13   Let ϕ, ψ be $\mathcal{PC}$ sentences. Suppose that ψ has one or more occurrences of constant κ and no occurrences of constant β. Suppose that ϕ is just like ψ except for having one or more occurrences of β where ψ has occurrences of κ, but not necessarily all occurrences of κ are replaced by β. Let **I**, **J** be any interpretations that are exactly alike except that **J** assigns the same domain element to β as it (and **I**) assigns to κ. Then the truth value of ϕ under **J** is the same as the truth value of ψ under **I**.

Proof. The proof is by CVI on $siz(\phi)$. The induction hypothesis is $\mathcal{I}(k)$: The statement of the theorem holds for all sentences ϕ of size less than k, where k is a positive integer.

Case 0. In this case, $siz(\phi) = 0$, so ϕ is atomic: It is a predicate followed by one or more constants, some of which are β. ψ is like ϕ except for having κ's in place of the β's in ϕ. Under their respective interpretations, they correspond to the same ordered k-tuple of domain elements, so they will have the same truth values.

Case k. Now we have $siz(\phi) = k$, and there are several subcases to consider, where ϕ

- has the form $\neg\phi_1$. Then ψ has the form $\neg\psi_1$, and ϕ_1 and ψ_1 are alike except for the stated relationships between occurrences of β and of κ. The size of ϕ_1 is less than that of ϕ, so the induction hypothesis applies to the subsentences. Hence, ϕ has the same value under **J** as ψ under **I**.

- has the form $(\forall\nu)\phi_1$. Then ψ has the form $(\forall\nu)\psi_1$ where ψ_1 is like ϕ_1 except for the occurrences of β and κ. Let η be a constant not in ϕ and not in ψ. Then ϕ is true under **J** iff $\phi_1\left[\nu/\eta\right]$ is true under every variant of **J** with respect to η, and similarly for ψ, $\psi_1\left[\nu/\eta\right]$, and variants of **I**.

The size of $\phi_1\left[{}^{\nu}/_{\eta}\right]$ is less than that of ϕ. Therefore, by induction, for any variant of **J** with respect to η, $\phi_1\left[{}^{\nu}/_{\eta}\right]$ has the same truth value as $\psi_1\left[{}^{\nu}/_{\eta}\right]$ under the corresponding variant of **I**. Hence, ϕ has the same truth value under **J** as ψ under **I**.

- has the form, $(\exists\nu)\phi_1$, and various other sentential forms. These remaining cases are proved by similar arguments and are left for practice.

$$\textbf{\textit{Q.E.D.}}$$

This result is now used to help prove the following:

Metatheorem 4-14 (**Instantiation Theorem**) Let Γ be a set of $\mathcal{PC}$ sentences.

$$\Gamma \vDash (\forall\nu)\phi \;\Rightarrow\; \Gamma \vDash \phi\left[{}^{\nu}/_{\kappa}\right].$$

Proof. Suppose that $\Gamma \vDash (\forall\nu)\phi$. Let **I** be any interpretation that satisfies Γ. Let β be a constant not in ϕ. Then $\phi\left[{}^{\nu}/_{\beta}\right]$ is true under every variant $\mathbf{I}_\beta$ of **I** with respect to β.

Now suppose for RAA that $\phi\left[{}^{\nu}/_{\kappa}\right]$ is false under **I**. Let $\mathbf{I}_\beta$ be the variant with $\mathbf{I}_{CONST}(\beta) = \mathbf{I}_{CONST}(\kappa)$. Then by the preceding theorem, $\phi\left[{}^{\nu}/_{\beta}\right]$ is false under $\mathbf{I}_\beta$. This contradicts the previous paragraph. **_Q.E.D._**

The Instantiation Theorem is used to justify Rule UI. The next theorem justifies Rule UG.

Metatheorem 4-15 (**Generalization Theorem**) Let $\phi\left[{}^{\nu}/_{\kappa}\right]$ be a sentence, and let Γ be a set of $\mathcal{PC}$ sentences. Provided that no sentences in Γ have occurrences of κ and also that κ does not occur in ϕ,

$$\Gamma \vDash \phi\left[{}^{\nu}/_{\kappa}\right] \;\Rightarrow\; \Gamma \vDash (\forall\nu)\phi.$$

Proof. Suppose that $\Gamma \vDash \phi\left[{}^{\nu}/_{\kappa}\right]$. For RAA, assume that it is not the case that $\Gamma \vDash (\forall\nu)\phi$. Then there is some interpretation **I** that satisfies Γ but not $(\forall\nu)\phi$. Since κ does not occur in ϕ, there is some variant $\mathbf{I}_\kappa$ of **I** with respect to κ such that $\phi\left[{}^{\nu}/_{\kappa}\right]$ is false under $\mathbf{I}_\kappa$. But also κ does not occur in any sentences of Γ. Since the only difference between **I** and $\mathbf{I}_\kappa$ is what they assign to κ, and since **I** satisfies Γ, therefore $\mathbf{I}_\kappa$ satisfies Γ. Thus, $\mathbf{I}_\kappa$ satisfies Γ but not $\phi\left[{}^{\nu}/_{\kappa}\right]$, which contradicts the hypothesis that $\Gamma \vDash \phi\left[{}^{\nu}/_{\kappa}\right]$. **_Q.E.D._**

It is easy to prove metatheorems corresponding to Rules QE and EG; they are stated without proofs.

Metatheorem 4-16 (**Quantifier Exchange Theorem**) Let Γ be a set of $\mathcal{PC}$ sentences, and let ϕ be a formula. Then

$$\Gamma \vDash \neg(\exists\nu)\phi \;\Leftrightarrow\; \Gamma \vDash (\forall\nu)\neg\phi.$$

Similar relationships hold for the pairs:

$(\exists\nu)\neg\phi$	$\neg(\forall\nu)\phi$
$\neg(\exists\nu)\neg\phi$	$(\forall\nu)\phi$
$(\exists\nu)\phi$	$\neg(\forall\nu)\neg\phi$

Proof. Left for practice.

Metatheorem 4-17 (**Existential Generalization Theorem**) Let Γ be a set of $\mathcal{PC}$ sentences and ϕ be a formula. Then

$$\Gamma \vDash \phi\,[^\nu/_\kappa] \;\Rightarrow\; \Gamma \vDash (\exists\nu)\phi.$$

Proof. Left for practice.

The only remaining quantifier rule is the EA Rule. It is what is called a *derived rule*; it is a direct consequence of the other rules. Any derivation using EA could be replaced by a derivation that does not use EA, so EA is redundant. It is useful because it typically leads to proofs that are shorter than the corresponding proofs that do not use it. For this reason, such derived rules are sometimes called *short-cut rules*. The following metatheorem is proved by considering the structure of derivations, rather than by semantical arguments. It justifies the use of EA, assuming that the other rules are justified. In addition, this form of proof shows that EA is a derived rule by demonstrating how any $\mathcal{PC}$ derivation using EA could be replaced by a derivation not using it. It is of interest that EG is also a derived rule; it can easily be derived from P, UI, C, TC, and QE. For practice, one should prove it both semantically (as in the previous metatheorem) and as a derived rule. Its proof as a derived rule is simpler than the following.

Metatheorem 4-18 (**Existential Antecedent Theorem**) Let ϕ be a formula, σ a sentence, and Γ a set of $\mathcal{PC}$ sentences. Suppose that

$$\Gamma \vdash \phi\,[^\nu/_\kappa] \to \sigma.$$

Provided that κ does not occur in $\phi \to \sigma$ nor in any sentence of Γ, then

$$\Gamma \vdash (\exists\nu)\phi \to \sigma.$$

Proof. Suppose that $\Gamma \vdash \phi[^\nu/_\kappa] \to \sigma$, i.e., there is a derivation of

$$\phi[^\nu/_\kappa] \to \sigma$$

from Γ. Suppose that $\phi[^\nu/_\kappa] \to \sigma$ is on line (k) of this derivation. Since premise names can be considered as abbreviations for the sentences they denote, we can let Γ be the set of premise names of line (k). The derivation can then be extended as shown in the following *proof schema*:

Γ	(k)	$\phi[^\nu/_\kappa] \to \sigma$	[comment]
$\{ Pr_{k+1} \}$	$(k{+}1)$	$\neg((\exists\nu)\phi \to \sigma)$	P (for RAA)
$\{ Pr_{k+1} \}$	$(k{+}2)$	$(\exists\nu)\phi$	TC $(k{+}1)$
$\{ Pr_{k+1} \}$	$(k{+}3)$	$\neg\sigma$	TC $(k{+}1)$
$\Gamma \cup \{ Pr_{k+1} \}$	$(k{+}4)$	$\neg\phi[\nu/\kappa]$	TC (k), $(k{+}3)$
$\Gamma \cup \{ Pr_{k+1} \}$	$(k{+}5)$	$(\forall\nu)\neg\phi$	UG $(k{+}4)$
$\Gamma \cup \{ Pr_{k+1} \}$	$(k{+}6)$	$\neg(\exists\nu)\phi$	QE $(k{+}5)$
$\Gamma \cup \{ Pr_{k+1} \}$	$(k{+}7)$	$(\exists\nu)\phi \to \sigma$	TC $(k{+}2)$, $(k{+}6)$
Γ	$(k{+}8)$	$\neg((\exists\nu)\phi \to \sigma) \to ((\exists\nu)\phi \to \sigma)$	C $(k{+}1)$, $(k{+}7)$
Γ	$(k{+}9)$	$(\exists\nu)\phi \to \sigma$	TC $(k{+}8)$

From this schema, it can be seen that invoking Rule EA is just shortcutting from line (k) to line $(k+9)$. Also, the provisos on Rule EA have their source in the provisos for Rule UG, which was used to obtain line $(k+5)$. The constant κ does not occur in

$$\phi \to \sigma,$$

so does not occur in line $(k+1)$. It also does not occur in any sentence of Γ.

Q.E.D.

The justification for Rule P was given in Section 1.6.1. Rule C was justified by Metatheorem 1-22. Rule TC is sound because it makes direct use of the tautological consequence relation. The previous metatheorems in this section justify the use of the remaining $\mathcal{PC}$ rules. Now that the rules are individually justified, their combined use can be justified by the $\mathcal{PC}$ Soundness Theorem, which is proved by course of values induction on the length of derivations. The *length* of a derivation is just the number of lines it has.

Metatheorem 4-19 (Soundness Theorem) Let ϕ be a $\mathcal{PC}$ sentence and Γ be a set of sentences. Then $\Gamma \vdash \phi \Rightarrow \Gamma \vDash \phi$.

Proof. We use course of values induction. Let the inductive predicate, $\mathcal{I}(k)$, state that, for all derivations of length k, $\Gamma \vdash \phi \Rightarrow \Gamma \vDash \phi$. The CVI induction hypothesis states that, for all $i < k$, $\mathcal{I}(i)$ is true.

Case $k = 1$. If a proof has only one line with the sentence ϕ, the only two ways ϕ can be obtained are by P using the set of premise names $\Gamma = \{Pr_1\}$, or by TC using $\Gamma = \varnothing$ (ϕ is a tautology). In either case, ϕ is a consequence of Γ, so $\mathcal{I}(1)$ is true, and all one-line proofs are sound.

Now assume the CVI hypothesis, which can be paraphrased: if ϕ is derivable from a set of premises by a proof of length $< k$, then ϕ is a consequence of these premises. Now suppose that ϕ is derivable from Γ in a proof of length k. By the definition of derivability from premises, this means that the premises of line (k) are all elements of Γ. Thus, ϕ is also derivable from exactly those premises of line (k). Therefore, we can assume, without loss of generality, that Γ is the set of premises on line (k). So, line (k) has premises Γ and sentence ϕ.

Case k. Suppose that ϕ is derivable from Γ in a proof of length k. Then line (k) has premises Γ and sentence ϕ.

If ϕ was obtained by P, then it is a consequence of $\{Pr_k\}$. If it was obtained by TC with $\Gamma = \varnothing$, then it is a tautology. In either case, as in a one line proof, ϕ is a consequence of Γ.

Otherwise, ϕ and Γ must have been obtained by applying one of the rules, C, TC, QE, UI, EG, UG, or EA, to earlier lines of the proof. By the induction hypothesis, each sentence on each of those earlier lines is a consequence of the premises of those earlier lines. Since each of these rules has been individually proved sound in previous metatheorems, it follows that ϕ is a consequence of the premises of line (k) that are determined by the rule that was used to obtain line (k). Hence, the set of premises, Γ, implies ϕ. **Q.E.D.**

The Soundness Theorem has an important corollary: every $\mathcal{PC}$ theorem is valid.

Metatheorem 4-20 Let ϕ be a $\mathcal{PC}$ sentence. If $\vdash \phi$, then ϕ is valid.

Proof. Suppose that $\vdash \phi$, so ϕ is $\mathcal{PC}$ derivable from the empty set. By the Soundness Theorem, $\varnothing \vDash \phi$. But (just as with sentential calculus) any interpretation satisfies $\varnothing$, so any interpretation must satisfy ϕ. Hence, ϕ is valid.

Q.E.D.

Although not proved here, our set of rules is also complete.

Metatheorem 4-21 (**Completeness Theorem**) Let ϕ be a $\mathcal{PC}$ sentence, and Γ be a set of sentences. Then $\Gamma \vDash \phi \Rightarrow \Gamma \vdash \phi$.

Proof. The proof is beyond the scope of this book. This theorem was first proved in 1930 by K. Gödel. It has since been proved by several different techniques. Proofs can be found in Mates (1972), in Mendelson (1987), and in most books on mathematical logic. These proofs depend on the particular set of rules under consideration, but essentially the same proof techniques can be adapted to any good system of rules.

The Completeness Theorem has a corollary that is the converse of Metatheorem 4-20: Every valid sentence is a theorem.

Metatheorem 4-22 Let ϕ be a $\mathcal{PC}$ sentence. If ϕ is valid, then $\vdash \phi$.

Proof. Suppose that ϕ is valid. Then it is true under every interpretation, so it is a consequence of $\varnothing$. Therefore, by the Completeness Theorem, ϕ is derivable from $\varnothing$. *Q.E.D.*

After the next set of exercises, the following section presents a small but typical application example; some additional metatheoretic considerations will be introduced for this example.

EXERCISES

General Instructions

Throughout this book, if an exercise is simply the statement of a theorem or a metatheorem, then the task is to prove this theorem or metatheorem. In your proof, you may use the definitions given in the text. Unless stated to the contrary, you may also use theorems proved in the text.

When using $\mathcal{PC}$ interpretations, be sure that the domain is specified, as well as the values or meanings of all relevant sentential letters, individual constants, and predicates. Give informal reasons that justify the truth values that the interpretation imposes on the $\mathcal{PC}$ sentences that are of concern.

Exercises for which answers are provided at the back of the book are marked with "Answer Provided."

Finally, please note that exercises that introduce important principles, useful applications, are especially difficult, or are special in some way, are indicated by bold print. In some cases, parenthetical comments describe the special features of the problem. If such comments are missing, the exercise is nonetheless of special value in introducing a new concept, principle, or technique.

E 4-37 Write a $\mathcal{PC}$ derivation of $\neg(\exists x)\neg Cx$ from the premises

$$\{\,(\forall z)(Cz \vee Mz),\ (\forall w)(Cw \vee \neg Mw)\,\}.$$

E 4-38 Write a $\mathcal{PC}$ derivation of $\neg(\exists x)(Ax \wedge Bx)$ from the premises

$$\{(\forall x)(\forall y)(\,(Ax \wedge By) \rightarrow Sxy),\ (\forall x)(Ax \rightarrow \neg Sxx)\}.$$

E 4-39 Write a $\mathcal{PC}$ derivation of

$$(\forall x)Fax \rightarrow (\forall x)(\exists y)Ryx$$

from the premise $(\forall y)((\exists x)Fxb \rightarrow Rby)$.

E 4-40 Give a $\mathcal{PC}$ derivation of $(\exists y)\neg Ryy$ from the premises

$$(\forall x)(Fx \wedge Gx) \text{ and } (\exists x)Rax \rightarrow \neg Fa.$$

E 4-41 Give a $\mathcal{PC}$ derivation of

$$\neg(\exists x)(\exists y)Sxy \rightarrow \neg(\exists x)(\exists y)Rxy$$

from the premise $(\forall x)((Pb \vee (\exists y)Rxy) \rightarrow (\exists y)Sxy)$.

E 4-42 Give a $\mathcal{PC}$ derivation of $(\forall x)(\forall y)(Cxy \vee Cyx)$ from the premises

$$(\forall x)(\forall y)(Rxy \vee Ryx) \text{ and } (\forall x)(\forall y)(Cxy \leftrightarrow Ryx).$$

E 4-43 (Answers Provided for 1 and 4) For each of the following arguments, give a $\mathcal{PC}$ derivation of the conclusion from the premise(s). The conclusion is indicated by $/\therefore$.

1. $(\forall x)(\forall y)(Rxy \rightarrow \neg Ryx)$
   $/\therefore\ (\forall x)\neg Rxx$
   (Hint: Think of the informal proof that an asymmetric relation is irreflexive.)

2. $(\forall x)(\forall y)Rxy$
   $/\therefore\ (\forall x)(\forall y)(Rxy \rightarrow Ryx)$
   (Hint: First derive Rba, then make an elegant use of Rule C.)

3. $(\forall x)(\forall y)((Dx \wedge My) \rightarrow Txy)$
   $/ \therefore (\forall z)((Dz \wedge Mz) \rightarrow Tzz)$
   (Compare with: Suppose that, for any x, y, if x is a demon and y is a monster, then x talks to y. Therefore: If anyone is both a demon and a monster, then he (or she) talks to himself.)

4. $(\forall x)(\forall y)((Gx \wedge Nxy) \rightarrow Ly)$
   $(\exists x)(Gx \wedge Nxa)$
   $/ \therefore (\exists z)Lz$

5. $(\forall x)((Cx \wedge \neg(\exists y)Pyx) \rightarrow Ox)$
   Ca
   $(\forall z)\neg Pza$
   $/ \therefore Oa$

6. $(\forall x)(\exists y)(Fx \wedge Gy)$
   $/ \therefore (\forall x)Fx$

7. $(\exists x)(Rax \wedge (Rxb \wedge Rba))$
   $(\forall x)(\forall y)(\forall z)((Rxy \wedge Ryz) \rightarrow Rxz)$
   $/ \therefore (\exists u)Ruu$
   (Hint: Sketch the graph of the relation R as determined by the premises. UI may be repeated as desired, instantiating to different constants.)

8. $(\forall x)(Fx \rightarrow \neg Gx)$
   $(\forall x)(\exists y)(Fx \wedge Gy)$
   $/ \therefore (\forall x)(Fx \wedge Gx)$
   (Hint: The premises are unsatisfiable.)

9. $(\forall x)(\forall y)(((Ax \wedge Ay) \wedge \neg Ixy) \rightarrow (Rxy \vee Ryx))$
   $(\forall x)(Bx \rightarrow Ax)$
   $/ \therefore (\forall x)(\forall y)(((Bx \wedge By) \wedge \neg Ixy) \rightarrow (Rxy \vee Ryx))$
   (Hint: Think of I as identity; then the first premise says that R is connected on A. The second premise says that B is a subset of A, and the conclusion says that R is connected on B. If $(Ba \wedge Bb) \wedge \neg Iab$ is assumed as an extra premise, one should be able to derive $Rab \vee Rba$. The rest is just conditionalization and universal generalization. An informal proof would use the same line of reasoning.)

428   Chapter 4 ■ Predicate Calculus

10. $(\forall x)(Px \rightarrow (\exists y)Rxy)$
    $/ \therefore (\forall x)(\exists y)(Px \rightarrow Rxy)$
    (Hint: Consider RAA; assume the negation of the conclusion.)

E 4-44 Write a $\mathcal{PC}$ derivation of $(\forall x)(Qc \rightarrow \neg Lxc)$ from the premises

$$\{(\exists x)Lxc \rightarrow Pc, \ (\forall x)(Px \rightarrow \neg Qx)\}.$$

E 4-45 Write a $\mathcal{PC}$ derivation of $(\forall t)(Bt \rightarrow \neg At)$ from the set of premises

$$\{(\forall x)((\exists z)Rxz \rightarrow \neg Ax), \ (\forall x)(Bx \rightarrow Rxa)\}.$$

E 4-46 Give a $\mathcal{PC}$ derivation of $(\exists x)Gxx$ from the premises

$$(\forall x)Lax, \ (\exists x)Lxx \rightarrow (\forall y)Ryb,$$
$$(\exists z)(\exists u)Rzu \rightarrow \neg(\exists u)Su, \ \neg Sc \rightarrow Gaa.$$

E 4-47 From the premises

$$(\forall x)(Rkx \rightarrow (Ax \rightarrow Bc)), \ (\exists x)Ax, \ (\forall x)Rkx,$$

give a $\mathcal{PC}$ proof of $(\exists x)Bx$.

E 4-48 Either show by an interpretation that Δ is satisfiable, or else derive a contradiction from Δ, where

$$\Delta = \{ (\exists x)Ax, \ (\forall x)(Ax \rightarrow (\exists y)(By \wedge Rxy)),$$
$$(\forall x)(\forall y)(Rxy \rightarrow Ay), \ \neg(\exists z)(Az \wedge Bz) \}.$$

E 4-49 Either prove by a $\mathcal{PC}$ derivation that

$$(\forall x)(\exists y)(Ax \rightarrow By) \leftrightarrow (\exists y)(\forall x)(Ax \rightarrow By)$$

is valid, or else give a counterexample interpretation.

E 4-50 (Illustrates a well-known form of argument involving relations.) Consider the argument: A lizard is a reptile. Therefore, a tail of a lizard is a tail of a reptile. This can be paraphrased: All lizards are reptiles. Therefore, any tail of a lizard is a tail of a reptile. Let Lx represent 'x is a lizard', Rx represent 'x is a reptile', and Txy represent 'x is a tail of y'. The sentence

$$(\exists y)(Ly \wedge Tcy)$$

means 'c is a tail of a lizard'. Symbolize the argument and derive the conclusion from the premise.

E 4-51 Write derivations of the two $\mathcal{PC}$ theorems

$$(\forall x)Ax \rightarrow (\exists x)Ax \text{ and } (\forall z)(Az \rightarrow (\exists y)Ay).$$

E 4-52 Prove the two $\mathcal{PC}$ theorems:

$$(\forall x)(Px \rightarrow (Px \lor Qx)) \text{ and } (\forall x)((Gx \rightarrow Hx) \rightarrow (\neg Gx \lor Hx)).$$

E 4-53 Prove the two $\mathcal{PC}$ theorems:

$$(\forall x)Ax \rightarrow (\forall y)Ay \text{ and } (\exists x)Ax \rightarrow (\exists y)Ay.$$

E 4-54

1. Prove that $(\exists x)(Px \rightarrow (\forall x)Px)$ is a theorem of predicate calculus. (Hint: Consider RAA.)

2. By considering possible interpretations, write a brief discussion of the difference in meanings between

$$(\exists x)(Px \rightarrow (\forall x)Px),$$

and

$$((\exists x)Px \rightarrow (\forall x)Px).$$

E 4-55 Exercise E 2-76 defined a kind of relational system called a *quasi order*, which is reflexive and transitive. In predicate calculus, we can write the assumptions, or axioms, about quasi orders as follows:

$$(\forall x)\,Qxx,$$

$$(\forall x)(\forall y)(\forall z)((Qxy \land Qyz) \rightarrow Qxz).$$

E 2-76 also added a *definition* of an *indifference relation*:

$$(\forall x)(\forall y)(Ixy \leftrightarrow (Qxy \land Qyx)).$$

Using these three sentences as premises, give three derivations to prove that I is an equivalence relation, i.e., prove in predicate calculus that I is reflexive, symmetric, and transitive. You will need to use predicate calculus definitions of these three terms. For instance, to state that I is reflexive, one writes $(\forall x)Ixx$. To avoid confusion, write three separate derivations. Notice which premises are actually *needed* in each of the derivations.

E 4-56 Either derive the conclusion from, or else show that it is not a consequence of, the premises of the following argument.

> A
> $(B \lor A) \to Pa$
> $Qb \lor \neg Pa$
> $/ \therefore \quad Qb \to (\exists x)(Cx \land Qx).$

E 4-57 By deriving a contradiction, show that the following set of sentences is unsatisfiable:

$$\Gamma = \{(\forall x)(\forall y)(\forall z)(Bxyz \leftrightarrow ((Oxy \land Ozx) \lor (Oxz \land Oyx))),$$

$$(\exists x)(\exists y)Bxyy, \ (\forall x)(\forall y)(Oxy \to \neg Oyx)\}.$$

E 4-58 Prove the predicate calculus theorems at the end of Section 4.3.2. (Remark. This entails a lot of work.)

E 4-59 Show that the converse of predicate calculus theorem number 9 (at the end of Section 4.3.2) is not valid.

E 4-60 Derive $(\exists x)((\exists x)Px \to Px)$ from $\varnothing$.

E 4-61 A person x is a *Great Joker* (Jx) iff, for any person y: x makes a fool of y iff y does not make a fool of himself (or herself). Let Fxy denote 'x makes a fool of y'. Then we can define x *is a Great Joker* by

$$(\forall x)(Jx \leftrightarrow (\forall y)(Fxy \leftrightarrow \neg Fyy)).$$

From this definition, prove that Great Jokers do not exist, i.e., derive $\neg(\exists x)Jx$. I suggest trying RAA.

E 4-62 (Answer Provided) Prove Metatheorem 4-16 about the quantifier exchanges.

E 4-63 Prove Metatheorem 4-17.

E 4-64 Prove the derivational counterpart of Metatheorem 4-17 by showing that the EG Rule is a derived rule. Use a proof schema (as in Metatheorem 4-18) that does not use EG to show that

$$\Gamma \vdash \phi\,[^{\nu}\!/_{\kappa}] \ \Rightarrow \ \Gamma \vdash (\exists \nu)\phi.$$

E 4-65 Let Γ be a set of $\mathcal{PC}$ sentences, such that no sentence in Γ has an occurrence of the constant a. Let σ be a sentence with no occurrences of a. Without using Rule EA, give a direct semantical proof of the following:

$$\text{If } \Gamma \vDash Fa \rightarrow \sigma,$$

$$\text{then } \Gamma \vDash (\exists x)Fx \rightarrow \sigma.$$

(Suggestions. Try RAA; assume that the antecedent of this conditional is true and the consequent is false. In order to have the latter, there must be an interpretation **I**, with domain **D**, that satisfies Γ, but makes $(\exists x)Fx \rightarrow \sigma$ false. Hence, $(\exists x)Fx$ is true under **I**, but σ is false under **I**. Let **F** be the subset of **D** assigned to F by this interpretation. Since $(\exists x)Fx$ is true, there is some element, $\mathbf{e} \in \mathbf{D}$, such that $\mathbf{e} \in \mathbf{F}$. Let $\mathbf{I}_a$ be a variant of **I** such that $\mathbf{I}_a(a) = \mathbf{e}$. Now show that $\mathbf{I}_a$ is a counterexample interpretation to the truth of the antecedent of the conditional. Work out the details and write the proof clearly and carefully.)

E 4-66 A corollary of the result of the previous exercise is this: If $Fa \rightarrow \sigma$ is a valid $\mathcal{PC}$ sentence, then $(\exists x)Fx \rightarrow \sigma$ is also valid. Yet Part 5 of E 4-24 shows that $(\exists x)Bx \rightarrow P$ is *not* a consequence of $Bc \rightarrow P$. Discuss the differences between these two situations, and explain why they are *both* true.

4.4 Application Example

Predicate calculus is a powerful and abstract formal language that enables us to represent information about an unlimited variety of domains. This section will exhibit one type of application and briefly relate it to computer logic programming. Before describing the application, I will define a new term and prove a metatheorem. This theorem is a metatheoretic generalization of $\mathcal{PC}$ Theorem 2 in Section 4.3.2. It is an example of a wide variety of metatheorems that can be proved for the predicate calculus. The proof uses a general schema for an infinite set of possible $\mathcal{PC}$ derivations that could be constructed. In this case, the proof schema used in the proof of the metatheorem is a generalization of the kind of $\mathcal{PC}$ derivation that is used to prove $\mathcal{PC}$ Theorem 2 in Section 4.3.2. It is helpful to write a derivation for $\mathcal{PC}$ Theorem 2 while studying the proof of the metatheorem.

Recall that the individual constants can be well-ordered, as was done in Section 4.2.1. We can similarly well-order the individual variables: $s, t, \cdots, z,$ $s_1, t_1, \cdots, z_1, s_2, \cdots$. I will call this the *standard ordering* and use it in the following:

Definition 4-23 Let ϕ be a predicate calculus formula. If ϕ is a sentence, then *the universal closure* of ϕ is just ϕ. If ϕ is not a sentence, then let all of the individual variables with free occurrences in ϕ be $\nu_1, \cdots, \nu_k$, where the ν_i are ordered left to right according to the standard ordering, i.e., of all the ν_i, ν_1 occurs first in the standard ordering, ν_2 occurs next, etc. Then *the universal closure* of ϕ is the sentence,

$$(\forall\nu_1)(\forall\nu_2) \cdots (\forall\nu_k)\phi. \tag{4.4}$$

For example, if ϕ is

$$Pax_2y \rightarrow (\exists w)Qz_2bw,$$

then the universal closure is

$$(\forall y)(\forall x_2)(\forall z_2)(Pax_2y \rightarrow (\exists w)Qz_2bw).$$

The only reason for ordering the variables is so that *the* universal closure is a specific sentence. If we permuted the quantifiers in (4.4), we would still have *a* universal closure of ϕ.

Metatheorem 4-24 Let ϕ, ψ be predicate calculus formulas and μ be a variable. If μ does not occur free in ψ, then the universal closure of

$$((\exists\mu)\phi \rightarrow \psi) \leftrightarrow (\forall\mu)(\phi \rightarrow \psi) \tag{4.5}$$

is a $\mathcal{PC}$ theorem.

Proof. The proof consists of showing that we can always construct a derivation of the universal closure of (4.5). I will do this by exhibiting a *schema* for such a derivation. First we instantiate the closure. Let $\kappa_1, \cdots, \kappa_k$ be individual constants not occurring in (4.5), and suppose they are ordered in the usual way. Let $\nu_1, \cdots, \nu_k$ be the variables occurring free in (4.5). For each i, uniformly substitute κ_i for the free occurrences of ν_i in (4.5). Let Φ and Ψ be the same as ϕ and ψ, except for having the constants substituted for the ν_i. If (4.5) has no free occurrences of variables, then Φ and Ψ are correspondingly the same as ϕ and ψ. From the hypothesis of the metatheorem, it follows that Ψ has no free occurrences of μ, but Φ may. To remind us that μ may be free in Φ, I rewrite Φ as $\Phi[\mu]$. $\Phi\left[\mu/\eta\right]$ is the result of replacing all free occurrences of μ in $\Phi[\mu]$ with occurrences of η. In this proof schema, η and δ are constants that do not occur in the proof until they are first introduced by UI on line (2) and by P on line (6), respectively.

$\{ Pr_1 \}$	(1)	$(\forall\mu)(\Phi[\mu] \to \Psi)$	P
$\{ Pr_1 \}$	(2)	$(\Phi\,[\mu/\eta] \to \Psi)$	UI (1)
$\{ Pr_1 \}$	(3)	$((\exists\mu)\Phi\,[\mu] \to \Psi)$	EA (2)
$\varnothing$	(4)	$(\forall\mu)(\Phi[\mu] \to \Psi) \to ((\exists\mu)\Phi\,[\mu] \to \Psi)$	C (1), (3)
$\{ Pr_5 \}$	(5)	$((\exists\mu)\Phi\,[\mu] \to \Psi)$	P
$\{ Pr_6 \}$	(6)	$\Phi\,[\mu/\delta]$	P (for CP)
$\{ Pr_6 \}$	(7)	$(\exists\mu)\Phi\,[\mu]$	EG (6)
$\{ Pr_5, Pr_6 \}$	(8)	Ψ	TC (5), (7)
$\{ Pr_5 \}$	(9)	$\Phi\,[\mu/\delta] \to \Psi$	C (6), (8)
$\{ Pr_5 \}$	(10)	$(\forall\mu)(\Phi[\mu] \to \Psi)$	UG (9)
$\varnothing$	(11)	$((\exists\mu)\Phi\,[\mu] \to \Psi) \to (\forall\mu)(\Phi[\mu] \to \Psi)$	C (5), (10)
$\varnothing$	(12)	$((\exists\mu)\Phi\,[\mu] \to \Psi) \leftrightarrow (\forall\mu)(\Phi[\mu] \to \Psi)$	TC (4), (11)

At line (12) the proof of the metatheorem is nearly finished.

If ϕ and ψ are the same as Φ and Ψ, we are done.

On the other hand, if any κ_i were substituted for ν_i, we can extend the proof schema by applying UG one or more times after line (12) to obtain a derivation of the universal closure of

$$((\exists\mu)\phi \to \psi) \leftrightarrow (\forall\mu)(\phi \to \psi)$$

from $\varnothing$. **Q.E.D.**

As previously mentioned, this metatheorem is a strong generalization of $\mathcal{PC}$ Theorem 2, in the list of theorems at the end of Section 4.3.2. The general technique used to prove this metatheorem, the use of a proof schema, was previously used in the metatheorem that justifies Rule EA. This proof technique has many applications to other metatheorems about types of $\mathcal{PC}$ derivations and theorems. If one proves a specific $\mathcal{PC}$ theorem, like Theorem 2, it is interesting to consider how that theorem might be generalized into a metatheorem. As the application example will demonstrate, it is useful to know metatheorems about general forms of $\mathcal{PC}$ theorems.

As previously mentioned, the application example will be briefly related to logic programming. I will state the logical formulas in a form that is similar to that used in the Prolog programming language. No knowledge of Prolog is needed. The general form of sentences that I will use is a form used in many logic programming languages. Even if Prolog becomes obsolete, the general techniques and formats will remain useful. An excellent treatment of representing information in the form used below is Kowalski (1979), which discusses many different types of applications of logic as well as general relationships between logic and programming.

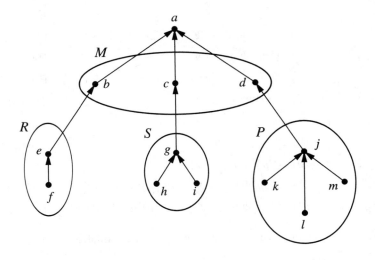

Figure 4.9. *The Omega Corporation*

A Prolog program consists largely of a set of predicate calculus sentences of the following two forms: atomic sentences and sentences of the form,

$$\phi \text{ if } (\psi_1 \text{ and } \psi_2 \text{ and } \cdots \text{ and } \psi_k),$$

in which ϕ and the ψ_i are atomic formulas. If there are free variables in the sentence, it is understood that they are all supposed to be universally quantified. In $\mathcal{PC}$, this sentence would be written as the (universal) closure of

$$(\psi_1 \wedge \psi_2 \wedge \cdots \wedge \psi_k) \rightarrow \phi.$$

We assume general associativity of conjunction, so omit the unnecessary parentheses. Prolog contains additional features, including some metapredicates, but these features are not needed for our example. There are many good books on Prolog, including Covington et al. (1988), Dodd (1990), and Sterling and Shapiro (1986).

Let us return to the Omega Corporation described in Chapter 2. Its organizational structure is shown again in Figure 4.9. We will construct a Prolog-like logical "knowledge base" that describes this organization. The individuals in the domain are $a, b, \cdots, m$. We use the following predicates with the specified interpretations:

Ox : x works in the Omega Corporation

Mx : x is in middle management

Rx : x is in the research and development department

Sx : x is in the shipping department

Px : x is in the production department

Hx : x is a department head

Dxy : y is directly superior to x

Hxy : y is up the chain of command from x

Axy : y is the annual salary (in dollars) of x

Bxy : y is the annual bonus (in dollars) of x

Let Ω be the set of sentences of the knowledge base. The atomic sentences (often called *facts*) in Ω are

$$Oa, \; Ob, \; Oc, \; Od, \; Oe, \; Of, \; Og, \; Oh, \; Oi, \; Oj, \; Ok, \; Ol, \; Om,$$

$$Mb, \; Mc, \; Md, \; Re, \; Rf, \; Sg, \; Sh, \; Si, \; Pj, \; Pk, \; Pl, \; Pm, \; He, \; Hg, \; Hj,$$

$$Dfe, \; Deb, \; Dba, \; Dhg, \; Dig, \; Dgc, \; Dca, \; Dkj, \; Dlj, \; Dmj, \; Djd, \; Dda.$$

The predicate D represents only the direct superior relation; Hxy represents the relationship that holds when either Dxy or there is a path (or chain) of people from x up to y such that each person in this path is a direct superior to the previous one in the path. We could state the connection between D and H by adding the following to Ω:

$$(\forall x)(\forall y)(Hxy \leftrightarrow (Dxy \vee (\exists u)(Dxu \wedge Huy))). \qquad (4.6)$$

This sentence is more complex than the two forms of Prolog sentences previously described. To achieve a Prolog-style knowledge base (KB), we express the connection in a different form. Let κ and δ be individual constants denoting individuals in the Omega Corporation. There are two sufficient conditions for $H\kappa\delta$: (i) $D\kappa\delta$, or (ii) there is some individual γ such that $D\kappa\gamma$ and $H\gamma\delta$. If either of these holds, then $H\kappa\delta$. Also, Sentence (4.6) says that the disjunction of these two conditions is necessary, so these are the only ways to have $H\kappa\delta$. Let us replace (4.6) with the closures of

$$Dxy \rightarrow Hxy, \qquad (4.7)$$

$$(\exists u)(Dxu \wedge Huy) \rightarrow Hxy. \qquad (4.8)$$

By virtue of the previous metatheorem, the universal closure of

$$((\exists u)(Dxu \wedge Huy) \rightarrow Hxy) \leftrightarrow (\forall u)((Dxu \wedge Huy) \rightarrow Hxy) \qquad (4.9)$$

is a $\mathcal{PC}$ theorem. If we now form the biconditional between the universal closure of the left side of (4.9) and the universal closure of its right side, then this biconditional is also a $\mathcal{PC}$ theorem. It follows by the Soundness Theorem that the closure of the left side of (4.9) is equivalent to the closure of the right side. Therefore, in the KB we can use the closure of

$$(Dxu \wedge Huy) \rightarrow Hxy \qquad (4.10)$$

instead of the closure of (4.8). The closure of (4.10) will have universal quantifiers for u, x, y; if we permute the order of these quantifiers, we still obtain logically equivalent sentences. Except for some syntactical differences, in Prolog one would use sentences (4.7) and (4.10) to express the sufficient conditions for $H\kappa\delta$. Notice that these two formulas provide a recursive definition of H, if we assume that the only pairs satisfying H are those implied by these two formulas. The purpose of all this is the following: We want a "smart" KB of information about the Omega Corporation. This KB should not only answer queries about atomic sentences stored in it, but it should also be able to use general 'if-then rules' to infer answers to other queries. A Prolog interpreter is programmed to search for answers to queries. If we have this KB in Prolog and ask it for all values of κ and δ such that $D\kappa\delta$, it will return to us all of the $< \kappa, \delta >$ pairs corresponding to the atomic sentences Dfe, Deb, Dba, Dhg, $\cdots$. If we ask it for all values of κ and δ such that $H\kappa\delta$, it will not find any atomic sentences of this form stored in the KB. But it will use (4.7) and (4.10) to infer answers like Hfe, Hfa, etc. It searches exhaustively for all true sufficient conditions for $H\kappa\delta$. Because any computer KB is finite, this search eventually ends. In using Prolog, and similar logic programming languages, it is assumed that the disjunction of all sufficient conditions in the KB is also necessary.

This example is intended to show that predicate calculus, even limited to the special forms of sentences in Prolog, is a useful language for representing information in a KB. It is interesting to note that (4.6), and hence also (4.7) and (4.10), are special forms of recursive definitions formulated in $\mathcal{PC}$. It takes considerable experience to become proficient in the use of $\mathcal{PC}$ for "knowledge representation." Let us enlarge the example a little to see how it can handle more complex deductions.

Extend the domain so that it includes the real numbers. Assume that the Omega Corporation pays everyone an annual salary according to his (or her) job classification. The boss, a, is an exception who is ignored here. For all others, the salary rules are the closures of

$$Mx \rightarrow Ax55000 \tag{4.11}$$

$$Rx \rightarrow Ax45000 \tag{4.12}$$

$$Sx \rightarrow Ax25000 \tag{4.13}$$

$$Px \rightarrow Ax35000 \tag{4.14}$$

Some individuals, who are department heads, receive annual bonuses of 5,000 dollars. The previous atomic sentences (facts) specify three department heads, e, g, j. The rule about their bonuses is added to Ω as follows:

$$Hx \rightarrow Bx5000 \tag{4.15}$$

Suppose that we are interested in the total compensation of the person e. We can make two specific queries: *Aey* and *Bez*. For the first, Prolog will search and find the solution, $y = 45000$, from the fact *Re* together with (4.12). Prolog will also find that *He* is true; using this, it will get the second solution, $z = 5000$, from (4.15). Searching for these solutions amounts to deriving *Ae* 45000 and *Be* 5000.

Although this example is simple, it suggests how predicate calculus can be used to represent complex information, perhaps with thousands of atomic sentences and hundreds of conditional sentences. A fast computer with a Prolog or similar interpreter can quickly respond to queries about this complex information by deductively inferring consequences from the KB.

EXERCISES

General Instructions

Throughout this book, if an exercise is simply the statement of a theorem or a metatheorem, then the task is to prove this theorem or metatheorem. In your proof, you may use the definitions given in the text. Unless stated to the contrary, you may also use theorems proved in the text.

When using $\mathcal{PC}$ interpretations, be sure that the domain is specified, as well as the values or meanings of all relevant sentential letters, individual constants, and predicates. Give informal reasons that justify the truth values that the interpretation imposes on the *PC* sentences that are of concern.

Exercises for which answers are provided at the back of the book are marked with "Answer Provided."

Finally, please note that exercises that introduce important principles, useful applications, are especially difficult, or are special in some way, are indicated by bold print. In some cases, parenthetical comments describe the special features of the problem. If such comments are missing, the exercise is nonetheless of special value in introducing a new concept, principle, or technique.

E 4-67 Construct a *PC* derivation showing that

$$(\forall x)(\forall y)(Rxy \leftrightarrow Lxy)$$

implies

$$(\forall x)(\forall y)Rxy \leftrightarrow (\forall x)(\forall y)Lxy.$$

Also, construct a counterexample to show that the converse implication does not hold.

E 4-68 For any persons x, y it is true that

1. x is an ancestor of y if x is the father of y.

2. x is an ancestor of y if x is the mother of y.

3. x is an ancestor of y if (there exists z such that z is the father of y and x is an ancestor of z).

4. x is an ancestor of y if (there exists z such that z is the mother of y and x is an ancestor of z).

Notice that the last two conditions are recursive in a manner similar to that used before in the characterization of Hxy. Let Axy represent 'x is an ancestor of y', Fxy represent 'x is the father of y', and Mxy represent 'x is the mother of y'. Suppose that $a, b, \cdots, i$ are people, and we have the following data:

$$Fba, \ Mca, \ Fjb, \ Meb, \ Ffc, \ Mgc, \ Fhe, \ Mie.$$

Make a the root node, and sketch this (family) tree. Now symbolize 1 through 4 in predicate calculus. Use these symbolized sentences as premises *together with* the family tree data. Give a predicate calculus derivation of Aia from all of these premises. The sketch of the family tree should help to guide you through the derivation.

E 4-69 These are problems pertaining to the Omega Corporation, as described in Section 4.4.

1. Use the closures of (4.7) and (4.10), plus other relevant data in the knowledge base Ω, to derive Hma.

2. Let S^3 be a predicate that applies to real numbers such that, for any reals, x, y, z,

$$S^3 xyz \Leftrightarrow z = x + y.$$

Suppose that S^3 is added to the set of predicates used to represent information about the Omega Corporation. Also, we add another predicate, C, such that Cxy holds iff x is a department head and y is the total annual compensation (in dollars) of x. We also add the closure of

$$(Hx \wedge (Axu \wedge (Bxv \wedge S^3 uvt))) \rightarrow Cxt.$$

Finally, assume that $S^3 45000\,5000\,50000$ is an additional premise obtained as the result of a numerical computation. Now recall that e is a member of the Omega Corporation. He (or she) receives both a salary and a bonus. Use the information in the text together with the preceding to prove that $Ce50000$. After this, you may want to try the next exercise.

E 4-70 (This is a somewhat complicated continuation of the previous exercise.) Notice that the above formula for Cxt applies only to department heads. Also, the sum sentence, $S^3 45000\,5000\,50000$, applies only to the particular person e. To compute the total compensation of the other two department heads, g, j, we need additional sum sentences. Also, suppose that we obtain additional information that the boss a receives a salary of 100,000 dollars and no bonus. Because there is exactly one boss, we can express this information by $Aa100000$. It would now be convenient to have the knowledge base imply the total annual compensation of any given person in the Omega Corporation. Add additional sentences to Ω so that this knowledge base has the following feature: Let κ be any one of the constants denoting the people in OmegaCorp. Let τ be the number corresponding to the total annual compensation of κ. Then, for any such κ and τ, if $C\kappa\tau$ is a true statement of the total annual compensation of κ, then $\Omega \vdash C\kappa\tau$. Thus, from your expanded knowledge base Ω, you should be able to derive $Ca100000$, $Ce50000$, $Ch25000$, etc. There are a number of ways to solve this problem; try to use no more sentences than are necessary.

4.5 Identity and Function Symbols

4.5.1 Extension of the Syntax

In this section, $\mathcal{PC}$ is extended by the addition of special symbols for *functions* and a special symbol for the *identity relation*. I will call this extended first-order predicate calculus $\mathcal{PCI}$.

An interpretation of $\mathcal{PC}$ includes a nonempty domain $\mathbf{D}$. Each n-ary predicate of $\mathcal{PC}$ is assigned to an n-ary relation on $\mathbf{D}$ (with 1-ary predicates being assigned to subsets of $\mathbf{D}$). Because a function from $\mathbf{D}^k$ into $\mathbf{D}$, for $k \geqslant 1$, is just a special kind of relation on $\mathbf{D}$, $\mathcal{PC}$ interpretations enable one to represent such functions with $\mathcal{PC}$ predicates. Nevertheless, it is usually more convenient to use a direct way to represent functions. Because a function from $\mathbf{D}^k$ into $\mathbf{D}$ is often called a k-ary *operation* on $\mathbf{D}$, *function symbols* are often called *operation symbols*. To make this extension even more useful, it is customary to introduce a special symbol for the *identity relation*. Any binary predicate could be interpreted as identity, but then it would be like any other predicate. But traditionally, identity has been considered to be a special *logical constant* that obeys its own special inference rules. In the following, identity will be treated in this special way. To begin, we extend the syntax of predicate calculus.

Definition 4-25 The syntax of the predicate calculus with identity and function symbols ($\mathcal{PCI}$) consists of *symbols* and *formulas* as follows:

Symbols:

> *parentheses and comma*: (,) and , (used for grouping and spacing)
>
> *sentential connectives*: $\neg$, $\vee$, $\wedge$, $\rightarrow$, $\leftrightarrow$
>
> *quantifiers*: $\forall, \exists$
>
> *identity symbol*: $=$
>
> The connectives, quantifiers, and identity symbol are called *logical constants*.
>
> *SC letters (sentential letters)*: Same as for the sentential calculus (and $\mathcal{PC}$)
>
> *predicate symbols*: Same as for $\mathcal{PC}$
>
> *individual constants*: Same as for $\mathcal{PC}$
>
> *individual variables*: Same as for $\mathcal{PC}$
>
> *function symbols*: lowercase letters $a, \cdots , z$ with positive integer superscripts (e.g., a^3), and with or without positive integer subscripts.

We now introduce a recursive definition of a new form of expression that is used in $\mathcal{PCI}$.

Terms:

> (i) An individual constant is a term.

(ii) An individual variable is a term.

(iii) If η^k is a function symbol, and $\tau_1, \cdots, \tau_k$ are terms, then an expression of the form $\eta^k(\tau_1, \cdots, \tau_k)$ is a term.

An expression is a term only by virtue of (i)–(iii).

Formulas: The set of all ($\mathcal{PCI}$) formulas is defined recursively, beginning with the atomic formulas.

> *Atomic Formula*: Any single $\mathcal{SC}$ letter, or an n-ary predicate followed by exactly n terms, or an expression of the form $\tau = \sigma$, where τ and σ are terms.

> *Formula*: Any atomic formula, or any expression obtained from atomic formulas by use of the same recursive construction rules that are used for $\mathcal{PC}$ formulas.

Here are some examples of terms:

$$a, \ b_7, \ x, \ f^3(a, x, b), \ f^3(h^1(a), \ f^3(m, n, a), \ h^1(z)).$$

For convenient abbreviations, we will usually omit the superscripts from function symbols, as is also done with predicates. We will also use the conventional expression, $\tau \neq \sigma$, as an abbreviation for $\neg \tau = \sigma$.

Definition 4-26 *Free* and *bound* occurrences of variables are defined just as for $\mathcal{PC}$; also, a *sentence* is a formula with no free occurrences of any variable. A term that has no occurrences of variables is called a *ground term* or *constant term*. A *ground sentence* is a formula with no occurrences of any variables, free or bound.

Examples. Here are some formulas (with superscripts omitted):

$$z = y, \ z = a, \ a = b, \ Sxyc, \ Sabc, \ Sxf(a, b)c,$$

$$(\forall x)(\forall y)(Gxa \rightarrow (\exists z)z = s(x, a, y)), \quad (Gah(a, b) \vee Rf(b, a)b).$$

Notice that a, b, c, $h(a, b)$, $f(a, b)$, and $f(b, a)$ are ground terms. Also, $a = b$, $Sabc$, $Gah(a, b)$, and $Rf(b, a)b$ are atomic sentences. These atomic sentences are also ground sentences, as is $(Gah(a, b) \vee Rf(b, a)b)$. ∎

4.5.2 Semantics of Predicate Calculus with Identity and Function Symbols

It is straightforward to extend the definitions of *interpretation* and *true under an interpretation* from $\mathcal{PC}$ to $\mathcal{PCI}$. These extensions will be sketched briefly. $\mathcal{SC}$ letters, individual constants, and predicates are interpreted in the same manner as for $\mathcal{PC}$. In addition, each k-ary function symbol η is mapped to some k-ary function from $\mathbf{D}^k$ to $\mathbf{D}$. I will use the notation $\mathbf{I}(\eta)$ to denote the function that is assigned to η. Finally, the identity symbol, $=$, is assigned to the identity relation on the domain $\mathbf{D}$, namely, $\{< x, x > \mid x \in \mathbf{D}\}$. We still need to specify how ground terms are interpreted.

Definition 4-27 The *value* or *referent* of a ground term τ under an interpretation $\mathbf{I}$ is denoted by $\mathbf{I}(\tau)$ and is determined recursively as follows:

(i) If τ is an individual constant, then its value is

$$\mathbf{I}(\tau) = \mathbf{I}_{CONST}(\tau),$$

the element of $\mathbf{D}$ that the interpretation assigns to τ.

(ii) If η is a function symbol and $\tau_1, \cdots, \tau_k$ are ground terms, let $\mathbf{I}(\eta) = \mathbf{f}$ and, for each i, let $\mathbf{I}(\tau_i) = \mathbf{t}_i$. Then the value of

$$\eta(\tau_1, \cdots, \tau_k)$$

under $\mathbf{I}$ is

$$\mathbf{f}(\mathbf{t}_1, \cdots, \mathbf{t}_k).$$

The definition of *truth* is modified in predictable ways.

Definition 4-28 The definition of $\mathbf{V_I}$ (and the expression '*true under* $\mathbf{I}$') is exactly the same as Definition 4-6 for $\mathcal{PC}$ except that Part 2 of Definition 4-6 (pertaining to atomic sentences that are not $\mathcal{SC}$ letters) is modified as follows:

2a. Let τ and σ be ground terms. Then an atomic sentence of the form $\tau = \sigma$ is *true under* $\mathbf{I}$ iff $\mathbf{I}(\tau)$ is the same element of $\mathbf{D}$ as $\mathbf{I}(\sigma)$.

2b. Suppose that ϕ is an atomic sentence of the form $\mathbb{R}\tau_1 \cdots \tau_n$, for predicate $\mathbb{R}$ and ground terms τ_i. Let

$$\mathbf{R} = \mathbf{I}_{PRED,\,n}(\mathbb{R}),$$

and let each τ_i be interpreted as $\mathbf{t}_i$, respectively. Then ϕ is *true under* $\mathbf{I}$ iff

$$\mathbf{R}\mathbf{t}_1 \cdots \mathbf{t}_n,$$

$$(\text{i.e., } < \mathbf{t}_1, \cdots, \mathbf{t}_n > \in \mathbf{R}).$$

The rest of the definition of truth for $\mathcal{PCI}$ is the same as that for $\mathcal{PC}$.

Example.

Let $\mathbf{D} = Nat$, and let $\mathbf{I}$ include the following assignments:

$$
\begin{aligned}
a &: \quad 0, \\
a_i &: \quad i, \text{ for any positive integer } i, \\
s(x, y) &: \quad \text{the sum of } x \text{ and } y, \\
p(x, y) &: \quad \text{the product of } x \text{ and } y, \\
Lxy &: \quad x \text{ is less than } y.
\end{aligned}
$$

Then the following are true:

$$Laa_2,$$

$$(\forall x)(\forall y)s(x, y) = s(y, x),$$

$$(\exists y)(\exists z)p(z, y) = a_{90},$$

but this is false:

$$(\forall u)(\forall v)Lp(u, v)s(u, v). \quad \blacksquare$$

4.5.3 Derivation Rules for Predicate Calculus with Identity and Function Symbols

Most of the previous rules stay the same; two require obvious modifications. To save space, the rules will be stated and used, but not justified by metatheorems. Rules P, TC, C, QE, UG, and EA are used exactly as they are for $\mathcal{PC}$. Rules UI and EG are revised as follows, changing from constants to terms.

RULE UI (Universal Instantiation). Let ν be a variable and τ be any ground term. If a sentence of the form $(\forall \nu)\phi$ is on a line of a proof, then $\phi\left[^{\nu}/_{\tau}\right]$ may be placed on a new line with the same set of premise names as $(\forall \nu)\phi$.

RULE EG (Existential Generalization). Let ν be a variable, τ be a ground term, and ϕ be a $\mathcal{PCI}$ formula. If $\phi\,[^\nu\!/_\tau]$ is on a line of a proof, then $(\exists\nu)\phi$ may be placed on a new line with the same set of premise names as $\phi\,[^\nu\!/_\tau]$.

We require one new rule for identity; it has two parts.

RULE I (Identity).

(i) Let ν be a variable. Then any sentence of the form $(\forall\nu)\nu = \nu$ may be entered on a line with the empty set of premise names.

(ii) Suppose that a sentence ϕ is on a line with premise names Γ. Also, suppose that the sentence $\tau = \sigma$, for ground terms τ, σ, is on a line with premise names Δ. Let ψ be like ϕ except for having occurrences of σ substituted for zero or more occurrences of τ in ϕ and having occurrences of τ substituted for zero or more occurrences of σ in ϕ. Then ψ may be placed on a new line with premise names $\Gamma \cup \Delta$.

Part (i) is called *the reflexivity property of identity* and part (ii) *the substitutivity of identity*. Note that (i) and (ii) in Rule I correspond respectively to I_1 and I_2 in Figure 0.1 of Section 0.3.3. We can now construct derivations in $\mathcal{PCI}$ using Rule I plus the previous $\mathcal{PC}$ rules, with UI and EG as just revised. Here are two simple examples, which are also important theorems about identity.

$\mathcal{PCI}$ **Theorem.**

$\varnothing$	(1) $(\forall x)x = x$	I
$\varnothing$	(2) $a = a$	UI (1)
$\{\,Pr_3\,\}$	(3) $a = b$	P
$\{\,Pr_3\,\}$	(4) $b = a$	I (2), (3)
$\varnothing$	(5) $a = b \rightarrow b = a$	C (3), (4)
$\varnothing$	(6) $(\forall y)(a = y \rightarrow y = a)$	UG (5)
$\varnothing$	(7) $(\forall x)(\forall y)(x = y \rightarrow y = x)$	UG (6)

$\mathcal{PCI}$ **Theorem.**

$\{\ Pr_1\ \}$	(1)	$a = b \land b = c$	P
$\{\ Pr_1\ \}$	(2)	$a = b$	TC (1)
$\{\ Pr_1\ \}$	(3)	$b = c$	TC (1)
$\{\ Pr_1\ \}$	(4)	$a = c$	I (2), (3)
$\varnothing$	(5)	$(a = b \land b = c) \to a = c$	C (1), (4)
$\varnothing$	(6)	$(\forall z)((a = b \land b = z) \to a = z)$	UG (5)
$\varnothing$	(7)	$(\forall y)(\forall z)((a = y \land y = z) \to a = z)$	UG (6)
$\varnothing$	(8)	$(\forall x)(\forall y)(\forall z)((x = y \land y = z) \to x = z)$	UG (7)

From Rule I (i), identity is reflexive. From the first theorem just presented, it is symmetric; and from the second theorem, it is transitive. Thus, identity is an equivalence relation (see Definition 2-22). From Chapter 2, we know that there are many equivalence relations other than identity. For instance, the relation of *similarity* on the set of all plane Euclidean triangles is an equivalence relation, but is certainly not the identity relation. It is interesting to note that we have proved symmetry and transitivity from the reflexivity and substitutivity properties of identity. The latter is a very powerful feature.

Rule I (i) states the reflexivity of identity in terms of universally quantified sentences, but this reflexivity is also manifested in instances. Let τ be any ground term. It is obvious that $\tau = \tau$ can be derived from $\varnothing$ by using UI together with I (i). The EG and EA Rules are useful derived rules; here is another one:

RULE I (Identity). (iii) Let τ be any ground term. Then the sentence $\tau = \tau$ may be placed on a line with the empty set of premise names.

The justification has already been sketched; here is a little more detail. Suppose that we wish to place $\tau = \tau$ on a line of a derivation. We can always do the following, using I (i) and UI.

$\cdots$	$\cdots$	$\cdots$	$\cdots$
$\varnothing$	(k)	$(\forall z)z = z$	I
$\varnothing$	$(k{+}1)$	$\tau = \tau$	UI (k)

Rule I (iii) simply omits line (k), which could always be included if we desired.

When commenting on uses of Rule I, we will just use 'I' alone, without reference to parts (i), (ii), or (iii).

4.5.4   Use of Identity in Representing Information

Recall that a *model* of a set Γ of $\mathcal{PC}$ sentences is an interpretation that satisfies Γ. We use the same terminology for $\mathcal{PCI}$. The *cardinality* or *size* of a model is the number of elements in its domain. Any model must have at least one element, and some models have an infinite number of elements. Suppose that Γ is a set of $\mathcal{PC}$ sentences that has a model with some finite number of elements, say, five. Then we can add additional elements to the domain of this model without specifying any features of these elements. Thus, Γ will also have models of any finite size and also an infinite model. There are other ways in which $\mathcal{PC}$ is limited in its ability to constrain the size of models of sets of sentences.

Identity extends the representational power of predicate calculus by enabling one to specify certain types of constraints on distinctness and certain forms of uniqueness. For instance, in $\mathcal{PCI}$ it is easy to specify 'at least three' by using

$$(\exists x)(\exists y)(\exists z)((x \neq y \wedge y \neq z) \wedge x \neq z).$$

For a given positive integer n, some $\mathcal{PC}$ sentences can require that their models have at least n elements, but such sentences tend to be more complex than the preceding. As noted before, no set of $\mathcal{PC}$ sentences with an n-element model can constrain its models to at most n elements, but $\mathcal{PCI}$ can specify 'at most n'. For example, here is 'at most two',

$$(\forall x)(\forall y)(\forall z)((x = y \vee y = z) \vee x = z).$$

Informally, let x, y, and z be assigned to any elements of the domain. Then at least one pair of these variables will be assigned to the same object. If the domain had three or more elements, this would not always be true. Thus, the domain has at most two elements. This sentence is equivalent to the negation of the previous $\mathcal{PCI}$ sentence. By using combinations of such forms, one can also specify 'exactly n', although the $\mathcal{PCI}$ sentences become lengthy and awkward.

Identity probably has greater practical use in the specification of uniqueness. Suppose we wish to say, 'Everest is the (unique) highest mountain', where it is understood that the domain is some set of geographical objects on earth. Let e denote Everest, Mx represent 'x is a mountain', and Huv represent v is higher than u. We can symbolize the English sentence by

$$Me \wedge (\forall x)((Mx \wedge x \neq e) \rightarrow Hxe).$$

In this example, it seems pretty clear that the English sentence implies that Everest is a mountain; hence the first conjunct. Some types of exceptions are also forms of uniqueness. For instance, consider 'Everyone perished except Ishmael'. We might symbolize this as

$$(\forall x)(x \neq i \rightarrow Px).$$

From this sentence, it follows that anyone other than Ishmael, e.g., Captain Ahab, perished. Yet the symbolization does not imply what happened to Ishmael. It might be argued that the use of *except* in the original English sentence also implies that Ishmael did not perish, and that this proposition should be added to the symbolization. Whether or not this proposition should be added, I would guess that most readers of the English sentence would infer, for lack of reason to the contrary, that Ishmael survived, although they might withdraw this inference in the face of new information. Also notice that this type of exception is a unique individual. There are other types of exceptions: 'No dogs are allowed in the building except guide dogs.' Here, the exception is an entire subset of the set of dogs. Most commonsense reasoning is entangled with various types of exceptions.

A familiar feature of everyday, commonsense reasoning is this: One often makes an inference that is justified on the basis of limited evidence that implies the conclusion only in "normal" situations. If additional evidence is revealed that makes the situation somehow "abnormal" or an exceptional case, then the inference is blocked, even though the original evidence is still accepted as true. Standard deductive logic does not behave this way: If $\Gamma \vDash \phi$, then also $\Gamma \cup \Delta \vDash \phi$, where Δ is any set of additional premises. Much current artificial intelligence research is concerned with the analysis of commonsense reasoning and how it may be related to deductive logic, inductive logic, and probability theory. A useful introduction to these issues is in Genesereth and Nilsson (1987); a large anthology of articles is Ginsberg (1987); and an approach that combines logical and epistemological considerations is presented in Causey (1991), (1994), and (2003). There is now a large body of literature in this field.

EXERCISES

General Instructions

When using $\mathcal{PC}$ interpretations, be sure that the domain is specified, as well as the values or meanings of all relevant sentential letters, individual constants, and predicates. Give informal reasons that justify the truth values that the interpretation imposes on the $\mathcal{PC}$ sentences that are of concern.

Exercises for which answers are provided at the back of the book are marked with "Answer Provided."

E 4-71 (Answer Provided for 1) For each of the following sets of $\mathcal{PCI}$ sentences, give a suitable interpretation to show that the set is satisfiable. Informally justify the truth values of the sentences under your interpretation. When interpreting a function symbol, be sure to specify a genuine function and show that it satisfies the conditions stated by the $\mathcal{PCI}$ sentences.

1. $\{(\forall x)(Ax \rightarrow x = a),\ \neg Ba,\ (\exists x)(\exists y)x \neq y\}$

2. $\{(\exists z)(\forall x)x = z,\ (\exists x)Ax,\ (\exists x)Bx\}$

3. $\{(\exists x)(\exists y)(\exists z)(x \neq y \wedge (x \neq z \wedge y \neq z)),$
   $\qquad (\forall x)f(x) \neq x,\ (\forall x)(\forall y)(f(x) = f(y) \rightarrow x = y)\}$

4. $\{(\exists x)Ax,\ (\forall x)(Ax \rightarrow \neg Bx),\ (\forall x)(Ax \rightarrow Bf(x)),\ (\forall x)(Bx \rightarrow Ag(x))\}$

5. $\{(\forall x)(\forall y)(f(x) = f(y) \rightarrow x = y),\ (\forall x)Rxf(x),$
   $\qquad (\forall x)(\forall y)(Rxy \rightarrow \neg Ryx),\ (\exists x)(\forall y)x \neq f(y)\}$

6. $\{Na,\ (\forall x)(\forall y)((Ny\ \wedge\ x = f(y)) \rightarrow Nx)\}$

7. $\{Na,\ (\forall x)(\forall y)((Ny \wedge x = f(y)) \rightarrow Nx),\ (\forall x)g(x, a) = x,$
   $\qquad (\forall x)(\forall y)g(x, f(y)) = f(g(x, y))\}$

E 4-72 (Answer Provided) Give a counterexample interpretation to show that the conclusion is not a consequence of the premises. The conclusion is indicated by $/\!\therefore$.

$(\exists x)Ax$

$(\forall x)(Ax \rightarrow \neg Bx)$

$(\forall x)(Ax \rightarrow Bf(x))$

$(\forall x)(Bx \rightarrow Ag(x))$

$/\!\therefore\ (\forall x)(Ax \rightarrow g(f(x)) = x).$

E 4-73 Give a counterexample interpretation to show that the conclusion is not a consequence of the premises. The conclusion is indicated by $/\!\therefore$.

$(\forall x)p(x,\ a) = x$

$(\forall x)(\forall y)p(x,\ s(y)) = s(p(x,\ y))$

$/\!\therefore\ (\forall y)(\forall z)(\exists x)p(x,\ y) = z.$

E 4-74 (Answer Provided) Give a derivation of this $\mathcal{PCI}$ theorem from $\varnothing$:

$$(\forall x)(\forall y)(x = y \rightarrow f(x) = f(y)).$$

E 4-75 Derive the following sentence from $\varnothing$:

$$(\forall x)((x = a \vee x = b) \rightarrow (f(x) = f(a) \vee f(x) = f(b))).$$

E 4-76 (Answer Provided for 2) For each of the following arguments, give a $\mathcal{PCI}$ derivation of the conclusion from the premise(s). The conclusion is indicated by $/\therefore$.

1. $(\forall x)\neg Rxx$

   $/\therefore$ $(\forall x)(\forall y)(Rxy \rightarrow x \neq y)$.

   (Hint: Try assuming $a = b$ as an extra premise.)

2. $(\exists x)Ax$

   $(\exists x)\neg Ax$

   $/\therefore$ $(\exists u)(\exists v)u \neq v$.

3. $(\exists u)(\forall v)(Av \leftrightarrow v = u)$

   $(\forall x)Ax$

   $/\therefore$ $(\forall x)(\forall y)x = y$.

   (Remark. It is a little tricky to satisfy the provisos of EA and UG.)

4. $(\forall x)(x = a \vee x = b)$

   $/\therefore$ $(\forall x)(f(x) = f(a) \vee f(x) = f(b))$.

5. $(\forall x)(\forall y)(g(y) = x \leftrightarrow y = f(x))$

   $/\therefore$ $(\forall x)g(f(x)) = x$.

   (Hint: Remember that UI has been extended to allow instantiation to any ground term.)

6. $(\forall x)(\forall y)(g(y) = x \leftrightarrow y = f(x))$

   $/\therefore$ $(\forall x)f(g(x)) = x$.

7. $(\forall u)(\neg(\exists v)Pvu \leftrightarrow u = o)$

   $(\forall y)(\neg Ey \rightarrow (\exists v)Pvy)$

   $/\therefore$ Eo.

E 4-77 From

$$(\forall x)(\forall y)(Rxy \rightarrow Ryx)$$

and

$$(\forall x)(\forall y)(Rxy \leftrightarrow Sf(x)f(y)),$$

derive

$$(\forall x)(\forall y)(Sf(x)f(y) \rightarrow Sf(y)f(x)).$$

E 4-78 Derive the following from ∅:

$$(\exists x)(\exists y)f(x) \neq f(y) \rightarrow (\exists x)(\exists y)x \neq y.$$

E 4-79 Let ϕ be the sentence

$$(\forall x)(\forall y)(Exy \leftrightarrow f(x) = f(y)).$$

Using ϕ as the only premise, give $\mathcal{PCI}$ derivations that establish that E is an equivalence relation, i.e., prove that E is reflexive, symmetric, and transitive.

E 4-80 The domain is the set of human beings. For any person x, let $m(x)$ denote the (biological) mother of x. Let Sxy represent 'x and y are sisters.' Let Bxy represent 'x and y are brothers.' Let Cxy represent 'x and y are matrilateral parallel cousins.' This means that $m(x)$ and $m(y)$ are sisters.

1. In $\mathcal{PCI}$, write a formal definition of Cxy in terms of m and S. This definition should have the form,

   $$(\forall x)(\forall y)(Cxy \leftrightarrow \phi(x, y)),$$

   where $\phi(x, y)$ is a formula using S and m.

2. Symbolize the sentence: For any x, y: If x and y are brothers, then they have the same mother.

3. Symbolize: For any x, y, z: If x and z are matrilateral parallel cousins, and x and y are brothers, then y and z are matrilateral parallel cousins.

4. Using the two $\mathcal{PCI}$ sentences from 1 and 2 as premises, give a formal derivation of the $\mathcal{PCI}$ sentence obtained from 3.

4.6 Formalized Theories

Symbolic logic is widely used as a precise language for the expression of formalized theories, especially mathematical theories. Some of the most important developments in twentieth-century foundations of mathematics pertain to formalized versions of Peano's Axioms for the natural numbers and to formalized set theory. This section is intended to introduce briefly some of the basic ideas and techniques connected with formalized theories. A basic understanding of these techniques is necessary for study of philosophy, science, and many other fields. These concepts are also helpful in organizing knowledge bases for artificial intelligence and deductive database programs. In addition, they have been used in the formal analysis of programming languages.

Definition 4-29 The *nonlogical symbols* of a formal language (such as $\mathcal{SC}$, $\mathcal{PC}$, and $\mathcal{PCI}$) are all of the symbols, except the logical constants, the variables, parentheses, and comma. If Γ is a nonempty set of sentences of a formal language, the set of all nonlogical symbols occurring in sentences of Γ is *the nonlogical vocabulary of* Γ, denoted by $VOC(\Gamma)$. If Θ is a set of nonlogical symbols and Γ is a set of sentences in which no nonlogical symbol occurs that is not in Θ, then Γ *is said to be formulated in terms of* Θ.

If $\Gamma = \{\phi\}$, then we may write $VOC(\phi)$ in place of $VOC(\{\phi\})$. If $\Gamma = \{\phi\}$, and Γ is formulated in terms of a set of nonlogical symbols Θ, then we may also say that ϕ is formulated in terms of Θ. In the case of $\mathcal{PCI}$, the nonlogical symbols are all of the individual constants, $\mathcal{SC}$ letters, predicates, and function symbols. These are the symbols that are assigned referents by an interpretation. Any set of sentences Γ is formulated in terms of its nonlogical vocabulary; but, in general, there will be additional sentences that can also be formulated in terms of this vocabulary that are not in Γ.

Definition 4-30 Let Γ be a nonempty set of sentences of $\mathcal{PCI}$. Then Γ is *deductively closed* iff: For any sentence ϕ formulated in terms of $VOC(\Gamma)$,

$$\phi \in \Gamma \;\Leftrightarrow\; \Gamma \vDash \phi.$$

A set Γ of $\mathcal{PCI}$ sentences is a *$\mathcal{PCI}$ theory* iff it is deductively closed.

This definition is stated in terms of $\mathcal{PCI}$, but similar definitions apply to $\mathcal{PC}$ and other formal languages. We will continue to use $\mathcal{PCI}$ unless otherwise stated. Also, unless otherwise specified, *logical consequence, satisfiability*, $\vDash$, $\vdash$, etc., will denote use of these terms for $\mathcal{PCI}$. Now let Γ be any set of sentences (of $\mathcal{PCI}$). It may or may not be deductively closed; if not, then it is a proper subset of a larger set that is.

Definition 4-31 Let Γ be a set of $\mathcal{PCI}$ sentences. The *deductive closure* of Γ is

$$DCLOSURE(\Gamma) =$$

$$\{\phi \mid \phi \text{ is formulated in terms of } VOC(\Gamma) \ \& \ \Gamma \vDash \phi\}.$$

If Γ happens to be a finite set, then we say that Γ is a *finite set of axioms* for $DCLOSURE(\Gamma)$, and $DCLOSURE(\Gamma)$ is said to be an *axiomatic theory* that is *generated from* Γ.

Any theory, such as $DCLOSURE(\Gamma)$, that is generated from a finite set of axioms is a *finitely axiomatizable theory*.

This definition can be extended to include infinite sets of axioms, provided there is an effective procedure to determine, in a finite number of steps, whether or not an arbitrary sentence in the vocabulary is an element of the set of axioms. The theory of such procedures is beyond the scope of this book. If we consider any infinite sets of axioms, they will be cases in which the existence of such a procedure is obvious.

When working with axiomatic theories, one is interested in consequences of the axioms. In any proof, any of the axioms could be introduced by Rule P. All proofs would then end with a line whose premise names correspond to a subset of the axioms. In practice, a different convention is normally used. Any axiom may be entered on a line with the empty set of premise names. We then seek proofs in which the last line has the empty set of premise names, with the understanding that the axioms of the theory have been used in the proof. If ϕ is on the last line of such a proof, then it is said to be a *theorem* of the axiomatic theory. Of course, theorems may then be entered on lines of additional proofs, using the empty set of premise names, for such theorems could always be reproved as subproofs of new proofs if one desired to do so much redundant work. These two operations are codified as rules that we will use in work with axiomatic theories.

RULE AX. Let Γ be a set of axioms of an axiomatic theory, and let $\phi \in \Gamma$. Then ϕ may be placed on a line of a derivation within this theory, using the empty set of premise names.

RULE THM. Let θ be a *theorem* of an axiomatic theory, i.e., with the help of Rule AX, θ can be derived from $\varnothing$. Then θ may be placed on a line of a new derivation within this theory, using the empty set of premise names.

These ideas will now be illustrated by the theory of groups. In Chapter 2, a group was defined as a special kind of relational system. Any system of the form

$$\langle \mathbf{G},\ \mathbf{i},\ * \rangle$$

satisfying G1–G3 of Definition 2-48 is a group. All of this is stated in the language of informal set theory, and I shall now use boldface print to emphasize that the relational system is being described in the metalanguage. Now we can axiomatize group theory in $\mathcal{PCI}$.

In Definition 2-48, $\mathbf{i}$ is a distinguished element of $\mathbf{G}$ and $*$ is a binary operation on $\mathbf{G}$. Correspondingly, in $\mathcal{PCI}$ we use an individual constant, say, i, for $\mathbf{i}$, and a function symbol, say, g^2, for $*$. However, instead of writing terms such as $g^2(x, y)$, I will write xy and consider this to abbreviate $g^2(x, y)$. This violates the original syntax rules for $\mathcal{PCI}$; we shall bend the rules for our convenience.

Axiom G1 states that the operation is associative; this is easy to formalize. Axiom G2 asserts the existence of at least one identity element, $\mathbf{i}$, which is later proved to be unique. The simplest way to handle this is to use the constant i in the statement of G2 and then again in the statement of G3.

Definition 4-32 The *theory of groups* is the deductive closure of the following axioms:

$$(\forall x)(\forall y)(\forall z)x(yz) = (xy)z. \tag{4.16}$$

$$(\forall x)xi = x. \tag{4.17}$$

$$(\forall x)(\exists z)xz = i. \tag{4.18}$$

Any interpretation of these axioms has a nonempty domain $\mathbf{G}$, an element $\mathbf{i}$ of $\mathbf{G}$, and a binary operation $*$. In other words, an interpretation, or model, of these three group axioms can be described as a relational system, $\langle \mathbf{G},\ \mathbf{i},\ * \rangle$, where $\mathbf{i}$ is the element of $\mathbf{G}$ assigned to i and $*$ is the binary operation on $\mathbf{G}$ that is assigned to g^2 by the interpretation. Thus, the set of all possible groups as defined in Definition 2-48 is just the set of all possible models of the three $\mathcal{PCI}$ axioms above. This can be stated as follows: A group is any model of the theory of groups. The theory is the deductive closure of the axioms, so the theory is a set of theorems that can be derived from these axioms. Any group is a model of the theory, so any theorem of group theory holds for any group.

As illustrated in Section 2.4, there are many different groups, so the theory has many models. In fact, because every set of permutation functions on a set S is a group, there is an infinite set of distinct groups. See the fourth example following Definition 2-48.

The theory of groups is mentioned here because it is a good illustration of a simple theory with a very wide variety of models. Many interesting theorems can be proved about groups. Theorem 2-49 stated several elementary lemmas about groups; these were proved informally in Chapter 2. I will now restate these lemmas in $\mathcal{PCI}$ form and present a couple of the proofs.

Theorem 4-33 The following are theorems of group theory:

(i) $(\forall x)(\forall y)(\forall z)(xz = yz \rightarrow x = y)$ (right cancellation).

(ii) $(\forall x)\ ix = x.$

(iii) $(\forall x)(\forall z)(xz = i \rightarrow zx = i).$

(iv) $(\forall x)(\forall y)(\forall z)(zx = zy \rightarrow x = y)$ (left cancellation).

(v) $(\forall y)((\forall x)xy = x \rightarrow y = i)$ (unique identity element).

(vi) $(\forall x)(\forall y)(\forall z)((xy = i \wedge xz = i) \rightarrow y = z).$

Proofs. To save space, I will combine some obvious multiple uses of UI and UG into single steps. Also, because any instance of an axiom or a theorem that can be obtained by UI is a theorem, such instances may be placed on lines by the THM Rule. I will also assume the theorem that identity is symmetric and combine symmetry operations with substitutions of identicals. This is a standard algebraic transformation of equations, and the theory of identity is part of $\mathcal{PCI}$ and therefore part of any theory formalized in $\mathcal{PCI}$.

(i)

$\{\ Pr_1\ \}$	(1)	$ac = bc$	P
$\varnothing$	(2)	$ai = a$	THM, instance of AX (4.17)
$\varnothing$	(3)	$(\exists z)cz = i$	THM, inst. of AX (4.18)
$\{\ Pr_4\ \}$	(4)	$cd = i$	P (ExEx, d an inverse of c)
$\{\ Pr_4\ \}$	(5)	$a = a(cd)$	I (2), (4)
$\varnothing$	(6)	$a(cd) = (ac)d$	THM, inst. of AX (4.16)
$\{\ Pr_4\ \}$	(7)	$a = (ac)d$	I (5), (6)
$\{\ Pr_1, Pr_4\ \}$	(8)	$a = (bc)d$	I (1), (7)

$\varnothing$	(9)	$b(cd) = (bc)d$	THM, inst. of AX (4.16)
$\{\ Pr_1,\ Pr_4\ \}$	(10)	$a = b(cd)$	I (8), (9)
$\{\ Pr_1,\ Pr_4\ \}$	(11)	$a = bi$	I (4), (10)
$\varnothing$	(12)	$bi = b$	THM, inst. of AX (4.17)
$\{\ Pr_1,\ Pr_4\ \}$	(13)	$a = b$	I, (11), (12)
$\{\ Pr_4\ \}$	(14)	$ac = bc \to a = b$	C (1), (13)
$\varnothing$	(15)	$cd = i \to$	
		$\qquad (ac = bc \to a = c)$	C (4), (14)
$\varnothing$	(16)	$(\exists z)\ cz = i \to$	
		$\qquad (ac = bc \to a = b)$	EA (15)
$\varnothing$	(17)	$(ac = bc \to a = b)$	TC (3), (16)
$\varnothing$	(18)	$(\forall x)(\forall y)(\forall z)(xz = yz \to$	
		$\qquad x = y)$	UG (17), 3 times

(ii)

$\{\ Pr_1\ \}$	(1)	$ab = i$	P (b is an inverse of a)
$\varnothing$	(2)	$ii = i$	THM, inst. of AX (4.17)
$\{\ Pr_1\ \}$	(3)	$i(ab) = i$	I (1), (2)
$\varnothing$	(4)	$i(ab) = (ia)b$	THM, inst. of AX (4.16)
$\{\ Pr_1\ \}$	(5)	$(ia)b = i$	I (3), (4)
$\{\ Pr_1\ \}$	(6)	$(ia)b = ab$	I (1), (5)
$\varnothing$	(7)	$(ia)b = ab \to ia = a$	THM, inst. of Theorem (i)
$\{\ Pr_1\ \}$	(8)	$ia = a$	TC (6), (7)
$\varnothing$	(9)	$ab = i \to ia = a$	C (1), (8)
$\varnothing$	(10)	$(\exists z)\ az = i \to ia = a$	EA (9)
$\varnothing$	(11)	$(\exists z)\ az = i$	THM, inst. of AX (4.18)
$\varnothing$	(12)	$ia = a$	TC (10), (11)
$\varnothing$	(13)	$(\forall x)\ ix = x$	UG (12)

The formal style of these derivations is tedious, but the basic steps in the proofs are essentially the same as those used in the informal proofs of Theorem 2-49. The only striking difference is that in the informal proofs, I used x^{-1} to denote any inverse element of the element x. Because the formal axiom,

$$(\forall x)(\exists z)\ xz = i,$$

is stated with an existential quantifier, in $\mathcal{PCI}$ the expression x^{-1} is not defined. Thus, it is necessary to use a constant (such as the b in line (1) of (ii)), to denote an inverse.

The remaining proofs are left as exercises. They are not difficult to construct if one is guided by the informal proofs of Theorem 2-49. **Q.E.D.**

Mathematicians usually give informal proofs. The purpose of writing a few formal proofs as exercises is to gain a better understanding of how the formal proofs are constructed. In the future, complex formal proofs will be checked, and at least partly constructed, by computers, just as we now use these machines for complex numerical and symbolic mathematical calculations.

It is annoying to introduce a new constant every time an inverse element is required, as is done in the preceding proofs. Fortunately, the last of the above propositions, (vi), establishes that every element x has a *unique* inverse. Once this is proved, one can extend the language of the theory by defining a new, unary operation as follows:

$$(\forall x)(\forall y)(f^1(x) = y \leftrightarrow xy = i).$$

We cannot introduce a new function symbol into the language of a theory unless we know that it denotes a genuine function. It is necessary to know that a term of the form, $f^1(\kappa)$, where κ is any constant, denotes a *unique* element of the domain of any interpretation. In the case of groups, we know that this is true because of (vi). Suppose we have $ab = i$ and $ac = i$. From (vi), we obtain $b = c$, and this type of implication holds for any constant in place of a. Therefore, the preceding universally generalized biconditional formula can be added to the formal theory of groups *as a definition of* f^1. The language of the theory has been expanded. In practice, instead of writing $f^1(x)$, one would follow tradition and write x^{-1}. The negative superscript just denotes the inversing operation, f^1, so x^{-1} denotes the unique inverse element of the element x.

Using definitions to add new predicates and function symbols to the language of a formal theory is a great convenience. It enables us to rewrite complex expressions in simpler forms. This is why mathematics, and other complex fields such as law and medicine, use so many specialized definitions. By using the defined expression, x^{-1}, we can, for example, rewrite the third axiom in the form,

$$(\forall x)\; xx^{-1} = i,$$

and can rewrite Theorem (iii) as

$$(\forall x)(xx^{-1} = i \rightarrow x^{-1}x = i).$$

5

From these two sentences, one immediately obtains

$$(\forall x)\ x^{-1}x = i.$$

After the introduction of x^{-1}, many formulas and proofs take on a simpler form, which is closer to the form used in informal proofs. It is possible to formalize the theory of groups by including the inversing operation *ab initio* in the axioms. I have started with a weaker set of axioms and proved this uniqueness as a theorem.

Although the formal derivations are tedious, their rigid structure has some advantages. For instance, given a formal proof like (i) or (ii), an appropriate computer program could check each line and determine whether or not the proof is indeed correct. It is much more difficult to write a program that constructs its own original proofs, but much progress has been made with this difficult endeavor. A good review of earlier work in this field is Bledsoe (1977); see also Boyer and Moore (1979) and (1998). There is now a large body of work, and continuing research, in the field of *automatic theorem proving*.

EXERCISES

E 4-81 Prove statements (iii) – (vi) of Theorem 4-33. It may be helpful, though it is not necessary, to review the informal proofs of Theorem 2-49.

Answers to Selected Exercises

Chapter 1 Answers

<u>E 1-3</u> (part 1) Using the following symbolizations:

M : the machine runs	S_1 : Switch1 is on
S_2 : Switch2 is on	P : the power cord is plugged in
F : a fuse is blown	C : the fuse is replaced,

we obtain

1. $(M \rightarrow (S_1 \wedge S_2))$

2. $(M \rightarrow P)$

3. $(F \rightarrow \neg M)$

4. $(((\neg M \wedge P) \wedge (S_1 \wedge S_2)) \rightarrow C)$

E 1-7   (parts 1 and 2)

1.

P	Q	$\neg((P \to Q) \to Q)$
T	T	F
T	F	F
F	T	F
F	F	T

2.

A	B	C	$((A \to (B \to C)) \leftrightarrow (A \wedge (B \wedge C)))$
T	T	T	T
T	T	F	T
T	F	T	F
T	F	F	F
F	T	T	F
F	T	F	F
F	F	T	F
F	F	F	F

E 1-16   We may use the following assignments: P: **F**, Q: **T**, and R: **T**. (This can also be stated as: $\mathbf{I}(P) = \mathbf{F}$, $\mathbf{I}(Q) = \mathbf{T}$, and $\mathbf{I}(R) = \mathbf{T}$.) Under this interpretation (of P, Q, R), we obtain the following values:

$$\mathbf{V_I}(P \to \neg Q) = (\mathbf{F} \to \neg \mathbf{T}) = \mathbf{T},$$

$$\mathbf{V_I}(Q \wedge R) = (\mathbf{T} \wedge \mathbf{T}) = \mathbf{T},$$

$$\mathbf{V_I}(\neg R \vee Q) = (\neg \mathbf{T} \vee \mathbf{T}) = \mathbf{T},$$

so **I** satisfies the given set of sentences.

E 1-33   *Metatheorem.* Let τ be a tautology and Γ be any set of $\mathcal{SC}$ sentences. Then $\Gamma \vDash_T \tau$.

Proof. Let **I** be any interpretation. Since τ is tautologous, it is true under **I**. Therefore, if **I** satisfies Γ, then **I** also satisfies τ. Thus, $\Gamma \vDash_T \tau$. **Q.E.D.**

<u>E 1-34</u>   *Metatheorem.* Let ρ be an $\mathcal{SC}$ sentence. Then $\varnothing \vDash_T \rho$ iff ρ is tautologous.

Proof. Suppose that $\varnothing \vDash_T \rho$, and let **I** be any interpretation. By Metatheorem 1-7, **I** satisfies $\varnothing$. Hence, **I** satisfies ρ. Hence, ρ is true under every interpretation, so it is tautologous.

Conversely, suppose that ρ is a tautology. Then it is true under every interpretation. Hence, any interpretation that satisfies $\varnothing$ also satisfies ρ, so $\varnothing \vDash_T \rho$.

Q.E.D.

<u>E 1-47</u>   (part 2)

  2.

$\{ Pr_1 \}$	(1)	$(A \vee B) \to C$	P
$\{ Pr_2 \}$	(2)	$R \to A$	P
$\{ Pr_3 \}$	(3)	R	P
$\{ Pr_2, Pr_3 \}$	(4)	A	MP (2), (3)
$\{ Pr_2, Pr_3 \}$	(5)	$A \vee B$	Add (4)
$\{ Pr_1, Pr_2, Pr_3 \}$	(6)	C	MP (1), (5)

<u>E 1-53</u>   (parts 1 and 4)

  1.

$\{ Pr_1 \}$	(1)	$A \vee (B \wedge C)$	P
$\{ Pr_1 \}$	(2)	$(A \vee B) \wedge (A \vee C)$	Dist (1)
$\{ Pr_1 \}$	(3)	$A \vee C$	Simp (2)
$\{ Pr_4 \}$	(4)	$A \to C$	P (for CP)
$\{ Pr_4 \}$	(5)	$\neg A \vee C$	CDis (4)
$\{ Pr_1, Pr_4 \}$	(6)	$C \vee C$	Cut (3), (5)
$\{ Pr_1, Pr_4 \}$	(7)	C	Taut (6)
$\{ Pr_1 \}$	(8)	$(A \to C) \to C$	C (4), (7)

4. This derivation uses RAA. Also try to prove it without using RAA.

{ Pr_1 }	(1)	$A \leftrightarrow \neg B$	P
{ Pr_2 }	(2)	$B \vee \neg C$	P
{ Pr_3 }	(3)	$C \rightarrow A$	P
{ Pr_4 }	(4)	$\neg\neg C$	P (for RAA)
{ Pr_1 }	(5)	$(A \rightarrow \neg B) \wedge (\neg B \rightarrow A)$	Equiv (1)
{ Pr_4 }	(6)	C	DN (4)
{ Pr_3, Pr_4 }	(7)	A	MP (3), (6)
{ Pr_1 }	(8)	$A \rightarrow \neg B$	Simp (5)
{ Pr_1, Pr_3, Pr_4 }	(9)	$\neg B$	MP (7), (8)
{ Pr_1, Pr_2, Pr_3, Pr_4 }	(10)	$\neg C$	DS (2), (9)
{ Pr_1, Pr_2, Pr_3 }	(11)	$\neg\neg C \rightarrow \neg C$	C (4), (10)
{ Pr_1, Pr_2, Pr_3 }	(12)	$\neg C$	Clav (11)

E 1-54   (part 3)

3. Let us try to construct a counterexample interpretation. To do this, we must have both C and P false. Hence, we need E false in order to make the third sentence true. Thus, we also need A true to make the first sentence true. But, with A true and E false, the antecedent of the fourth sentence is true, and this requires that F be true. Since A is true, we need the consequent of the second sentence to be true. This requires that Q be false. These considerations lead to the assignments:

$$A: \mathbf{T}, C: \mathbf{F}, E: \mathbf{F}, F: \mathbf{T}, P: \mathbf{F}, Q: \mathbf{F}.$$

To show that it really is a counterexample, we calculate the values of the sentences.

$$\mathbf{V_I}(A \vee E) = (\mathbf{T} \vee \mathbf{F}) = \mathbf{T},$$

$$\mathbf{V_I}(A \rightarrow ((P \vee Q) \rightarrow C)) = (\mathbf{T} \rightarrow ((\mathbf{F} \vee \mathbf{F}) \rightarrow \mathbf{F})) = \mathbf{T},$$

$$\mathbf{V_I}(\neg E \vee C) = (\neg \mathbf{F} \vee \mathbf{F}) = \mathbf{T},$$

$$\mathbf{V_I}((A \wedge \neg E) \rightarrow F) = ((\mathbf{T} \wedge \neg \mathbf{F}) \rightarrow \mathbf{T}) = \mathbf{T},$$

$$\mathbf{V_I}(C \vee P) = (\mathbf{F} \vee \mathbf{F}) = \mathbf{F}.$$

Thus, $\mathbf{I}$ satisfies the premises but not the conclusion, so it is a counterexample.

<u>E 1-61</u>   Proof of the first $\mathcal{SC}$ theorem:

1.

$\{\,Pr_1\,\}$	(1) $A \wedge B$	P (for CP)
$\{\,Pr_1\,\}$	(2) $(A \wedge B) \vee (\neg A \wedge \neg B)$	Add (1)
$\{\,Pr_1\,\}$	(3) $A \leftrightarrow B$	Equiv (2)
$\varnothing$	(4) $(A \wedge B) \rightarrow (A \leftrightarrow B)$	C (1), (3)

<u>E 1-64</u>   (part 1)

1.

$$(A \rightarrow B) \rightarrow (C \wedge A)$$

$$\neg(\neg A \vee B) \vee (C \wedge A)$$

$$(\neg\neg A \wedge \neg B) \vee (C \wedge A)$$

$$(A \wedge \neg B) \vee (C \wedge A)$$

$$((A \wedge \neg B) \vee C) \wedge ((A \wedge \neg B) \vee A)$$

$$(C \vee (A \wedge \neg B)) \wedge (A \vee (A \wedge \neg B))$$

$$((C \vee A) \wedge (C \vee \neg B)) \wedge ((A \vee A) \wedge (A \vee \neg B))$$

Since $(A \vee A)$ is equivalent to A, we can further simplify to

$$((C \vee A) \wedge (C \vee \neg B)) \wedge (A \wedge (A \vee \neg B)).$$

<u>E 1-67</u>   In CNF, the premises are $\neg A \vee (C \vee D)$, $\neg B \vee A$, $\neg D \vee C$, and B. In CNF, the negation of the conclusion is $\neg B \vee \neg C$. Here is the Resolution Derivation.

$\{\,Pr_1\,\}$	(1) $\neg A \vee (C \vee D)$	P
$\{\,Pr_2\,\}$	(2) $\neg B \vee A$	P
$\{\,Pr_3\,\}$	(3) $\neg D \vee C$	P

{ Pr_4 }	(4)	B	P
{ Pr_5 }	(5)	$\neg B \vee \neg C$	P
{ Pr_1, Pr_2 }	(6)	$\neg B \vee (C \vee D)$	Cut (1), (2)
{ Pr_1, Pr_2, Pr_4 }	(7)	$C \vee D$	DS (4), (6)
{ Pr_4, Pr_5 }	(8)	$\neg C$	DS (4), (5)
{ Pr_1, Pr_2, Pr_4, Pr_5 }	(9)	D	DS (8), (7)
{ Pr_3, Pr_4, Pr_5 }	(10)	$\neg D$	DS (3), (8)

There is a contradiction between lines (9) and (10).

Chapter 2 Answers

<u>E 2-3</u>   *Theorem.* For any sets, A, B, C, D,

$$\text{if } A \subseteq B \ \& \ C \subseteq D, \text{ then } A \cup C \subseteq B \cup D.$$

Proof. Let $A \subseteq B$ and $C \subseteq D$. Now let $x \in A \cup C$. Then, by the definition of $\cup$, $x \in A$ or $x \in C$. Since $A \subseteq B$, by the definition of $\subseteq$, if $x \in A$, then $x \in B$. Similarly, since $C \subseteq D$, if $x \in C$, then $x \in D$. Hence, by the $\mathcal{SC}$ CD rule, $x \in B$ or $x \in D$. Thus, $x \in B \cup D$. Hence, $x \in A \cup C$ implies $x \in B \cup D$. Therefore,

$$A \cup C \subseteq B \cup D. \quad \textbf{\textit{Q.E.D.}}$$

<u>E 2-6</u>   *Theorem.* For any sets, A, B,

$$A = B \ \Leftrightarrow \ (A \cup B) \subseteq (A \cap B).$$

Proof.

1. Suppose that $A = B$. Then, using substitutivity of identity, we have

$$A \cup B = A \cup A.$$

Also, for any x, by the $\mathcal{SC}$ Taut rule, and the definitions of union and intersection,

$$x \in A \cup A \ \Leftrightarrow \ (x \in A \text{ or } x \in A) \ \Leftrightarrow \ x \in A,$$

$$x \in A \cup A \ \Leftrightarrow \ x \in A \ \Leftrightarrow \ (x \in A \ \& \ x \in A) \ \Leftrightarrow \ x \in A \cap A.$$

Hence, by Extensionality, $A \cup A = A \cap A$, and thus $A \cup B = A \cap A$. Since, $A = B$, we have $A \cup B = A \cap B$. This stronger result implies what the theorem states, namely, $A \cup B \subseteq A \cap B$.

2. Conversely, suppose that $(A \cup B) \subseteq (A \cap B)$. Let $x \in A$. Then, by the Add rule,

$$x \in A \ \text{ or } \ x \in B,$$

so, by the definition of $\cup$,

$$x \in A \cup B.$$

But it is given that $(A \cup B) \subseteq (A \cap B)$, so by using the HS rule and the definition of subset, we have

$$x \in A \cap B.$$

By the definition of $\cap$, we also have, $x \in B$. This proves that $A \subseteq B$. Similarly, $B \subseteq A$, so by Theorem 2-4, $A = B$. **Q.E.D.**

E 2-33   *Theorem.* For any sets, A, B, $\mathcal{P}(A \cap B) = \mathcal{P}(A) \cap \mathcal{P}(B)$.

It will help to prove a lemma first.

Lemma. For any sets, A, B, S,

$$S \subseteq A \cap B \;\Leftrightarrow\; (S \subseteq A \;\&\; S \subseteq B).$$

Proof. Suppose that $S \subseteq A \cap B$. Let $x \in S$. Then, by the definition of subset, $x \in A \cap B$; so, by the definition of intersection, $x \in A \;\&\; x \in B$. Hence, $S \subseteq A \;\&\; S \subseteq B$.

Conversely, suppose that $(S \subseteq A \;\&\; S \subseteq B)$. Let $x \in S$. Then $x \in A \;\&\; x \in B$, so $x \in A \cap B$. Hence, $S \subseteq A \cap B$. This completes the proof of the lemma. (Remark. This lemma can also be proved by a sequence of biconditional statements. This should be tried for practice.)

Proof of the Theorem. Let S be any set. Then

$$
\begin{array}{llll}
S \in \mathcal{P}(A \cap B) & \Leftrightarrow & S \subseteq A \cap B & \text{Def. of } \mathcal{P} \\
 & \Leftrightarrow & (S \subseteq A \;\&\; S \subseteq B) & \text{by the lemma} \\
 & \Leftrightarrow & S \in \mathcal{P}(A) \;\&\; S \in \mathcal{P}(B) & \text{Def. of } \mathcal{P} \\
 & \Leftrightarrow & S \in \mathcal{P}(A) \cap \mathcal{P}(B) & \text{Def. } \cap
\end{array}
$$

Thus, by Extensionality, $\mathcal{P}(A \cap B) = \mathcal{P}(A) \cap \mathcal{P}(B)$. **Q.E.D.**

E 2-42   *Theorem.* Let A, B, C be nonempty sets. Then

$$A \times (B \cap C) = (A \times B) \cap (A \times C).$$

Proof. $A \times (B \cap C)$ is a set of ordered pairs, so any element of this set has the form $< x, y >$, where $x \in A$ and $y \in B \cap C$. We have

$$
\begin{aligned}
< x, y > \in A \times (B \cap C) \quad &\Leftrightarrow \quad x \in A \And y \in B \cap C && \text{Def. of } \times \\
&\Leftrightarrow \quad x \in A \And (y \in B \And y \in C) && \text{Def. of } \cap \\
&\Leftrightarrow \quad (x \in A \And x \in A) \And \\
&\qquad\qquad (y \in B \And y \in C) && \mathcal{SC} \text{ Taut} \\
&\Leftrightarrow \quad (x \in A \And y \in B) \And \\
&\qquad\qquad (x \in A \And y \in C) && \mathcal{SC} \text{ Assoc, Comm} \\
&\Leftrightarrow \quad < x, y > \in A \times B \And \\
&\qquad\qquad < x, y > \in A \times C && \text{Def. of } \times \\
&\Leftrightarrow \quad < x, y > \in \\
&\qquad\qquad (A \times B) \cap (A \times C) && \text{Def. } \cap
\end{aligned}
$$

The equation follows by Extensionality. ***Q.E.D.***

E 2-49   (part 1)

Theorem. Let R be a binary relation on a nonempty set A, and let R^{-1} be the converse of R. Then

1. If R is asymmetric on A, then R^{-1} is asymmetric on A.

Proof. Let x, $y \in A$ and suppose that $R^{-1}xy$. Then, by the definition of converse relation, we have Ryx. Since R is asymmetric, it follows that not Rxy. Hence, by the definition of converse, not $R^{-1}yx$. Therefore,

$$
R^{-1}xy \;\Rightarrow\; \text{not } R^{-1}yx,
$$

so R^{-1} is asymmetric. ***Q.E.D.***

E 2-56   Let Γ be the set of *all* $\mathcal{SC}$ sentences. For any ϕ, $\psi \in \Gamma$, define $T\phi\psi$ iff ϕ and ψ are tautologically equivalent. Then T is an equivalence relation on Γ.

Proof. For any $\mathcal{SC}$ interpretation **I**, we have

$$
\mathbf{V_I}(\phi) = \mathbf{V_I}(\phi),
$$

so by Metatheorem 1–16, ϕ is tautologically equivalent to ϕ. Hence, $T\phi\phi$, so T is reflexive.

Also, if T$\phi\psi$, then ϕ and ψ are tautologically equivalent. For any $\mathcal{SC}$ interpretation,

$$\mathbf{V_I}(\phi) = \mathbf{V_I}(\psi).$$

Hence,

$$\mathbf{V_I}(\psi) = \mathbf{V_I}(\phi).$$

Thus, $T\psi\phi$, so T is symmetric.

Finally, let ϕ, ψ, $\rho \in \Gamma$, and suppose that $T\phi\psi$ & $T\psi\rho$. Then ϕ and ψ are tautologically equivalent, and so are ψ and ρ. Let $\mathbf{I}$ be any $\mathcal{SC}$ interpretation. Then

$$\mathbf{V_I}(\phi) = \mathbf{V_I}(\psi)$$

and

$$\mathbf{V_I}(\psi) = \mathbf{V_I}(\rho).$$

Whence, $\mathbf{V_I}(\phi) = \mathbf{V_I}(\rho)$, so $T\phi\rho$. Thus, T is transitive. **Q.E.D.**

(Remark. This proof used Metatheorem 1–16 together with the fact that $=$ is an equivalence relation. From Metatheorem 1–15, it follows that T$\phi\psi$ iff $\phi \leftrightarrow \psi$ is tautologous. It is easy to prove that $\leftrightarrow$ is reflexive, symmetric, and transitive. Using these facts, one can construct a slightly different proof of the preceding result.)

<u>E 2-71</u>   *Theorem.* Let $\langle A, L\rangle$ be a strict simple order. Then

1. L is irreflexive on A.

2. For any x, $y \in A$, one and only one of the following holds: $x = y$, Lxy, Lyx.

Proof.

1. Since L is a strict simple order, L is asymmetric by definition. Moreover, any asymmetric relation is irreflexive, as this shows: By asymmetry, for any x, $y \in A$,

$$Lxy \Rightarrow \text{not } Lyx.$$

Since x and y can be any elements, they can be the same; so from the previous formula we get,

$$Lxx \Rightarrow \text{not } Lxx.$$

Hence, by the Clavius rule, not Lxx. Thus, L is irreflexive.

2. By definition, L is connected, so for any x, $y \in A$,

$$\text{if } x \neq y, \text{ then } (Lxy \text{ or } Lyx),$$

so

$$x = y \text{ or } (Lxy \text{ or } Lyx).$$

Now if $x = y$, then not Lxy and not Lyx, because L is irreflexive.

Suppose that Lxy. Since L is asymmetric, not Lyx. Also, if $x = y$, then Lxy would contradict the fact that L is irreflexive. Thus, if Lxy, then not Lyx and not $x = y$.

Similarly, if Lyx, then not Lxy and not $x = y$.

Therefore, exactly one of the following holds: $x = y$, Lxy, Lyx. Notice that this proof did not use the fact that L is transitive. **Q.E.D.**

<u>E 2-77</u>   The relation f on Re^+ is given by $f = \{\ < x, y > \mid xy = 7\ \}$.

1. We first show that f satisfies the definition of a function.

   (i) Let $x \in Re^+$. Then $y = (7/x) \in Re^+$, and $xy = x(7/x) = 7$. Thus, for any $x \in Re^+$, there exists $y \in Re^+$, such that $< x, y > \in f$, i.e., fxy.

   (ii) Suppose that fxy and fxz. Then $xy = 7$ and $xz = 7$, so

   $$xy = xz.$$

   Since $x \neq 0$, $y = z$ and f is a function from Re^+ to Re^+.

2. Let x_1, $x_2 \in Re^+$, and suppose that $f(x_1) = f(x_2) = y$. Then

   $$x_1 y = 7 = x_2 y.$$

   Since $y \neq 0$, $x_1 = x_2$, so f is one–one. Hence, it also has a unique inverse, which is its converse relation, by Theorem 2-43.

3. From 1, we have $f(x) = 7/x$. It is easy to guess that $f^{-1}(x) = 7/x$, and this can also be easily verified. Let $x \in Re^+$. Then

$$f^{-1}(f(x)) = f^{-1}(7/x) = 7/(7/x) = x.$$

(Remark. This is a special kind of one–one function for which $f = f^{-1}$. Notice that f is a symmetric relation.)

<u>E 2-87</u> *Theorem.* Let A, B, C be nonempty sets and suppose that

$$f : (A \cup B) \longrightarrow C.$$

Then

$$Img(f, (A \cup B)) = Img(f, A) \cup Img(f, B).$$

Proof.

1. Let $y \in Img(f, (A \cup B))$. By the definition of image, there exists $x \in A \cup B$ such that $y = f(x)$. Using the definition of union, if $x \in A \cup B$, then $x \in A$ or $x \in B$. Therefore, $y \in Img(f, A)$ or $y \in Img(f, B)$, so

$$y \in Img(f, A) \cup Img(f, B).$$

2. Conversely, let

$$y \in Img(f, A) \cup Img(f, B).$$

Then $y \in Img(f, A)$ or $y \in Img(f, B)$. In the first case, there is $x_1 \in A$ such that $y = f(x_1)$. In the second case, there is $x_2 \in B$ such that $y = f(x_2)$. Hence, there is some $x \in A$ or some $x \in B$ such that $y = f(x)$. Therefore, there is some $x \in A \cup B$ such that $y = f(x)$. Thus,

$$y \in Img(f, (A \cup B)).$$

This proves the converse.

We now have

$$y \in Img(f, (A \cup B)) \iff y \in Img(f, A) \cup Img(f, B),$$

so

$$Img(f, (A \cup B)) = Img(f, A) \cup Img(f, B).$$

$$\boldsymbol{Q.E.D.}$$

E 2-96   *Theorem.* Let $\langle A, L \rangle$ be a relational system such that, for any x, $y \in A$, one and only one of the following holds: $x = y$, Lxy, Lyx. Let

$$f : A \longrightarrow A$$

such that, for any x, $y \in A$,

$$Lxy \;\Rightarrow\; Lf(x)f(y).$$

Then f maps A one–one to A.

Proof. Let x, $y \in A$. Exactly one of the following holds: $x = y$, Lxy, Lyx. Also, since f maps A to A, $f(x)$, $f(y) \in A$. Therefore, exactly one of the following holds: $f(x) = f(y)$, $Lf(x)f(y)$, $Lf(y)f(x)$.

Suppose that $f(x) = f(y)$; then not $Lf(x)f(y)$ and also not $Lf(y)f(x)$. To prove that f is one-one, we need $x = y$. But we have exactly one of $x = y$, Lxy, or Lyx. If Lxy, then by the given assumption about f, we would have $Lf(x)f(y)$, which is already ruled out. Similarly, Lyx is impossible. Therefore, $x = y$, so f is one–one. *Q.E.D.*

Chapter 3 Answers

<u>E 3-1</u>   Let $M = \{\, a,\, b,\, c\,\}$, where a, b, c are alphabetic characters. Let

$$F : M \longrightarrow M,$$

with the values $F(a) = b$, $F(b) = c$, $F(c) = a$. The desired function is defined by

$$(1)\ r(0) = a$$

and

$$(2)\ \text{for every } k \in \textit{Nat},\ r(s(k)) = F(r(k)).$$

The requested values are

k	$r(k)$
$s(0)$	$F(r(0)) = F(a) = b,$
$s(s(0))$	$F(r(s(0))) = F(b) = c,$
$s(s(s(0)))$	$F(r(s(s(0)))) = F(c) = a,$
$s(s(s(s(0))))$	$F(r(s(s(s(0))))) = F(a) = b,$
$s(s(s(s(s(0)))))$	$F(r(s(s(s(s(0)))))) = F(b) = c.$

<u>E 3-10</u>   Complete the proof of Theorem 3-8, using results (in the text) that precede it.

Proof of the second equation. Let $\phi(k)$ be $0k = 0$, and use induction on k.

Case 0. By the definition of *times*, $0(0) = \textit{times}(0,\, 0) = 0$, so $\phi(0)$ is true.

Case $k + 1$. Assume $\phi(k)$. Using the definition of *times* and the induction hypothesis, we have

$$0(k + 1) = \textit{times}(0, k + 1) = 0 + \textit{times}(0, k) = 0 + 0k = 0 + 0 = 0,$$

where the last equality results from the definition of addition. This proves that $\phi(k)$ implies $\phi(k + 1)$.

The two cases together with IP imply that, for any $k \in Nat$, $0k = 0$.

For the next two proofs, recall that the text has shown, in Equation (3.1), that $s(k) = k + 1$, where $k \in Nat$.

Proof of the third equation. Let $\phi(k)$ be $1 + k = s(k)$.

Case 0. Using the definitions of addition and of 1,

$$1 + 0 = plus(1,0) = 1 = s(0),$$

which establishes $\phi(0)$.

Case $k + 1$. Assume $\phi(k)$. Using Equation (3.1) and the definition of addition,

$$1 + (k + 1) = plus(1, k + 1) = plus(1, s(k)) = s(plus(1, k)) = s(1 + k).$$

Using the induction hypothesis $\phi(k)$ and (3.1), we obtain

$$1 + (k + 1) = s(s(k)) = s(k + 1).$$

Hence, $\phi(k)$ implies $\phi(k + 1)$.

Proof of the fourth equation. Let $\phi(k)$ be $1k = k$.

Case 0. Using the definition of multiplication, $1k = 1(0) = 0 = k$.

Case $k + 1$. Assume $\phi(k)$. Then

$$1(k + 1) = times(1, k + 1) = 1 + times(1, k) = 1 + 1k.$$

Applying the induction hypothesis,

$$1(k + 1) = 1 + 1k = 1 + k.$$

By the previously proved equation, $1 + k = s(k)$. Also by (3.1), $s(k) = k + 1$, so

$$1(k + 1) = 1 + k = k + 1,$$

which completes the induction step. ***Q.E.D.***

E 3-25   It is easy to guess that we have a sequence of squares. With a little more work, one can construct the table of values given below.

k	r_k
0	0
1	$1 = (0+1)^2 = (r_0 + 1)^2$
2	$4 = (1+1)^2 = (r_1 + 1)^2$
3	$25 = (r_2 + 1)^2$
4	$676 = (r_3 + 1)^2$
5	$458329 = (r_4 + 1)^2$
6	$210066388900 = (r_5 + 1)^2$
7	$44127887745906175987801 = (r_6 + 1)^2$

The table suggests the recursion equation, $r_{k+1} = (r_k + 1)^2$, for $k \in Nat$, and the definition,

$$r_k = r(k) = [\text{ if } k = 0 \text{ then } 0 \text{ else } (r(k-1) + 1)^2] .$$

It is easy to check that this definition returns the values in the table.

Theorem. Let r_k be defined as above. Then, for any $k \in Nat$,

$$r_{k+1} - r_k = r_k^2 + r_k + 1.$$

Proof. Directly from the definition of r_k, we have

$$r_{k+1} = (r_k + 1)^2 = r_k^2 + 2r_k + 1.$$

Hence,

$$r_{k+1} - r_k = r_k^2 + r_k + 1. \textbf{ Q.E.D.}$$

E 3-46   *Lemma.* For any $k \in Nat$, $0 < k!$.

Proof. Let $k \in Nat$ and let $\phi(k)$ be $0 < k!$. Use induction on k.
Case 0. By definition, $0! = 1 > 0$.

Case $k + 1$. From the definition, $(k + 1)! = (k + 1)k!$. From the induction hypothesis, $0 < k!$. Also, $0 < k + 1$. It follows from Equation (3.3) in Theorem 3-12 that the product of nonzero numbers is nonzero. Hence, it further follows that in Nat, the product of numbers greater than zero is greater than zero.

Therefore, $0 < (k+1)k! = (k+1)!$. **Q.E.D.**

Theorem. Let $k \in Nat$ and $k \geqslant 4$. Then $2^k < k!$.

Proof. Let $k \geqslant 4$ and let $\phi(k)$ be: $2^k < k!$. I use induction on k, with the base case of 4.

Case $k = 4$. By direct calculation, we have $2^4 = 16$ and $4! = 24$, so $2^4 < 4!$.

Case $k+1$, where $k \geqslant 4$. By the induction hypothesis, $2^k < k!$. Also, $0 < (k+1)$, so by using Lemma 3-30, and the definition of factorial, we have

$$(k+1)2^k < (k+1)k! = (k+1)!.$$

Since $k \geqslant 4$, $k+1 > 4 > 2$. We are assuming the laws of exponents, from which it follows that $0 < 2^k$, so using Lemma 3-30 again, we obtain

$$2^{k+1} = (2^k)(2) < (2^k)(k+1) = (k+1)2^k.$$

By the transitivity of $<$,

$$2^{k+1} < (k+1)!,$$

which is $\phi(k+1)$. This completes the induction step. **Q.E.D.**

E 3-63 *Metatheorem.* Let τ be a tautology, i, j, $m \in Nat$, and ϕ, ψ, χ be $\mathcal{SC}$ sentences. Define a sequence of sets of sentences as follows:

$$\Gamma_0 = \{\tau\},$$

and, for any $m > 0$,

$$\Gamma_m = \{(\phi \wedge \psi) \mid \text{there exist } i, j < m \text{ such that } \phi \in \Gamma_i \,\&\, \psi \in \Gamma_j\}.$$

Now let

$$\Delta = \bigcup\{\Gamma_k \mid k \in Nat\}.$$

Then for any $\chi \in \Delta$, χ is tautologous.

Proof. Let $\chi \in \Delta$. Then, since Δ is the union of all Γ_k, it follows that, for some $k \in Nat$, $\chi \in \Gamma_k$. It therefore suffices to prove that, for all $k \in Nat$, any sentence in Γ_k is tautologous. This will be done by CVI. Let $\phi(t)$ be: Any sentence in Γ_t is tautologous.

Case 0. The only sentence in Γ_0 is τ, which is given to be a tautology. Therefore, any sentence in Γ_0 is tautologous.

Case $k > 0$. Assume the CVI hypothesis: For all $t < k$, any sentence in Γ_t is tautologous. Let $\chi \in \Gamma_k$. By the recursive definition, there exist $\phi \in \Gamma_i$ and $\psi \in \Gamma_j$, with $i, j < k$, such that χ is $(\phi \wedge \psi)$. By the CVI hypothesis, both ϕ and ψ are tautologies, so are true under every interpretation. Hence, their conjunction χ is true under every interpretation and is therefore also a tautology. This shows that any sentence in Γ_k is tautologous and completes the induction step. **Q.E.D.**

E 3-69   Prove that, for any $k \in Nat$, $3|k(k+1)(k+2)$.

Proof. By the Division Algorithm Theorem, there exist unique q, r such that $r < 3$ and $k = 3q + r$. We consider cases.

Case $r = 0$. Then $k = 3q$, so $3|k$.

Case $r = 1$. Then $k = 3q + 1$, so
$k + 2 = 3q + 3 = 3(q+1)$, and $3|(k+2)$.

Case $r = 2$. Then $k = 3q + 2$, so $k + 1 = 3q + 3 = 3(q+1)$,
and $3|(k+1)$.

In each case, 3 divides one of the factors of $k(k+l)(k+2)$. Since divisibility is transitive, in each case, $3|k(k+1)(k+2)$. **Q.E.D.**

E 3-75   Let $\langle A, R \rangle$ be a relational system in which A has $n \neq 0$ elements. Let R be connected, symmetric, and irreflexive on A, and let $p(n)$ be the number of elements of R when A has n elements. Then $p(n) = n(n-1)$.

1. *Definition.*
If $n = 1$, then $p(n) = 0$. Suppose that we have n elements in A and $p(n)$ in R. Let a_i be any element in A. If we add a new element b to A, then R acquires all new pairs of the forms, $< a_i, b >$ and $< b, a_i >$, since R is connected and symmetric. There are $2n$ pairs with these forms. Thus, after adding b to A, R now has $p(n+1) = p(n) + 2n$ pairs, so we obtain the definition

$$p(1) = 0,$$

and, for $n \in Nat^+$,

$$p(n+1) = p(n) + 2n.$$

2. *Theorem.* For every $n \in Nat^+$, $p(n) = n(n-1)$.

Proof. Let $\phi(k)$ be $p(k) = k(k-1)$, for $0 < k$.

Case $k = 1$. $p(k) = p(1) = 0$. Also, $k(k-1) = 1(0) = 0$. Hence, in this case, $p(k) = k(k-1)$.

Case $k + 1$. From the recursive definition, we have

$$p(k+1) = p(k) + 2k.$$

and from the induction hypothesis, we have

$$p(k) = k(k-1).$$

Substituting the latter into the former,

$$p(k+1) = k(k-1) + 2k = k^2 - k + 2k = k^2 + k = (k+1)k.$$

Hence,

$$p(k+1) = (k+1)k = (k+1)((k+1) - 1),$$

which is $\phi(k+1)$. This completes the induction step. ***Q.E.D.***

E 3-77   *Theorem.* For any strings, σ, γ,

$$length_s(append_s(\sigma, \gamma)) = length_s(\sigma) + length_s(\gamma).$$

Proof. Let $k \in Nat^+$. We use induction on the length k of σ. The induction hypothesis is $\phi(k)$: For any string γ and any string σ of length k,

$$length_s(append_s(\sigma, \gamma)) = length_s(\sigma) + length_s(\gamma).$$

Case $k = 1$. If the length of σ is 1, then σ is a character. By the definition of $append_s$ (Definition 3-44),

$$append_s(\sigma, \gamma) = prfx(\sigma, \gamma),$$

and by the definition of $length_s$ (Definition 3-43),

$$length_s(prfx(\sigma, \gamma)) = 1 + length_s(\gamma).$$

Hence,

$$length_s(append_s(\sigma, \gamma)) = length_s(\sigma) + length_s(\gamma).$$

Case $k + 1$. Let $length_s(\sigma) = k + 1$. By Definition 3-44,

$$append_s(\sigma, \gamma) = prfx(head_s(\sigma), append_s(tail_s(\sigma), \gamma)),$$

and

$$length_s(prfx(head_s(\sigma), append_s(tail_s(\sigma), \gamma))) =$$

$$1 + length_s(append_s(tail_s(\sigma), \gamma)).$$

But $length_s(\sigma) = k + 1$, so $length_s(tail_s(\sigma)) = k$. Therefore, by the induction hypothesis we have

$$length_s(append_s(tail_s(\sigma), \gamma)) = length_s(tail_s(\sigma)) + length_s(\gamma).$$

Therefore,

$$length_s(append_s(\sigma, \gamma)) = 1 + length_s(tail_s(\sigma)) + length_s(\gamma),$$

so

$$length_s(append_s(\sigma, \gamma)) = length_s(\sigma) + length_s(\gamma),$$

which establishes $\phi(k + 1)$. **Q.E.D.**

E 3-82   Let λ be any list with length $n \geqslant 1$. Let $1 \leqslant k \leqslant n$. Define the function f as follows:

$$f(k, \lambda) = [\text{ if } k = 1 \text{ then } tail(\lambda) \text{ else } tail(f(k - 1, \lambda))].$$

Prove that

$$k + length(f(k, \lambda)) = length(\lambda).$$

Proof. Let $k \in Nat^+$. I use the induction hypothesis $\phi(k)$: For any list λ with $k \leqslant length(\lambda)$,

$$k + length(f(k, \lambda)) = length(\lambda).$$

Case $k = 1$. In this case, we have $f(1, \lambda) = tail(\lambda)$. Also,

$$length(tail(\lambda)) = length(\lambda) - 1,$$

so

$$k + length(f(k, \lambda)) = 1 + length(f(1, \lambda)) = 1 + length(tail(\lambda)),$$

so

$$k + length(f(k, \lambda)) = length(\lambda).$$

This establishes the base case.

Case $k + 1$. Assume $\phi(k)$, and let λ be a list with length not less than $k + 1$. We have

$$f(k + 1, \lambda) = tail(f(k, \lambda)).$$

$$length(f(k + 1, \lambda)) = length(tail(f(k, \lambda))) = length(f(k, \lambda)) - 1.$$

$$1 + length(f(k + 1, \lambda)) = length(f(k, \lambda)).$$

But $length(\lambda) > k$, so by $\phi(k)$,

$$length(f(k, \lambda)) = length(\lambda) - k,$$

so

$$1 + length(f(k + 1, \lambda)) = length(\lambda) - k,$$

or

$$(k + 1) + length(f(k + 1, \lambda)) = length(\lambda).$$

By the last equation, we have established $\phi(k + 1)$, which completes the induction step. ***Q.E.D.***

Chapter 4 Answers

<u>E 4-2</u>   (parts 1–5)

1. Formula: $P_3 \rightarrow Rxy$.

2. Not a formula: P_3^1 must be followed by one individual constant or variable.

3. Not a formula: b is not a formula, so it may not be conjoined with anything else.

4. Formula: $Raxr \lor (\forall x)Qxy$.

5. Formula: $(\forall x)(\exists y)Rxy$.

<u>E 4-5</u>   (parts 1–5)

1. $(\exists x)Wxc$. (Remark. Any other variable could have been used in place of x. For instance, we may write, $(\exists z)Wzc$.)

2. $(\exists x)(Wrx \land Mx)$.

3. $(\exists x)(Wrx \land Mx) \rightarrow (\exists y)(Wry \land Py)$. (Remark. Since the antecedent and consequent are separate sentences, we may also use x as the variable in the consequent instead of y.)

4. $(\exists z)Lzf$.

5. $(\forall x)(F_1 x \rightarrow (\exists y)Lyx)$. (Remarks. The order of occurrence of the two variables after L must agree with the meaning of the English sentence, the specified interpretation of L, and with the variable after F_1. Although it is not as similar in structure to the English sentence, we may also use the symbolization,

$$(\forall x)(\exists y)(F_1 x \rightarrow Lyx).$$

Later developments in this chapter enable one to prove that these two sentences are logically equivalent. However, in general, changing the position of quantifiers in a sentence produces a nonequivalent sentence, or a nonsentence. Such changes should not be made without justification by the logical theory.)

NOTE: In the following, if $\mathbf{I}$ is an interpretation, I will use $\mathbf{I}$ as an abbreviation for $\mathbf{I}_{CONST}$, $\mathbf{I}_{SC}$, $\mathbf{I}_{PRED,k}$ according to the type of argument that is passed to it.

E 4-11   (parts 1 and 3)

1. We may use the interpretation $\mathbf{I}$ with domain $\mathbf{D} = \{0, 1\}$, and

$\quad A \;:\; \{0, 1\}$

$\quad B \;:\; \{0\}$

$\quad C \;:\; \{0\}$

$\quad b \;:\; 0$

Under this interpretation, Ab, Bb, Cb are all true, since $\mathbf{I}(b) = 0$ and 0 is an element of each of $\mathbf{I}(A)$, $\mathbf{I}(B)$, $\mathbf{I}(C)$.

Also, $(\exists x)(Ax \wedge \neg(Bx \vee Cx))$ is true under $\mathbf{I}$, since $1 \in \mathbf{I}(A)$, but it is not the case that 1 is in either $\mathbf{I}(B)$ or in $\mathbf{I}(C)$. This can be stated more simply by writing

$$1 \in \mathbf{I}(A) \cap (\mathbf{I}(B) \cup \mathbf{I}(C))'.$$

Finally, the only element in $\mathbf{I}(C)$ is 0, which is also an element of $\mathbf{I}(A)$, so $(\forall x)(Cx \rightarrow Ax)$ is true under $\mathbf{I}$. The truth of this sentence follows from the fact that $\mathbf{I}(C) \subseteq \mathbf{I}(A)$.

3. Let $\mathbf{I}$ have domain $\mathbf{D} = \{1, 2, 3\}$ with

$\quad A \;:\; \{1\}$

$\quad B \;:\; \{2, 3\}$

$\quad R \;:\; \{<1, 2>, <2, 1>, <2, 3>, <3, 2>\}$

Both $\mathbf{I}(A)$ and $\mathbf{I}(B)$ are nonempty, so

$$(\exists x)Ax \wedge (\exists x)Bx$$

is true under $\mathbf{I}$.

Since $\mathbf{I}(A) \cap \mathbf{I}(B) = \varnothing$, the sentence $\neg(\exists x)(Ax \wedge Bx)$ is true under $\mathbf{I}$. By inspection of the ordered pairs in $\mathbf{I}(R)$, one sees that this relation on $\mathbf{D}$ is irreflexive and symmetric, so the sentences

$$(\forall x)\neg Rxx, \ (\forall x)(\forall y)(Rxy \rightarrow Ryx)$$

are true.

Consider any element in $\mathbf{I}(A)$. There is only one, namely, 1. We have $< 1, 2 > \in \mathbf{I}(R)$, and $2 \in \mathbf{I}(B)$. Thus, any element in $\mathbf{I}(A)$ is related (by $\mathbf{I}(R)$) to some element in $\mathbf{I}(B)$. Hence, $(\forall x)(Ax \rightarrow (\exists y)(By \wedge Rxy))$ is true.

Finally, consider any element in $\mathbf{I}(B)$. There are two such elements, 2 and 3. Each of these is related to the other. Therefore, any element in $\mathbf{I}(B)$ is related to some element of $\mathbf{I}(B)$; hence,

$$(\forall x)(Bx \rightarrow (\exists y)(By \wedge Rxy))$$

is true under $\mathbf{I}$.

<u>E 4-17</u>   To motivate the solution, consider this: A, B, C must be interpreted as subsets of the domain, say, $\mathbf{A}$, $\mathbf{B}$, $\mathbf{C}$, respectively. The sentence

$$(\forall x)(Ax \rightarrow (Bx \vee \neg Cx))$$

asserts that $\mathbf{A} \subseteq (\mathbf{B} \cup \mathbf{C}')$. Thus, to make a counterexample, we need an element in $\mathbf{A}$ that is neither in $\mathbf{B}$ nor in $\mathbf{C}'$. Here is such an interpretation.

Let $\mathbf{I}$ have domain $\mathbf{D} = \{0, 1\}$ with

$A \ : \ \{0\} = \mathbf{A}$

$B \ : \ \{1\} = \mathbf{B}$

$C \ : \ \{0\} = \mathbf{C}$

Then we have $0 \in \mathbf{A}$. But $0 \notin \mathbf{B}$ and (since $0 \in \mathbf{C}$) $0 \notin \mathbf{C}'$. The sentence is false under this interpretation, so it is not valid.

E 4-24   (part 1)

1. Let **I** have domain **D** $= Nat$ with

$$A : \{x \mid x \text{ is even}\} = \mathbf{A}.$$

Then $(\exists y)Ay$ is true under **I** since 2 is an even number. But $(\forall x)Ax$ is false under **I** since not all natural numbers are even. In particular, 3 is not even. (Remark. Clearly, many other interpretations would have worked just as well with these two sentences.)

E 4-43   (parts 1 and 4)

1.

$\{\ Pr_1\ \}$	(1)	$(\forall x)(\forall y)(Rxy \to \neg Ryx)$	P
$\{\ Pr_1\ \}$	(2)	$(\forall y)(Ray \to \neg Rya)$	UI (1)
$\{\ Pr_1\ \}$	(3)	$Raa \to \neg Raa$	UI (2)
$\{\ Pr_1\ \}$	(4)	$\neg Raa$	TC (3)
$\{\ Pr_1\ \}$	(5)	$(\forall x)\neg Rxx$	UG (4)

4.

$\{\ Pr_1\ \}$	(1)	$(\forall x)(\forall y)((Gx \wedge Nxy) \to Ly)$	P
$\{\ Pr_2\ \}$	(2)	$(\exists x)(Gx \wedge Nxa)$	P
$\{\ Pr_3\ \}$	(3)	$Gb \wedge Nba$	P (for ExEx)
$\{\ Pr_1\ \}$	(4)	$(\forall y)((Gb \wedge Nby) \to Ly)$	UI (1)
$\{\ Pr_1\ \}$	(5)	$(Gb \wedge Nba) \to La$	UI (4)
$\{\ Pr_1, Pr_3\ \}$	(6)	La	TC (3), (5)
$\{\ Pr_1, Pr_3\ \}$	(7)	$(\exists z)Lz$	EG (6)
$\{\ Pr_1\ \}$	(8)	$(Gb \wedge Nba) \to (\exists z)Lz$	C (3), (7)
$\{\ Pr_1\ \}$	(9)	$(\exists x)(Gx \wedge Nxa) \to (\exists z)Lz$	EA (8)
$\{\ Pr_1, Pr_2\ \}$	(10)	$(\exists z)Lz$	TC (2), (9)

E 4-62   *Metatheorem* (Quantifier Exchange). Let Γ be a set of $\mathcal{PC}$ sentences, and let ϕ be a formula. Then

$$\Gamma \vDash \neg(\exists\nu)\phi \;\Leftrightarrow\; \Gamma \vDash (\forall\nu)\neg\phi.$$

Similar relationships hold for the pairs:

$(\exists\nu)\neg\phi$	$\neg(\forall\nu)\phi$
$\neg(\exists\nu)\neg\phi$	$(\forall\nu)\phi$
$(\exists\nu)\phi$	$\neg(\forall\nu)\neg\phi$

Proof. Suppose that $\Gamma \vDash \neg(\exists\nu)\phi$. Let $\mathbf{I}$ be any interpretation that satisfies Γ. For RAA, suppose that $\mathbf{I}$ does not satisfy $(\forall\nu)\neg\phi$. Let κ be a constant that does not occur in $\neg\phi$. Then there is a variant $\mathbf{I}_\kappa$ of $\mathbf{I}$ with respect to κ such that $\neg\phi\,[^\nu/_\kappa]$ is false under $\mathbf{I}_\kappa$. Hence, $\phi\,[^\nu/_\kappa]$ is true under $\mathbf{I}_\kappa$. Thus, $(\exists\nu)\phi$ is true under $\mathbf{I}$, so $\neg(\exists\nu)\phi$ is false under $\mathbf{I}$. This contradicts the assumption that $\Gamma \vDash \neg(\exists\nu)\phi$.

Conversely, suppose that $\Gamma \vDash (\forall\nu)\neg\phi$. Let $\mathbf{I}$ be any interpretation that satisfies Γ. Then $(\forall\nu)\neg\phi$ is true under $\mathbf{I}$. Let κ be a constant not in $\neg\phi$. Then $\neg\phi\,[^\nu/_\kappa]$ is true under every variant of $\mathbf{I}$ with respect to κ. Hence, $\phi\,[^\nu/_\kappa]$ is false under every such variant of $\mathbf{I}$. Thus, there is no variant of $\mathbf{I}$ with respect to κ under which $\phi\,[^\nu/_\kappa]$ is true, from which $(\exists\nu)\phi$ is false under $\mathbf{I}$. Therefore, if $\mathbf{I}$ is any interpretation satisfying Γ, then $\neg(\exists\nu)\phi$ is true under $\mathbf{I}$. Therefore, $\Gamma \vDash \neg(\exists\nu)\phi$.

The other pairs can be proved equivalent by similar arguments. *Q.E.D.*

E 4-71   (part 1)

1. Let $\mathbf{I}$ have domain $\mathbf{D} = \{0, 1\}$ with

$$A \,:\, \{\,0\,\} = \mathbf{A}, \quad B \,:\, \{1\} = \mathbf{B}, \quad a \,:\, 0.$$

The only element in $\mathbf{A}$ is 0, and $\mathbf{I}(a) = 0$, so

$$(\forall x)(Ax \to x = a)$$

is true under $\mathbf{I}$. This can be stated more precisely as follows: Consider

$$Ab \to b = a.$$

Let $\mathbf{I}_b$ be any variant of $\mathbf{I}$ with respect to b. There are only two elements that can be assigned to b, namely, 0 and 1. If we let $\mathbf{I}_b(b) = 1$, then Ab is false, so

$Ab \rightarrow b = a$ is true. If we let $\mathbf{I}_b(b) = 0$, then Ab is true, but also $b = a$ is true, so again $Ab \rightarrow b = a$ is true. Since $Ab \rightarrow b = a$ is true under every variant of $\mathbf{I}$ with respect to b,

$$(\forall x)(Ax \rightarrow x = a)$$

is true under $\mathbf{I}$.

The sentence $\neg Ba$ is true under $\mathbf{I}$, since 0 is not in $\mathbf{B}$.

The sentence $(\exists x)(\exists y)x \neq y$ is true under the interpretation because there are at least two elements in $\mathbf{D}$, namely, 0 and 1.

E 4-72   Let $\mathbf{I}$ have domain $\mathbf{D} = Nat$ with

$$A : \{\, x \mid x \text{ is even} \,\} = \mathbf{A}, \quad B : \{\, x \mid x \text{ is odd} \,\} = \mathbf{B},$$

$$f : \text{ the successor function on } Nat, \quad g : \text{ the successor function on } Nat.$$

We have $(\exists x)Ax$ true under $\mathbf{I}$, since $\mathbf{A}$ is nonempty.

Since any even number is not odd, $(\forall x)(Ax \rightarrow \neg Bx)$ is true.

If a number n is even, then its successor $n + 1$ is odd, so $(\forall x)(Ax \rightarrow Bf(x))$ is true.

Since the successor of an odd number is even,

$$(\forall x)(Bx \rightarrow Ag(x))$$

is also true.

Thus, all four premises are true under $\mathbf{I}$. Now notice that $2 \in \mathbf{A}$. But the successor of the successor of 2 is 4. Since $4 \neq 2$, the conclusion,

$$(\forall x)(Ax \rightarrow g(f(x)) = x),$$

is false under the interpretation. Hence, the argument is unsound.

E 4-74

$\{ Pr_1 \}$	(1)	$a = b$	P
$\varnothing$	(2)	$f(a) = f(a)$	I
$\{ Pr_1 \}$	(3)	$f(a) = f(b)$	I (1), (2)
$\varnothing$	(4)	$a = b \to f(a) = f(b)$	C (1), (3)
$\varnothing$	(5)	$(\forall y)(a = y \to f(a) = f(y))$	UG (4)
$\varnothing$	(6)	$(\forall x)(\forall y)(x = y \to f(x) = f(y))$	UG (5)

E 4-76   (part 2)

2.

$\{ Pr_1 \}$	(1)	$(\exists x)Ax$	P
$\{ Pr_2 \}$	(2)	$(\exists x)\neg Ax$	P
$\{ Pr_3 \}$	(3)	Aa	P (for ExEx)
$\{ Pr_4 \}$	(4)	$\neg Ab$	P (for ExEx)
$\{ Pr_5 \}$	(5)	$a = b$	P (for RAA)
$\{ Pr_4, Pr_5 \}$	(6)	$\neg Aa$	I (4), (5)
$\{ Pr_3, Pr_4, Pr_5 \}$	(7)	$a \neq b$	TC (3), (6)
$\{ Pr_3, Pr_4 \}$	(8)	$a = b \to a \neq b$	C (5), (7)
$\{ Pr_3, Pr_4 \}$	(9)	$a \neq b$	TC (8)
$\{ Pr_3, Pr_4 \}$	(10)	$(\exists v)a \neq v$	EG (9)
$\{ Pr_3, Pr_4 \}$	(11)	$(\exists u)(\exists v)u \neq v$	EG (10)
$\{ Pr_3 \}$	(12)	$\neg Ab \to (\exists u)(\exists v)u \neq v$	C (4), (11)
$\{ Pr_3 \}$	(13)	$(\exists x)\neg Ax \to (\exists u)(\exists v)u \neq v$	EA (12)
$\{ Pr_3, Pr_2 \}$	(14)	$(\exists u)(\exists v)u \neq v$	TC (2), (13)
$\{ Pr_2 \}$	(15)	$Aa \to (\exists u)(\exists v)u \neq v$	C (3), (14)
$\{ Pr_2 \}$	(16)	$(\exists x)Ax \to (\exists u)(\exists v)u \neq v$	EA (15)
$\{ Pr_2, Pr_1 \}$	(17)	$(\exists u)(\exists v)u \neq v$	TC (1), (16)

The Greek Alphabet

NAME	UPPER CASE	LOWER CASE
Alpha	A	α
Beta	B	β
Gamma	Γ	γ
Delta	Δ	δ
Epsilon	E	ε
Zeta	Z	ζ
Eta	H	η
Theta	Θ	θ
Iota	I	ι
Kappa	K	κ
Lambda	Λ	λ
Mu	M	μ
Nu	N	ν
Xi	Ξ	ξ
Omicron	O	o
Pi	Π	π
Rho	P	ρ
Sigma	Σ	σ
Tau	T	τ
Upsilon	Υ	υ
Phi	Φ	ϕ
Chi	X	χ
Psi	Ψ	ψ
Omega	Ω	ω

Glossary of Symbols

References

Abelson, H., and G. Sussman, 1996. *The Structure and Interpretation of Computer Programs*, 2nd ed., MIT Press, Cambridge, MA. (Introduction to programming theory and techniques; uses the Scheme version of LISP.)

Benacerraf, P., and H. Putnam, eds., 1983. *Philosophy of Mathematics: Selected Readings*, 2nd ed., Cambridge University Press, New York. (Anthology of important articles.)

Berge, C., 1962. *The Theory of Graphs and Its Applications*, translated by Alison Doig, John Wiley, New York. (Develops basic graph theory and applies the results to many interesting, and historically significant, problems.)

Beth, E. W., 1966. *The Foundations of Mathematics*, Harper & Row, New York. (Advanced book on logic, paradoxes, and philosophy of mathematics.)

Birkhoff, G., and S. MacLane, 1953. *A Survey of Modern Algebra*, revised ed., Macmillan, New York. (Introduction to algebraic problems and systems.)

Bledsoe, W. W., 1977. "Non-Resolution Theorem Proving," *Artificial Intelligence*, Vol. 9, No. 1, pp. 1–35. (Reprinted in Webber and Nilsson (1981).)

Bonevac, D., 1987. *Deduction*, Mayfield Publishing Co., Palo Alto, CA.

Boyer, R. S., and J. S. Moore, 1979. *A Computational Logic*, Academic Press, New York.

Boyer, R. S., and J. S. Moore, 1998. *A Computational Logic Handbook*, 2nd ed., Academic Press, New York.

Burstall, R. M., 1969. "Proving Properties of Programs by Structural Induction," *The Computer Journal*, Vol. 12, No. 1, pp. 41–48.

Causey, R. L., 1977. *Unity of Science*, D. Reidel, Dordrecht and Boston.

Causey, R. L., 1991. "The Epistemic Basis of Defeasible Reasoning," *Minds and Machines*, Vol. 1, No. 4, pp. 437–458.

Causey, R. L., 1994. "EVID: A System for Interactive Defeasible Reasoning," *Decision Support Systems*, Vol. 11, Issue 2, pp. 103–131.

Causey, R. L., 2003. "Computational Dialogic Defeasible Reasoning," *Argumentation*, Vol. 17, No. 4, pp. 421–450.

Chandy, K. M., and S. Taylor, 1992. *An Introduction to Parallel Programming*, Jones and Bartlett, Boston.

Covington, M. A., *et al.*, 1988. *Prolog Programming in Depth*, Scott, Foresman and Co., Glenview, IL.

Dalen, D. van, and A. F. Monna, 1972. *Sets and Integration: An Outline of the Development*, Wolters-Noordhoff Publishing, Groningen, Holland. (Historical material.)

Dalen, D. van, H. C. Doets, and H. de Swart, 1978. *Sets: Naive, Axiomatic and Applied*, Pergamon Press, Oxford. (Extensive development of set theory with some advanced mathematical applications.)

Dodd, T., 1990. *Prolog: A Logical Approach*, Oxford University Press, Oxford.

Dijkstra, E. W., and C. S. Scholten, 1990. *Predicate Calculus and Program Semantics*, Springer-Verlag, New York.

Enderton, H. B., 2001. *A Mathematical Introduction to Logic*, 2nd ed., Academic Press, New York. (Mathematical logic, including some model theory, undecidability results, and higher order logic.)

Escher, M. C., 1971. *The Graphic Work of M. C. Escher*, Ballantine Books, New York.

Feferman, S., 1991. "Proofs of Termination and the '91' Function," in Lifschitz (1991), pp. 47–63.

Gardner, M., 1982. *Logic Machines and Diagrams*, University of Chicago Press, Chicago. (Historical discussions and portrayals.)

Genesereth, M. R., and N. J. Nilsson, 1987. *Logical Foundations of Artificial Intelligence*, Morgan Kaufmann, San Mateo, CA.

Ginsberg, M. L., ed., 1987. *Readings in Nonmonotonic Reasoning*, Morgan Kaufmann, San Mateo, CA.

Gries, D., 1981. *The Science of Programming*, Springer-Verlag, New York.

Halmos, P. R., 1974. *Naive Set Theory*, Springer-Verlag, New York. (Systematic introduction to set theory.)

Harel, D., 1992. *Algorithmics: The Spirit of Computing*, 2nd ed., Addison-Wesley, Reading, MA.

Hartshorne, C., and P. Weiss, eds., 1960. *Collected Papers of Charles Sanders Peirce*, Vols. III–IV, Belknap Press, Cambridge, MA.

Heath, T. L., ed., 1956. *The Thirteen Books of Euclid's Elements*, 2nd ed., Dover Publications, New York. (In three volumes.)

Hoare, C. A. R., and J. C. Shepherdson, eds., 1985. *Mathematical Logic and Programming Languages*. Prentice-Hall International, Englewood Cliffs, NJ. (Collection of research articles.)

Hochberg, H., 1981. "The Wiener–Kuratowski Procedure and the Analysis of Order," *Analysis*, Vol. 41, No. 4, pp. 161–163.

Hofstadter, D. R., 1979. *Gödel, Escher, Bach: An Eternal Golden Braid*, Basic Books, New York. (A charming, eclectic book. Among other things, it contains many illustrations of mathematical ideas, including the use of recursion, in art and music.)

Horowitz, E., and S. Sahni, 1978. *Fundamentals of Computer Algorithms*, Computer Science Press, Rockville, MD.

Horowitz, E. and S. Sahni, 1982. *Fundamentals of Data Structures*, Computer Science Press, Rockville, MD.

Kalish, D., R. Montague, and G. Mar, 1980. *Logic: Techniques of Formal Reasoning*, Harcourt Brace Jovanovich, New York.

Knuth, D. E., 1991. "Textbook Examples of Recursion," in Lifschitz (1991), pp. 207–229.

Kowalski, R., 1979. *Logic for Problem Solving*, North Holland, New York. (Applications of logic treated with techniques suitable for logic programming.)

LeVeque, W. J., 1990. *Elementary Theory of Numbers*, Dover Publications, New York.

Lewis, H. R. and C. H. Papadimitriou, 1998. *Elements of the Theory of Computation*, 2nd ed., Prentice-Hall, Upper Saddle River, NJ. (Detailed text on automata theory and formal languages.)

Lifschitz, Vladimir, ed., 1991. *Artificial Intelligence and Mathematical Theory of Computation*, Academic Press, Boston.

Manna, Z., 1974. *Mathematical Theory of Computation*, McGraw-Hill, New York.

Manna, Z., and R. Waldinger, 1985. *The Logical Basis for Computer Programming*, Vol. I: Deductive Reasoning, Addison-Wesley, Reading, MA.

Manna, Z. and R. Waldinger, 1990. *The Logical Basis for Computer Programming*, Vol. II: Deductive Systems, Addison-Wesley, Reading, MA.

Martin, J. J., 1986. *Data Types and Data Structures*, Prentice-Hall, Englewood Cliffs, New Jersey.

Martin, N. M., 1989. *Systems of Logic*, Cambridge University Press, Cambridge, UK.

Mates, B., 1972. *Elementary Logic*, 2nd ed., Oxford University Press, New York.

Mendelson, E., 1987. *Introduction to Mathematical Logic*, 3rd ed., Wadsworth & Brooks/Cole, Monterey, CA.

Peirce, C. S., 1881. "On The Logic of Number," *American Journal of Mathematics*, Vol. 4, pp. 85–95. (Reprinted in Hartshorne and Weiss (1960), pp. 158–170.)

Piff, M., 1991. *Discrete Mathematics: An Introduction for Software Engineers*, Cambridge University Press, Cambridge, UK.

Pollock, J., 1990. *Technical Methods in Philosophy*, Westview Press, Boulder, CO. (Survey of introductory set theory, recursion, and logic.)

Prior, A. N., 1962. *Formal Logic*, 2nd ed., Oxford University Press, Oxford. (Surveys various logical systems, with many interesting historical references. Uses the so-called "Polish notation," in which all of the sentential connectives are prefix operators and parentheses are not required for grouping.)

Quine, W. V. O., 1960. *Word and Object*, MIT Press, Cambridge, MA. (A well-known work in the philosophy of language.)

Roberts, E. S., 1986. *Thinking Recursively*, John Wiley & Sons, New York. (Introduction to the use of recursion in programming with examples in Pascal.)

Roberts, F. S., 1979. *Measurement Theory with Applications to Decision-making, Utility, and the Social Sciences*, Addison-Wesley, Reading, MA. (Largely concerned with theorems about mappings of various kinds of non-numerical relational systems into numerical systems. These mappings provide quantitative and qualitative measurements used in natural and social sciences.)

Robinson, J. A., 1992. "Logic and Logic Programming," *Communications of the Association for Computing Machinery*, Vol. 35, No. 3, pp. 40–65.

Sainsbury, M., 1988. *Paradoxes*, Cambridge University Press, Cambridge, UK.

Sainsbury, M., 1991. *Logical Forms: An Introduction to Philosophical Logic*, Basil Blackwell, Oxford.

Schagrin, M. L., *et al.*, 1985. *Logic: A Computer Approach*, McGraw-Hill, New York. (An introduction to logic, with emphasis on sentential calculus. Describes many procedures for checking syntax, constructing and checking derivations, etc.)

Sterling, L., and E. Shapiro, 1986. *The Art of Prolog*, MIT Press, Cambridge, MA.

Stewart, I., and D. Tall, 1977. *The Foundations of Mathematics*, Oxford University Press, Oxford. (Survey of various number systems, elementary set theory, and methods of proof.)

Stoll, R. R., 1963. *Set Theory and Logic*, Dover Publications, New York. (Extensive development of set theory and mathematical logic.)

Suppes, P., 1972. *Axiomatic Set Theory*, Dover Publications, New York. (Rigorous development of set theory.)

Tanimoto, S. L., 1987. *The Elements of Artificial Intelligence*, Computer Science Press, Rockville, MD.

Wang, H., 1957. "The Axiomatization of Arithmetic," *Journal of Symbolic Logic*, Vol. 22, No. 2, pp. 145–158.

Warner, S., 1990. *Modern Algebra*, Dover Publications, New York

Webber, B. L., and N. J. Nilsson, eds., 1981. *Readings in Artificial Intelligence*, Morgan Kaufmann, San Mateo, CA.

Weyl, H., 1983. *Symmetry*, Princeton University Press, Princeton, NJ.

Index

nonlogical 451
symmetric relation 161
symmetry operations 223
syntax 20
  PC 363
  PCI 440
  predicate calculus 363
  SC 20
  sentential calculus 20

tail 337
tail$_s$ 329
Tanimoto, S. L. 412
Tarski, A. 378
tautological consequence 46, 63, 391
Tautological Consequences (TC)
  401, 443
Tautological Equivalence rules 69, 70
Tautological Implication rules 65, 66
tautology 43
tautologous
  PC 389
  SC 43
Tautology Laws (Taut) 70
Taylor, S. 312fn.
term 440
  constant 441
  ground 441
  referent of 442
  value of 442
theorem
  CVI (Course of Values
    Induction) 303
  De Morgan's 137
  Division Algorithm 309
  existence 203, 211, 237, 238
  Euclidean Algorithm 313
  of axiomatic theory 452
  PC 400, 419
  predicate calculus 400, 419
  Prime Divisor 298, 305, 323
  Recursion 237, 240, 280
  SC 77, 98
  sentential calculus 77, 98

uniqueness 237–238
  See also metatheorem
theory
  axiomatic 452
  of groups 221, 453
  of sets 2, 115
  PCI 451
  proof 62
THM 452
times 253
top 346
transitive
  closure 279
  extension 276
  relation 161
tree 23, 34
trichotomy law 188
true under an interpretation 43,
  379, 381, 442
truth 43
  logical 381
  table 35
  value 35, 378, 442
truth functional structure 389

Unification Algorithm 412
union 7, 129, 139
  generalized 139
unique readability 24, 34
universal closure 431
universal generalization 367
Universal Generalization (UG)
  405, 443
Universal Instantiation (UI)
  402, 443
universal specification 403
universe of discourse 121

valid 381
valuation function 34, 378, 442
variable 116
  bound occurrence 365
  free occurrence 365
  individual 363, 364